Vehicle-to-Vehicle and Vehicle-to-Infrastructure Communications

A Technical Approach

Vehicle-to-Vehicle and Vehicle-to-Infrastructure Communications

A Technical Approach

Edited by
Fei Hu

CRC Press
Taylor & Francis Group
Boca Raton London New York

CRC Press is an imprint of the
Taylor & Francis Group, an **informa** business

CRC Press
Taylor & Francis Group
6000 Broken Sound Parkway NW, Suite 300
Boca Raton, FL 33487-2742

International Standard Book Number-13: 978-1-138-70683-5 (Hardback)

Library of Congress Cataloging-in-Publication Data

Names: Hu, Fei, 1972- editor.
Title: Vehicle-to-vehicle and vehicle-to-infrastructure communications : a technical approach / edited by Fei Hu.
Description: Boca Raton : Taylor & Francis, CRC Press, [2018] | Includes bibliographical references.
Identifiers: LCCN 2017044927 | ISBN 9781138706835 (hb : alk. paper)
Subjects: LCSH: Vehicular ad hoc networks (Computer networks) | Vehicle-infrastructure integration.
Classification: LCC TE228.37 .V385 2018 | DDC 629.2/72--dc23
LC record available at https://lccn.loc.gov/2017044927

Visit the Taylor & Francis Web site at
http://www.taylorandfrancis.com

and the CRC Press Web site at
http://www.crcpress.com

Dedicated to my lovely wife, *Fang Yang*, the mother of three beautiful kids (Gloria, Edward, and Edwin).

Contents

Preface

The concept of a smart city has attracted attention from many countries. Intelligent transportation is the critical component in a smart city. To achieve intelligent transportation, all vehicles must be aware of the local and nearby traffic situations to avoid traffic jams. This requires that a vehicle should keep real-time, continuous communications with nearby vehicles as well as the road traffic base stations that connect to the city infrastructure.

Up to this point there is no book to cover the engineering designs of vehicle-to-vehicle and vehicle-to-infrastructure communications. Those existing books just cover general vehicle communication issues from the city infrastructure or commercial perspectives. This will be the first book to cover the communication hardware and software details as well as system integration models for two important vehicle communication types: vehicle-to-vehicle (V2V) and vehicle-to-infrastructure (V2I) communications. We use V2X to refer to both network types.

The targeted audience includes academic researchers, graduate students, company R&D people, government decision-makers, and other interested readers.

The book is organized as follows:

Section I. Overview: In this part, we summarize the technical challenges of V2X communications, network architectures, protocol stack, and how V2X systems can be used to detect congestion locations and levels.

Section II. V2X Routing: First, we introduce a comprehensive and comparative study in a city environment of representative routing protocols developed for inter-vehicular networks and wireless mobile ad hoc networks. The strengths and weaknesses of these techniques are discussed based on various network scenarios with regard to their support for highly mobile nodes. Second, we present a service architecture that exercises a software-defined network (SDN) concept to enable efficient data services in vehicular networks by exploiting the synergy between V2V and V2I communications. SDN is an emergent paradigm which brings flexibility and programmability to networks.

Section III. V2X Networking Schemes: There are many different network protocols and standards that can be used to achieve V2X communications. Here we select three important network types: (1) Dedicated short-range communication (DSRC): The digital tachograph (DT) is a device installed in commercial vehicles above 3.5 tonnes in Europe to monitor the driving time of commercial drivers. We will analyze the application of DSRC to the DT, discuss the benefits for law enforcers and propose a feasibility study to evaluate the transmission and reception of data from the DT. (2) In-cabin Wi-Fi: Automobile manufacturers are delivering a new generation of connected vehicles with in-cabin Wi-Fi devices, enabling advances in a wide range of in-vehicle communications and infotainment capabilities. We will answer the following question: if every running vehicle is equipped with an in-cabin Wi-Fi, how will the communication performance be affected by the varied number of surrounding vehicles, the transmission power, and the data rate? (3) Vehicle light communications: Considering the wide deployment of light emitting diodes (LEDs) both in automotive lighting and roadside infrastructure, visible light communication (VLC) has emerged as a potential complementary technology for V2X communications.

Section IV: V2X Applications: In this part, we will introduce some interesting applications that are based on a V2X platform. For example, we can build a cyber-physical system for platoon-aware traffic light rescheduling in the smart cities (SC) of tomorrow. It can enhance urban mobility by exploring methods and management strategies that increase system efficiency and improve individual mobility through information sharing, avoid congestion of key transportation corridors through cooperative navigation systems, and handle nonrecurring congestion. As another application, automated vehicle technologies have been implemented in practice progressively for at least a decade in passenger cars and commercial vehicles. A predecessor technology for automated vehicles is cruise control (CC), which regulates the vehicle speed to the driver's desired/selected speed regardless of traffic immediately in front, so the driver needs to adjust the set speed for a proper following distance. CC can utilize V2X for a smooth operation.

Section V: Antennas for V2X: Multiple input multiple output (MIMO) antenna systems offer high data rates (gigabit wireless link) and good quality of service (QoS) in non-line-of-sight (NLOS) environments. The applicability of MIMO techniques in V2V communications is examined in this part. MIMO helps to meet safety requirements and acceptable user experience for infotainment applications. We will show the effect of antenna beam direction, orientation, and position on the MIMO system performance through site-specific numerical simulations. Directive antenna performance is studied and compared with omnidirectional antennas. This part will also expound on (a) the electromagnetic waves as the energy departs the antenna, (b) the antenna patterns, (c) significant characteristics in antenna patterns, (d) common antenna patterns, (e) free space antenna path loss, and (f) the antenna location on a vehicle and a roadside unit. Finally, we introduce V2V channel measurements and modeling, which are important for future V2V system designs.

Section VI. Physical Layer Technologies: We will then move to the lowest network stack—the physical layer of V2X platforms. The combination of high data transfer rates, high performance, and multipath attenuation strength renders orthogonal frequency-division multiplexing (OFDM) ideal for current and future communication applications, allowing a response to high service quality requirements. The system presented in this part is prototyped based on the IEEE 802.11p standard, customized for vehicular ad hoc networks, and the signals are transmitted and received using a bandwidth of 10 MHz. The OFDM transceiver implementation is based on field-programmable gate array (FPGA) technology and the very high speed integrated circuit hardware description language (VHDL). In addition, any different decoding techniques have been tried and tested for their performance under a vehicular noisy channel by computer-based simulations under modulation schemes and transmitted through additive white Gaussian noise (AWGN) and Rayleigh fading channel models.

Section VII. Future Developments: In this part, we will introduce some future technologies that can advance V2X designs. First, we will introduce a simulation methodology that can implement SDN-based vehicle networks. Second, we will explain the concept of World Wide Wheels (WWW). We will explain how XG enables the high throughput of V2X.

Finally, we thank all V2X experts for their excellent work in the chapter writing. Due to the time limitation, this book might have some issues. We have tried to give credit to each cited reference in the chapters. If you see any problems in this book, please let us know. Thank you for reading this V2X book!

Editor

Dr. Fei Hu is currently a full professor in the Department of Electrical and Computer Engineering at the University of Alabama, Tuscaloosa, Alabama, USA. He obtained his PhD degrees from Tongji University (Shanghai, China) in the field of signal processing (in 1999) and from Clarkson University (New York, USA) in electrical and computer engineering (in 2002). He has published over 200 journal/conference papers and books. Dr. Hu's research has been supported by the US National Science Foundation, Cisco, Sprint, and other sources. His research expertise can be summarized as 3S: security, signals, and sensors. (1) *Security*: This is about how to overcome different cyberattacks in a complex wireless or wired network. Recently he focused on cyber-physical system security and medical security issues. (2) *Signals*: This mainly refers to intelligent signal processing, that is, using machine learning algorithms to process sensing signals in a smart way in order to extract patterns (i.e., pattern recognition). (3) *Sensors*: This includes microsensor design and wireless sensor networking issues.

Contributors

Nischal Adhikari
Department of Electrical Engineering
University of North Dakota
Grand Forks, North Dakota

Osman D. Altan
Federal Highway Administration (FHWA)
Turner Fairbank Highway Research Center
McLean, Virginia

Muhammed Ali Aydin
Department of Computer Engineering
Istanbul University
Istanbul, Turkey

Gianmarco Baldini
European Commission
Joint Research Centre (JRC)
Ispra, Italy

Ke Bao
Department of Electrical and Computer
    Engineering
University of Alabama
Tuscaloosa, Alabama

Ali Boyaci
Department of Computer Engineering
Istanbul Commerce University
Istanbul, Turkey

Jiannong Cao
Department of Computing
Hong Kong Polytechnic University
Hung Hom, Hong Kong, China

Xi Chen
Cyber-Physical Systems Laboratory
School of Computer Science
McGill University
Montreal, Canada

Xiang Chen
School of Information Science and Technology
Sun Yat-sen University
Guangzhou, China

and

Research Institute of Information Technology
    (RIIT) of Tsinghua University
Beijing, China

Christos N. Efrem
School of Electrical and Computer Engineering
National Technical University of Athens
    (NTUA)
Athens, Greece

Ali Riza Ekti
Department of Electrical-Electronics Engineering
Balikesir University
Istanbul, Turkey

Sinem Coleri Ergen
Department of Electrical and Electronics
    Engineering
Koç University
Istanbul, Turkey

Ryan Florin
Old Dominion University
Norfolk, Virginia

Ruisi He
Beijing Jiaotong University
Beijing, China

Yao-Hua (Danny) Ho
Network and Systems Laboratory (NSL)
Department of Computer Science and
    Information Engineering (CSIE)
National Taiwan Normal University
    (NTNU)
Taipei, Taiwan Republic of China

Fei Hu
Department of Electrical and Computer
    Engineering
University of Alabama
Tuscaloosa, Alabama

Aravind Kailas
Volvo Trucks
Greensboro, North Carolina

Sithamparanathan Kandeepan
School of Electrical and Computer
    Engineering
RMIT University
Melbourne, Australia

George Kiokes
Hellenic Air Force Academy
Dekeleia, Greece

Victor C. S. Lee
Computer Science Department
City University of Hong Kong
Kowloon Tong, Hong Kong, China

Chunhai Li
The Key Laboratory of Cognitive Radio and
    Information Processing
Guilin University of Electronic Technology
Ministry of Education
Guilin, China

Xiaohuan Li
Beihang University
Beijing, China

and

The Key Laboratory of Cognitive Radio and
    Information Processing
Guilin University of Electronic Technology
Ministry of Education
Guilin, China

Kai Liu
College of Computer Science
Chongqing University
Chongqing, China

Xiao-Yun Lu
PATH Program
Institute of Transportation Studies
University of California, Berkeley
Berkeley, California

Dan C. Marinescu
University of Central Florida
Orlando, Florida

Milad Mirzaee
Department of Antenna and Applied
    Electromagnetics
University of North Dakota
Grand Forks, North Dakota

Omer Narmanlioglu
Department of Electrical and Electronics
    Engineering
Ozyegin University
Istanbul, Turkey

Joseph K. Y. Ng
Computer Science Department
Hong Kong Baptist University
Kowloon Tong, Hong Kong, China

Sima Noghanian
Department of Antenna and Applied
    Electromagnetics
University of North Dakota
Grand Forks, North Dakota

Stephan Olariu
Old Dominion University
Norfolk, Virginia

Athanasios D. Panagopoulos
School of Electrical and Computer Engineering
National Technical University of Athens (NTUA)
Athens, Greece

Steven E. Shladover
PATH Program
Institute of Transportation Studies
University of California
Berkeley, California

Sang H. Son
Department of Information and
    Communication Engineering
Daegu Gyeongbuk Institute of Science and
    Technology (DGIST)
Daegu, Korea

Xin Tang
The Key Laboratory of Cognitive Radio and
    Information Processing
Guilin University of Electronic Technology
Ministry of Education
Guilin, China

Bugra Turan
Department of Electrical and Electronics
    Engineering
Koç University
Istanbul, Turkey

Murat Uysal
Department of Electrical and Electronics
    Engineering
Ozyegin University
Istanbul, Turkey

Jonathan B. Walker
U.S. Department of Transportation (U.S.
    DOT)
Federal Highway Administration (FHWA)
Washington, DC

Myounggyu Won
South Dakota State University
Brookings, South Dakota

Rongbin Yao
The Key Laboratory of Cognitive Radio and
    Information Processing
Guilin University of Electronic Technology
Ministry of Education
Guilin, China

Serhan Yarkan
Center for Applied Research on Informatics
    Technologies
Istanbul Commerce University
Istanbul, Turkey

Hongwei Zhang
Department of Electrical and Computer
    Engineering
Iowa State University
Ames, Iowa

Erietta Zountouridou
I-SENSE Group
Institute of Communication and Computer
    Systems (ICCS)
Athens, Greece

OVERVIEW

I

Chapter 1

A Review on V2V Communication for Traffic Jam Management

Myounggyu Won

South Dakota State University
Brookings, South Dakota

Contents

1.1 Introduction

Traffic demand has ever increased over the past decades as the population of city areas has grown substantially, which led to generation of severe traffic congestions all around the globe [1]. Traffic jams incur a huge amount of economic losses, environmental pollution, and even driver stress; in the United States, more than 5.5 billion hours were dissipated, which is commensurate with 2.9 billion gallons of fuel costing more than $121 billion [2]. The traditional approach to reduce traffic jams by building more road networks is no longer a feasible solution as the capacity has reached the limit in many countries [3]. Reducing traffic jams remains a significant research challenge for the academic and industry communities [4].

Vehicle-to-vehicle (V2V) communication, also known as the connected car technology, is becoming reality—the US Department of Transportation has started to deploy the connected vehicle technology throughout the US light-duty vehicles [5]. The key enabling technology for V2V communication is the wireless communication standard called dedicated short-range

3

communications (DSRC) and wireless access for vehicular environments (WAVE) that utilizes a region of unlicensed 5.9 GHz band, for which the specifications are defined by IEEE 802.11p [6]. The V2V technology enables vehicles to collect previously unobtainable high-fidelity traffic information such as surrounding vehicles' speed, positions, acceleration, destinations, maneuvers, and more depending on the protocol design. Utilizing the rich traffic data, vehicles are allowed to make rapid and accurate assessment of traffic conditions, sparking the development of various V2V-based approaches to reduce traffic jams.

In this chapter, we provide a comprehensive review on V2V-based solutions for reducing traffic jams. We categorize those solutions based on three key aspects: traffic information dissemination, traffic jam detection and classification, and traffic jam reduction. Traffic information dissemination refers to the schemes that are designed to allow vehicles to share with other vehicles various kinds of traffic information such as vehicle speed, position, and acceleration. Utilizing such shared traffic information, traffic jam detection and classification techniques are developed to effectively identify a traffic jam and to classify the intensity of a traffic jam. Once a traffic jam is detected traffic jam reduction methods are employed to allow vehicles to take appropriate actions to effectively alleviate traffic jams.

Figure 1.1 displays an overview of our classification of V2V-based solutions designed to reduce traffic jams. As shown, those solutions are classified based on the three major components, that is, traffic information propagation, traffic jam detection/classification, and traffic jam reduction. Various approaches have been used to detect/classify traffic jams ranging from a simple threshold-based method (*e.g.,* comparison of the average vehicle speed with a predefined threshold to detect a traffic jam) to a more advanced neural network-based mechanism. Effective dissemination of traffic information such as the vehicle speed and location is critical in successfully detecting traffic jams. Such traffic information is propagated in various ways, for example, disseminating the information to all vehicles (broadcasting), or to some vehicles in a certain region of interest (geocasting). Once a traffic jam is detected, various methods are applied to reduce traffic jams. For example, vehicles are advised to adjust their speed or headway distance to the preceding vehicle, or vehicles are provided with alternative routes. This chapter, based on this classification method, provides a review on a variety of techniques developed to effectively disseminate traffic information, detect/classify traffic jams, and alleviate them.

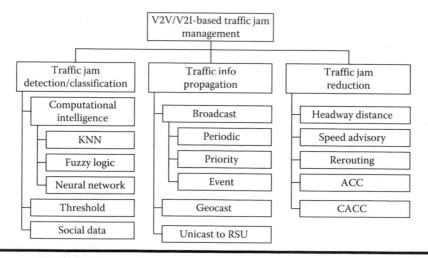

Figure 1.1   Classification of V2V/V2I-based traffic jam reduction schemes (ACC: adaptive cruise control, CACC: cooperative adaptive cruise control).

Table 1.1   Summary of Traffic Information Propagation Techniques

Propagation Mechanism	References
Periodic broadcast	[3,7–11,12–18]
Priority-based broadcast	[19]
Event-driven broadcast	[20,21]
Event-driven geocast	[22–25]
Periodic unitcast to RSU (V2I)	[11,26–30]

This chapter is organized as follows. In Section 1.2, techniques to propagate traffic information are reviewed, which is followed by the discussion on the methods for traffic jam detection and classification in Section 1.3. Various traffic jam reduction schemes are reviewed in Section 1.4. This chapter is concluded in Section 1.5.

1.2 Traffic Information Propagation Techniques

Dissemination of traffic information is an essential part of most V2V-based solutions for alleviating traffic jams because vehicles, roadside units (RSU), and centralized systems identify traffic jams based on the traffic information received via communication, and any driving advisory in the form of recommended speed and alternative routes need to be propagated to all or some vehicles. While most traffic jam reduction approaches rely on communication of common traffic information such as the vehicle speed, location, and acceleration, the way that traffic information is propagated is quite diverse.

Table 1.1 summarizes various techniques used to propagate traffic information. Periodic broadcasting also called as the beaconing process is the most widely adopted method as it is defined in the DSRC/WAVE standard. While DSRC/WAVE defines the single-hop data communication, epidemic routing is assumed in many solutions to assure propagation of traffic information over multihops to cover much larger areas based on random pair-wise message exchanges, and to handle frequent network partitions caused by random positions of vehicles [31]. In periodic broadcasting, vehicle information (*e.g.*, vehicle speed, location, and acceleration) is contained in a message, but some protocols use a message that also contains the congestion information (*e.g.*, the location of a congested area and the timestamp of congestion occurrence [7]).

Broadcasting incurs high network traffic causing contention, collision, and redundancy because all nodes continuously transmit data, which is known as the broadcast storm problem [32]. This problem exacerbates when a traffic jam occurs since there are a large number of vehicles in a limited space performing the beaconing process. It is thus of importance to prioritize different types of messages so that important messages can be transmitted with a higher priority [19]. Also, the ACK message can be selectively sent for lightweight messages while preventing the sending of an ACK message for heavyweight messages, and the frequency of message transmission can be dynamically adjusted to reduce the network traffic [9]. To alleviate the broadcast storm problem, some approaches allow for beaconing only when a certain event occurs, for example, when the vehicle velocity becomes less than a threshold [20,21].

When a traffic jam is identified, some approaches ensure that not all vehicles are notified of the traffic jam information, that is, only a subset of vehicles take actions to reduce traffic jams, for example, by reducing speed, or by taking alternative routes. To make sure that only some of the vehicles receive the message, geocasting is utilized [33]. In geocasting, a message is sent to nodes in a certain geographical area only [23,24], for example, vehicles on a certain road segment. Similar to the event-driven broadcasting, geocasting may be triggered by an event; for example, when a traffic accident occurs, a message that contains the information about the location and the time of the traffic accident may be geocast [22]. Some approaches determine retransmissions based on the level of congestion to further reduce the network traffic [24].

For effective propagation of traffic information especially when transportation infrastructure such as the RSUs are available, vehicle-to-infrastructure (V2I) may be utilized. More specifically, vehicles periodically transmit traffic information to nearby RSUs when they pass the RSUs [26]. These RSUs are also used as computational resources for detecting and classifying traffic jams, and for computing alternative routes. Usually, V2I is used together with V2V, especially to propagate driving recommendations to vehicles that are far away from RSUs [11].

1.3 Traffic Jam Detection and Classification Techniques

Traffic information received via V2V or V2I such as the vehicle speed and location are used to detect traffic jams and to classify the intensity of traffic jams. Various techniques have been developed for detection and classification of traffic jams; those techniques are categorized and summarized in Table 1.2.

As shown in the table, a majority of traffic reduction mechanisms adopt the threshold-based traffic jam detection methods. The threshold-based techniques specify that a traffic jam is detected when a certain traffic parameter is smaller/greater than a predefined threshold. Various kinds of traffic parameters are considered. For example, the vehicle speed being smaller than a minimum expected value indicates the existence of a traffic jam [3]. The average speed of surrounding vehicles is calculated based on received traffic information of the vehicle speed of neighboring vehicles, and if the average speed is smaller than a threshold, a traffic jam is detected [7]. The vehicle density is used to detect a traffic jam as well [12]. More specifically, if the difference between the vehicle arrival rate and the vehicle departure rate in a given region is greater than a predefined threshold,

Table 1.2 Summary of Traffic Jam Detection and Classification Techniques

Detection Mechanism	References
Model-based	[19,25,29,34]
Threshold-based	[7–8,10,12–13,20–21,35]
Fuzzy-logic-based	[9,16–18]
k-NN-based	[26,30]
Event-based	[22]
Social data-based	[36]
Neural network-based	[24]

a traffic jam is detected. In other approaches, the traffic density K_i on road segment i can be computed via V2V and compared with the traffic density under congestion denoted by K_{jam}, that is, if $K_i/K_{jam} > \delta$, where $\delta \in [0,1]$, a traffic jam is detected [13]. Detection of a traffic jam often triggers broadcasting a message to notify other vehicles of the existence of the traffic jam [20,21].

A challenge of the threshold-based methods is to select appropriate thresholds. If the threshold is too small, unnecessary driving advisory may be provided even though a traffic jam does not exist. On the other hand, if the threshold is too large, traffic jams may not be handled properly resulting in exacerbation of the traffic jam. Determining an appropriate threshold is, however, a hard task since the traffic environment dynamically changes and thus requires the thresholds to be selected adaptively.

The model-based methods are another widely adopted technique to detect and classify traffic jams. Those methods use obtained traffic information via V2V and V2I such as vehicle speed, acceleration, and location of neighboring vehicles as input to a model that generates a decision on whether a traffic jam exists or not. For example, the congestion index CI is defined and used as a model to determine a traffic jam, which is given as follows [25].

$$CI = \frac{T - T_0}{T_0},$$

where T is the actual travel time and T_0 is the travel time when there is no congestion. This model specifies the existence of a traffic jam as well as the intensity of the traffic jam as a numerical value.

The time-varying characteristics of traffic jams due to random driving behavior and dynamically changing traffic environments make it difficult to accurately classify the intensity of traffic jams. To address this challenge, computational intelligence such as fuzzy logic, k-NN, or neural networks has been adopted for detection and classification of traffic jams. Fuzzy logic is well studied in the literature. Linguistic variables are defined to represent different degrees of traffic parameters such as vehicle speed and density [9]. More specifically, membership functions for traffic parameters are defined to map a numerical value for a traffic parameter to a linguistic variable, which in turn is provided as input to a fuzzy inference system to classify the intensity of a traffic jam. For example, membership functions for vehicle speed μ_v distance to a critical section μ_s (*i.e.*, a road segment where a traffic jam occurred) and elapsed time after the emergence of a traffic jam μ_t are defined and used to determine the intensity of a traffic jam [16–18]. Figure 1.2 illustrates an example of the membership functions that generate an input to a fuzzy inference system, which in turn outputs the driving advisory as a linguistic variable based on the membership function μ_p.

An artificial neural network (ANN) is adopted by Forster et al. [24] to detect and classify traffic jams. ANN is a powerful tool due to its self-learning capabilities and adaptability to new situations. The overview of the approach proposed in Ref. [24] is shown in Figure 1.3. Vehicle speed and density are used as input to ANN. As shown in the figure, two neurons with the vehicle speed and density are used to define the input layer; four neurons are used in the hidden layer to represent the learning ability; and the output layer with a single neuron represents the congestion level, which is sent to other vehicles and also used to derive an alternative route.

As mentioned in this section, the commonly used traffic information to detect and classify traffic jams are vehicle speed and density. On the other hand, there are some works that attempt to use different types of information to detect traffic jams. For example, social data such as mobile phone data are used to predict the traffic jam [36]. The basic idea is to utilize mobile phone use log, that is, the time and cell tower location, of users to estimate the travel demand that can be used to detect and classify potential traffic jams.

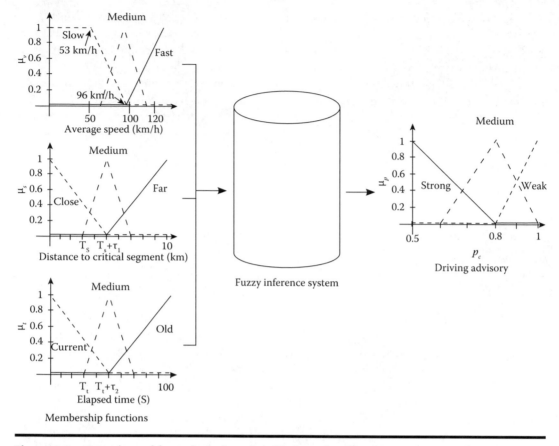

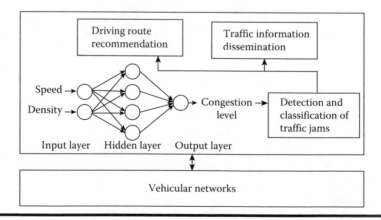

Figure 1.2   Overview of fuzzy logic-based traffic jam classification. T_s, τ_s, T_t, and τ_t are system parameters, and p_c represents a metric to adjust the inter-vehicle distance.

Figure 1.3   Overview of neural network-based approach for classifying traffic jams.

1.4 Traffic Jam Reduction Techniques

Various techniques have been developed to reduce traffic jams (Table 1.3). We categorize those techniques largely into two classes: microscopic methods and macroscopic approaches. The microscopic methods reduce traffic jams by adjusting the vehicle speed, or the distance to the preceding vehicle. These techniques are useful when the traffic density is not too high; in particular, traffic jams caused by minor perturbations even under moderate traffic density are often referred to as phantom jams [18]. On the contrary, if the traffic density is high, these microscopic methods may not be effective in reducing traffic jams. In contrast, macroscopic methods effectively alleviate traffic jams by reducing the traffic volume through rerouting vehicles from the congested area. There is a large body of research on reducing traffic jams through recommendation of alternative routes. Various methods are developed to calculate new routes that minimize different objectives such as the length of route or the travel time while avoiding traffic congestion [9,22]. However, these methods do not take into account the possibility of getting traffic congestion in other areas due to the rerouted traffic. To minimize the chance of getting such new traffic jams, different techniques have been investigated. For example, route changes are localized within the coverage of an RSU, and if the vehicle reaches the RSU, an updated route recommendation is generated [26]. Another approach defines a metric D that includes the congestion index CI, and depending on the new congestion level reflected in the index, a new route is determined adaptively. More specifically, the metric is defined as follows.

$$D = \alpha \cdot d + \beta \cdot CI.$$

Here the parameter β represents the significance of "least-congestness," and $\beta = 1 - \alpha$.

User preference can be incorporated in determining alternative routes. To represent the user preference, a new metric called the preference utility index is defined and driving routes are ordered according to this metric [37]. More specifically, preference utility weight p_w is defined to classify routes according to user preference.

$$p_w = a \cdot w_1 + c \cdot w_2 + (1 - i) \cdot w_3 + t \cdot w_4 + d \cdot w_5 + f \cdot w_6.$$

Table 1.3 Summary of Traffic Jam Reduction Techniques

Reduction Mechanism	References
Rerouting	[9–10,13–14,19,22–28, 30,34,36–37]
Speed adjustment	[3,8,11,20–21,29,35,38]
Inter-vehicle distance adjustment	[7,15–18,39]
Lane change	[11,38]
Cooperative adaptive cruise control	[40–42]

Here $w_1, \ldots, w_6$ are weight factors for the parameters defined in Table 1.4 [37]. Now the preference utility weights are classified into six preference bands, between 0 and 1 at intervals of 0.25 with the sixth band value of –1. Consequently, the preference utility index is defined as follows.

$$p_i = 0.25 \cdot ((5 - rank) + (5 - band) + (1 - alt) + dist).$$

Here *rank* represents the rank of the route according to the user preference ("1" meaning the most preferred route), *band* is the band number, *alt* is the number of alternative routes in that band, and *dist* specifies the difference in the band number of the best route and that of the next best route. Consequently, a route with the best preference utility index is recommended to driver.

While the macroscopic methods are useful when the traffic density is high, those approaches cause inconvenience to drivers since they need to drive on a new route that may be unfamiliar for

Table 1.4   Parameters for Calculation of Preference Utility Weight.

Parameters	Comments
Choice of Alternative Route a → *Value* = 1: if alternative route exists; *Value* = –1: if alternative route does not exist.	If a = –1, none of the remaining parameter values are calculated and p_w equals –1, indicating that the route is invalid for that vehicle as it does not lead to its destination.
Previous Compliance c → Value = 0: if driver complied more times than the threshold; *Value* = 1: if driver complied fewer times than the threshold.	Ensures that a vehicle is not made to compromise on its route choice repeatedly, as this may lead to driver noncompliance. In this work, the threshold is assumed to be 2.
Deviation Index i → Value range: 0–1.	Calculated as the ratio of the number of times a vehicle has deviated from its primary preference route to the total number of journeys undertaken. The smaller the index, the more flexible is the vehicle with its route choice.
Time Tolerance t → Value = 0: if current route travel time exceeds the threshold; *Value* = 1: if current route travel time is less than the threshold.	The threshold is calculated as the average of excess travel times encountered (over all previous route traversals) as compared to the "Shortest Time" route.
Distance Tolerance d → Value = 0: if travel distance exceeds the threshold; *Value* = 1: if travel distance is less than the threshold.	The threshold is calculated as the average of excess travel distance encountered (over all the previous route traversals) as compared to the "Shortest Distance" route.
Familiarity Index f → Value range: 0–1.	Calculated as the ratio of the number of times a particular route is taken to the total number of journeys undertaken on it and any of its alternate routes.

Source: US DOT Briefing Room, https://www.transportation.gov/briefing-room (Accessed 01 April 2017).

them, or they may have to drive a longer distance. On the contrary, if the traffic density is moderate to high, microscopic methods can be applied to effectively reduce traffic jams. Vehicle speed adjustment is a well-known technique of the microscopic solutions. The basic idea is that when the vehicle speed is appropriately reduced, the possibility of sudden acceleration and deceleration, one of the main causes for generating a shockwave (*i.e.*, a congested region that propagates upstream), is decreased, thereby reducing the chance of exacerbating the traffic jam [28]. A simple way of implementing vehicle speed adjustment is to predefine the vehicle speed, and when a traffic jam is detected, all affected vehicles reduce their speed to the predefined one [38]. However, this deterministic method does not cope well with varying traffic density. To address this challenge, a linear function is defined to map the vehicle density with the advised vehicle speed [8]. A more advanced car-following model is used to derive more effective recommendation of vehicle speed. For example, the Kraus car-following model [43] is adopted to compute the maximum safe velocity for the duration of τ to avoid the propagation of the shockwave [20]:

$$v_{safe} = v_j + \frac{d_{i,j}^* - v_j T}{\tau},$$

where $d_{i,j}^*$ represents the distance between the end of the shockwave and vehicle i (which can be determined based on the traffic information received via V2V), v_j is the velocity of vehicle j ahead of vehicle i that sent the V2V message, and T is the minimum time headway between i and j. A similar approach based on the model predictive car-following algorithm is proposed [29], that is, the recommended speed in terms of the desired time gap t_d is defined as follows.

$$t_d = t_{d,0} + \frac{s}{s_f}(t_{d,m} - t_{d,0}),$$

where $t_{d,0}$ is the minimum desired time gap, s is the distance between two vehicles, s_f is the distance threshold to apply this model, and $t_{d,m}$ is the user-defined maximum time gap.

Inter-vehicle distance adjustment methods are basically similar to the speed adjustment mechanisms, that is, an additional gap between the two adjacent vehicles is used to decrease the probability of the over deceleration effect which is one of the main factors that causes the shockwave [7,18]. While the speed adjustment techniques and distance adjustment techniques provide recommended speed and/or distance to human drivers, more advanced systems like cooperative adaptive cruise control systems are implemented to automate the speed/distance adjustment process [40,41].

1.5 Conclusions

This chapter presents a comprehensive review of V2V/V2I-based traffic jam management mechanisms. Although numerous approaches have been developed to reduce traffic jams, there remain a number of challenges to be addressed due to the inherent complexity of traffic jams. A congested area in which an excessive number of vehicles are squeezed in a limited space poses a new challenge to develop a method to efficiently disseminate the traffic information. In addition, real-time adaptive algorithms are needed to appropriately analyze the traffic jam and reduce the traffic jam considering the dynamic characteristics of traffic jams. We hope that this chapter will be a beneficial asset for beginning scientists and engineers in developing V2V-based solutions for reducing traffic jams.

References

1. P. Wang, T. Hunter, A. M. Bayen, K. Schechtner, and M. C. González, Understanding road usage patterns in urban areas, Scientific Reports vol. 2, no. 1101, pp. 1–6, 2012.
2. D. Schrank, B. Eisele, and T. Lomax, *TTIS 2012 urban mobility report*, Texas A&M Transportation Institute. Texas A&M University located in College Station, Texas. 2012.
3. H. -P. Wu, W. -H. Shen, Y. -L. Wei, H. -M. Tsai, and Q. Xie, Traffic shockwave mitigation with human-driven vehicles: Is it feasible? in *Proceedings of the First ACM International Workshop on Smart, Autonomous, and Connected Vehicular Systems and Services*, New York City, NY: ACM, 2016, pp. 38–43, October 03– 07, 2016.
4. Z. Zhang, H. Fujii, and S. Managi, How does commuting behavior change due to incentives? An empirical study of the Beijing subway system, *Transportation Research Part F: Traffic Psychology and Behaviour*, vol. 24, pp. 17–26, 2014.
5. US DOT Briefing Room, https://www.transportation.gov/briefing-room (Accessed 01 April 2017).
6. Q. Xu, T. Mak, J. Ko, and R. Sengupta, Vehicle-to-vehicle safety messaging in DSRC, in *Proceedings of the 1st ACM international workshop on Vehicular ad hoc networks*, Philadelphia, PA: ACM, pp. 19–28, October 01, 2004.
7. F. Knorr, D. Baselt, M. Schreckenberg, and M. Mauve, Reducing traffic jams via vanets, *IEEE Transactions on Vehicular Technology*, vol. 61, no. 8, pp. 3490–3498, 2012.
8. A. Chen, B. Khorashadi, C. -N. Chuah, D. Ghosal, and M. Zhang, *Smoothing vehicular traffic flow using vehicular-based ad hoc networking & computing grid (VGrid)*, in *Intelligent Transportation Systems Conference*, 2006. ITSC'06. IEEE. Toronto, Canada: IEEE, pp. 349–354, September 17–20, 2006.
9. G. B. Aráujo, M. M. Queiroz, F. de LP Duarte-Figueiredo, A. I. Tostes, and A. A. Loureiro, Cartim: A proposal toward identification and minimization of vehicular traffic congestion for VANET, in *2014 IEEE Symposium on Computers and Communication (ISCC)*, Madeira, Portugal: IEEE, 2014, pp. 1–6.
10. Z. Jiang, J. Wu, and P. Sabatino, Gui: GPS-less traffic congestion avoidance in urban areas with inter-vehicular communication, in *2014 IEEE 11th International Conference on Mobile Ad Hoc and Sensor Systems (MASS)*, Philadeliphia: IEEE, pp. 19–27, October 28–30, 2014.
11. S. Djahel, N. Jabeur, R. Barrett, and J. Murphy, Toward v2i communication technology-based solution for reducing road traffic congestion in smart cities, in *2015 International Symposium on Networks, computers and communications (ISNCC)*, Yasmine Hammamet - Tunisia: IEEE, pp. 1–6, May 13–15, 2015.
12. M. F. Fahmy and D. Ranasinghe, Discovering automobile congestion and volume using VANETs, in *8th International Conference on ITS Telecommunications, 2008. ITST 2008*, Phuket, Thailand: IEEE, pp. 367–372, October 24, 2008.
13. J. Pan, M. A. Khan, I. S. Popa, K. Zeitouni, and C. Borcea, Proactive vehicle rerouting strategies for congestion avoidance, in *2012 IEEE 8th International Conference on Distributed Computing in Sensor Systems (DCOSS)*, Hangzhou, Zhejiang, China: IEEE, pp. 265–272, May 16–18, 2012.
14. R. Doolan and G. -M. Muntean, VANET-enabled eco-friendly road characteristics-aware routing for vehicular traffic, in *2013 IEEE 77th Vehicular Technology Conference (VTC Spring)*, Dresden, Germany: IEEE, pp. 1–5, June 02–05, 2013.
15. F. Knorr and M. Schreckenberg, Influence of inter-vehicle communication on peak hour traffic flow, *Physica A: Statistical Mechanics and its Applications*, vol. 391, no. 6, pp. 2225–2231, 2012.
16. M. Won, T. Park, and S. H. Son, Fuzzyjam: Reducing traffic jams using a fusion of fuzzy logic and vehicular networks, in *2014 IEEE 17th International Conference on Intelligent Transportation Systems (ITSC)*, Qingdao, China: IEEE, pp. 1869–1875, October 8–11, 2014.
17. M. Won, T. Park, and S. H. Son, Lane-level traffic jam control using vehicle-to-Vehicle communications, in *2014 IEEE 17th International Conference on Intelligent Transportation Systems (ITSC)*, Qingdao, China: IEEE, pp. 2068–2074, October 8–11, 2014.
18. M. Won, T. Park, and S. H. Son, Toward mitigating phantom jam using vehicle-to-vehicle communication, IEEE Transactions on Intelligent Transportation Systems vol. 18, no. 5, pp. 1313–1324, 2017.

19. L. -D. Chou, D. C. Li, and H. -W. Chao, Mitigate traffic congestion with virtual data sink based information dissemination in intelligent transportation system, in *2011 Third International Conference on Ubiquitous and Future Networks (ICUFN)*, Dalian, China: IEEE, pp. 37–42, June 15–17, 2011.
20. M. Forster, R. Frank, M. Gerla, and T. Engel, A cooperative advanced driver assistance system to mitigate vehicular traffic shock waves, in *2014 Proceedings IEEE INFOCOM*, Toronto, Canada: IEEE, pp. 1968–1976, April 27 – May 02, 2014.
21. M. Forster, R. Frank, and T. Engel, An event-driven inter-Vehicle communication protocol to attenuate vehicular shock waves, in *2014 International Conference on Connected Vehicles and Expo (ICCVE)*, IEEE, 2014, pp. 540–545.
22. A. M. de Souza, R. S. Yokoyama, G. Maia, A. A. Loureiro, and L. A. Villas, Minimizing traffic jams in urban centers using vehicular ad hoc networks, in *2015 7th International Conference on New Technologies, Mobility and Security (NTMS)*, Paris, FR: IEEE, pp. 1–5, June 26–29, 2015.
23. A. Lakas and M. Cheqfah, Detection and dissipation of road traffic congestion using vehicular communication, in *2009 Mediterranean Microwave Symposium (MMS)*, Tangiers, Morocco: IEEE, pp. 1–6, November 15–17, 2009.
24. R. I. Meneguette, P. Geraldo Filho, D. L. Guidoni, G. Pessin, L. A. Villas, and J. Ueyama, Increasing intelligence in inter-vehicle communications to reduce traffic congestions: Experiments in urban and highway environments, *PLoS One*, vol. 11, no. 8, p. e0159110, 2016.
25. A. Lakas and M. Chaqfeh, A novel method for reducing road traffic congestion using vehicular communication, in *Proceedings of the 6th International Wireless Communications and Mobile Computing Conference*, Caen, France: ACM, pp. 16–20, June 28 – July 02, 2010.
26. A. M. de Souza, R. S. Yokoyama, G. Maia, A. Loureiro, and L. Villas, Real-time path planning to prevent traffic jam through an intelligent transportation system, in *2016 IEEE Symposium on Computers and Communication (ISCC)*, Messina, Italy: IEEE, pp. 726–731, Jun 27–30, 2016.
27. M. R. Jabbarpour, A. Jalooli, E. Shaghaghi, R. M. Noor, L. Rothkrantz, R. H. Khokhar, and N. B. Anuar, Ant-based vehicle congestion avoidance system using vehicular net1.5. Conclusions 19 works, *Engineering Applications of Artificial Intelligence*, vol. 36, pp. 303–319, 2014.
28. S. Wang, S. Djahel, Z. Zhang, and J. McManis, Next road rerouting: A multiagent system for mitigating unexpected urban traffic congestion, *IEEE Transactions on Intelligent Transportation Systems*, vol. 17, no. 10, pp. 2888–2899, 2016.
29. M. Wang, W. Daamen, S.P. Hoogendoorn, and B. van Arem, Connected variable speed limits control and car-following control with vehicle-Infrastructure communication to resolve stop-and-Go waves, *Journal of Intelligent Transportation Systems*, vol. 20, no. 6, pp. 559–572, 2016.
30. A. M. de Souza, R. S. Yokoyama, L. C. Botega, R. I. Meneguette, and L. A. Villas, Scorpion: A solution using cooperative rerouting to prevent congestion and improve traffic condition, in *2015 IEEE International Conference on Computer and Information Technology; Ubiquitous Computing and Communications; Dependable, Autonomic and Secure Computing; Pervasive Intelligence and Computing (CIT/IUCC/DASC/PICOM)*, Liverpool, UK: IEEE, pp. 497–503, October 26–28, 2015.
31. A. Vahdat and D. Becker, Epidemic routing for partially connected ad hoc networks, 2000. Technical Report CS-2000-06, Duke University, 2010.
32. Y. -C. Tseng, S.-Y. Ni, Y. -S. Chen, and J. -P. Sheu, The broadcast storm problem in a mobile ad hoc network, *Wireless Networks*, vol. 8, no. 2/3, pp. 153–167, 2002.
33. J. C. Navas and T. Imielinski, Geocast geographic addressing and routing, in *Proceedings of the 3rd Annual ACM/IEEE International Conference on Mobile Computing and Networking*, Budapest, Hungary: ACM, pp. 66–76, September 26–30, 1997.
34. D. H. Stolfi and E. Alba, Red swarm: Reducing travel times in smart cities by using bio-inspired algorithms, *Applied Soft Computing*, vol. 24, pp. 181–195, 2014.
35. A. Hegyi, B. Netten, M. Wang, W. Schakel, T. Schreiter, Y. Yuan, B. van Arem, and T. Alkim, A cooperative system based variable speed limit control algorithm against jam waves—An extension of the specialist algorithm, in *2013 16th International IEEE Conference on Intelligent Transportation Systems—(ITSC)*, The Hague, Netherlands: IEEE, pp. 973–978, October 06–09, 2013.

36. K. He, Z. Xu, P. Wang, L. Deng, and L. Tu, Congestion avoidance routing based on large-scale social signals, *IEEE Transactions on Intelligent Transportation Systems*, vol. 17, no. 9, pp. 2613–2626, 2016.
37. P. Desai, S. W. Loke, A. Desai, and J. Singh, Caravan: Congestion avoidance and route allocation using virtual agent negotiation, *IEEE Transactions on Intelligent Transportation Systems*, vol. 14, no. 3, pp. 1197–1207, 2013.
38. T. D. Hewer, M. Nekovee, and P. V. Coveney, Reducing congestion in obstructed highways with traffic data dissemination using ad hoc vehicular networks, *EURASIP Journal on Advances in Signal Processing*, vol. 2010, no. 1, p. 169503, 2010.
39. A. Kesting, M. Treiber, M. Schönhof, and D. Helbing, Adaptive cruise control design for active congestion avoidance, *Transportation Research Part C: Emerging Technologies*, vol. 16, no. 6, pp. 668–683, 2008.
40. J. Ploeg, A. F. Serrarens, and G. J. Heijenk, Connect & drive: Design and evaluation of cooperative adaptive cruise control for congestion reduction, *Journal of Modern Transportation*, vol. 19, no. 3, pp. 207–213, 2011.
41. B. Van Arem, C.J. Van Driel, and R. Visser, The impact of cooperative adaptive cruise control on traffic-flow characteristics, *IEEE Transactions on Intelligent Transportation Systems*, vol. 7, no. 4, pp. 429–436, 2006.
42. M. A. S. Kamal, J.-I. Imura, T. Hayakawa, A. Ohata, and K. Aihara, Smart driving of a vehicle using model predictive control for improving traffic flow, *IEEE Transactions on Intelligent Transportation Systems*, vol. 15, no. 2, pp. 878–888, 2014.
43. S. Krauss, P. Wagner, and C. Gawron, Metastable states in a microscopic model of traffic flow, *Physical Review E*, vol. 55, no. 5, p. 5597, 1997.

V2X ROUTING

II

Chapter 2

Routing Protocols in V2V Communication

Yao-Hua (Danny) Ho

National Taiwan Normal University (NTNU)

Taipei, Taiwan Republic of China

Contents

2.1 Introduction

Vehicular networks have become increasingly popular in recent years. There are two variations of vehicular networks: *infrastructure* networks and *infrastructureless* networks. The latter, also known as a vehicular ad hoc network (VANET), has no fixed routers for data exchange. This is on the same principles of a mobile ad hoc network (MANET) where the mobile nodes themselves function as routers which discover and maintain communication connections. Thus, VANET and MANET are a self-organizing multihop wireless network where all nodes participate in the routing and data forwarding process. Such a network can be easily deployed in situations where no base station is available, and a network must be built spontaneous. As examples, for applications such as vehicle-to-vehicle communications, battlefield communications, national crises, and disaster recovery, a wired network is not available and ad hoc networks provide the only feasible means of communications and information access.

There are many routing protocols designed to relay data in MANETs such as:

- ad hoc on-demand distance vector (AODV) [1]
- asymptotically optimal geometric routing [2]
- cluster gateway switch routing (CGSR) [3]
- connectionless approach (CLA) [4]
- contention-based forwarding (CBF) [5]
- distance routing effect algorithm for mobility (DREAM) [6]
- dynamic destination sequenced distance vector routing (DSDV) [7]
- dynamic source routing (DSR) [8]
- greedy perimeter stateless routing (GPSR) [9]
- location-aware routing protocol (such as GRID, which the geographic area is treated as a number of logical grid) [10]
- global state routing (GSR) [11]
- location-aided routing (LAR) [12]
- on-demand multicast routing [13]
- trajectory-based forwarding (TBF) [14]
- trigger-based distributed quality of service (QoS) routing [15]
- location-based routing (Terminode Routing, TMNR) [16]
- wireless routing protocol (WRP) [17]
- zone routing protocol (ZRP) [18]

However, most of them are not designed for high mobility and street environments.

Earlier protocols such as DSR [8], AODV [1], and LAR [12] require a source to use a route request to establish a hop-by-hop route between itself and a destination before sending data. In the street environments, however, obstacles and fast-moving nodes result in a very short window of communication between nodes on different streets. The established route expires quickly and the source needs to reissue another expensive network-wide route request after sending only a few data packets via the previous route. These protocols, when applied in the street environments, will incur a high control overhead in terms of route request packets.

To overcome the fragility of multiple-hop routes, one-hop-based approaches, such as TBF [14], let each forwarder select the next forwarding node by comparing the positions of all its neighbors with the trajectory defined by a source. This position information is obtained through periodic broadcasts from neighboring nodes. The short window of communication in the street environments, however, means that the nodes need to broadcast more frequently in order to maintain up-to-date location information. This strategy incurs a high control overhead in terms of frequent beaconing packets that also congest the wireless medium. The above protocols are defined as a *connection-oriented approach* because each link a packet traverses must first be established through a network-wide route request or location information exchange among all the nodes in the network.

Rather than using the expensive control overhead to preestablish each link, the *CLA* allows a node to dynamically participate in a forwarding of data by comparing its current location with headers of the data, which contain location information of a source, a destination, and a previous relayed node. Existing connectionless techniques such as CBF [5], beaconless routing (BLR) [19], and CLA to MANET [4,37] only allow nodes that have the shortest distance to the destination or are on the shortest geographic path (*i.e.*, a straight line) between the source and destination to relay data. When applying these CLA to the street environments, nodes that can relay data around obstacles often do not get to relay the data because they are farther from the destination than the previous relayed node or are not on the forwarding path of the data. Thus, these approaches cannot be applied directly to the street environments.

To cope with the obstacles in a city environment, a modified CBF technique [20] has been proposed to address the problems. This scheme allows a source to specify a forwarding path as a list of junctions, and applies CBF between consecutive junctions. This solution requires at least one node at each turning junction, which is often difficult to achieve over an extended communication period. To overcome this drawback, another method (*i.e.*, CLA-S) [21] adapts the CLA to a street environment to utilize multiple communication paths.

Performance comparisons of these ad hoc routing protocols have been studied using different simulation models. In Refs. [22–24], the ns-2 simulator [25] is used to study communications among 50–100 mobile nodes. Their moving speeds are limited to 20 m/s, with relativity long pause time ranging from 0 to 900 seconds. It is not clear if the simulation results are still valid under a high-mobility environment such as vehicle-to-vehicle communications and advanced battlefield applications. Other studies [26–29] focus on scalability by varying the network size and the number of nodes in the simulations. In Ref. [27], up to 400 nodes are studied, with node speed and pause time fixed at 14 m/s and 0 s, respectively. Up to 50 mobile nodes are considered in Ref. [26]. Their pause time is fixed at 30 seconds while node speed varies up to 30 m/s. The mobility assumed in these two studies, however, is too slow to reveal the impact of high mobility on performance. Large networks with up to 50,000 nodes are studied in Ref. [29]. In Ref. [4], street environments with the high-mobility scenarios considered are 14 m/s (32 mph) with zero pause time and 30 m/s (67 mph) with a 30-second pause time are studied. Since either a relatively low node speed or a long pause time is assumed, the effective node mobility is still too low to capture the characteristics of high-mobility environments such as vehicle-to-vehicle communications (*i.e.*, high node speed with zero pause time).

Many more advanced routing techniques have been developed for MANETs and vehicular networks in recent years. In this chapter, eight representative routing protocols (AODV, DSR, LAR, GRID, TMNR, GPSR, CBF-street version, and CLA-street version) are compared under a street environment with detailed simulation results. In particular, this chapter focuses on high mobility (*e.g.*, vehicles) and street environments, as many important emergent applications of this technology involve high-mobility nodes. As an example, cars in a vehicle-to-vehicle network are typically moving at speeds exceeding 30 mph in a city environment with large obstacles (*e.g.*, office buildings). Very little is known about how existing routing methods perform relative to each other in such street environments. The impact of high mobility and large obstacles on different routing protocols are investigated under various scenarios in this chapter. The results provide insight into routing protocols and offer guidelines for inter-vehicle network applications.

2.2 Routing Protocols Review

Early-generation routing protocols, such as DSDV, WRP, and GSR, establish communication links by maintaining routing information in a routing table at each node. A drawback to this solution is that every node needs to update its routing table and propagate the update, as the network topology changes, in order to maintain a consistent view of the network. This operation incurs excessive network traffic and computation overhead. Later techniques, such as DSR and AODV, attempt to reduce unnecessary network traffic by initiating route request on demand. This type of routing protocols establishes communication links by flooding the network to find a route to the destination node. This strategy is simple and robust; however, it is not energy efficient and can cause severe media congestion.

Another approach to reduce network flooding is to leverage location information obtained from a global positioning system (GPS) [30] or other location services [31,32]. For instance, LAR uses location information to limit the area of flooding, thereby reducing the number of route request messages. These schemes result in better power conservation and improve network scalability.

Some other techniques reduce not only the number of route requests but also route maintenance costs. This type of approach, such as GRID and CGSR, organizes mobile nodes into clusters. Each cluster has a cluster head and a number of gateways. Two clusters communicate via a gateway node within their communication range. An obvious advantage of this environment is that only cluster-heads and gateway nodes need to rebroadcast messages. However, the network throughput can be limited by the number of gateway nodes. Furthermore, cluster management incurs overhead.

To address mobility issues, one-hop approaches, such as TBF, TMNR, and GPSR, have been proposed. In these schemes, instead of the need to establish a complete connection from the source to the destination, the node only needs to establish the connection to the next hop (*i.e.*, one hop) and forward the data. To determine the next hop, a node compares the distances of its neighbor nodes to the destination node (*i.e.*, GPSR), the next waypoint (*i.e.*, TMNR), or a trajectory (*i.e.*, TBF).

CBF and CLA are more recent techniques developed for routing in MANETs. While GPSR, TMNR, and TBF need to maintain (proactively or reactively) neighbor nodes location information and establish a connection to the next hop before forwarding a data packet, CBF and CLA simply forward data packets without first establishing the link to the next node. Any node that happens to be in the general direction toward the destination node can compete for the "right" to forward data packets. Since there is no need to preestablish multihops or one-hop connections, these protocols, that is, CBF and CLA, are classified as a connectionless-based approach. The advantages of this approach are twofold. First, there is no need to maintain communication connections; and second, nodes do not need to maintain information about their neighboring nodes. Recently, CBF and CLA have been extended to address the obstacle problems in a city environment. In the modified CBF technique, a source specifies a forwarding path as a list of junctions and applies CBF between consecutive junctions. In the modified CLA (*i.e.*, CLA-S), a source specifies a forwarding zone as an area of streets and intersections, and applies a CLA approach.

In the remainder of this chapter, the eight selected techniques (AODV, DSR, LAR, GRID, TMNR, GPSR, CBF-street, and CLA-street) are introduced and explained in greater detail for the study. As illustrated in Figure 2.1, the selected techniques are grouped into four categories—multihop based, cluster based, one-hop based, and connectionless based. The shaded boxes represent the eight routing methods selected for the comparative study. These techniques are selected as representatives of their category. Proactive techniques are not included in this chapter because it would not be feasible to maintain up-to-date routing tables in a high-mobility environment where the network topology changes rapidly.

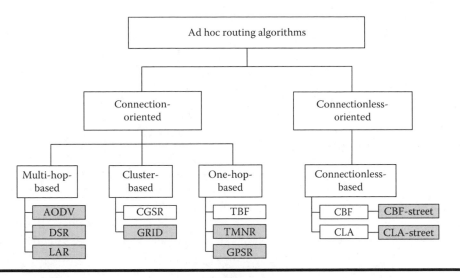

Figure 2.1   Classification of some existing routing protocols.

2.2.1 *Dynamic Source Routing*

DSR [8] is a simple and efficient routing protocol. It allows nodes to dynamically discover a multihop route across the MANET to any destination node. To do so, a source node S first sends out a ROUTE REQUEST message with the request ID to all of its neighbors. If the nodes receiving this message have not seen it before, they add themselves to the route and forward the message to their neighbors. If a receiving node D is the destination or has information on a route to the destination, D sends a REPLY message containing the full route. D may send the REPLY along the route. After receiving one or several routes, the source node picks the best route, stores it, and starts to send messages along this route. If the network topology changes due to a broken communication connection along the route, source node S can attempt to use any other route S happens to know about, or can issue another ROUTE REQUEST to find a new route. This is referred to as route maintenance.

DSR is perhaps the most cited routing protocol for MANETs. We use this routing protocol as a base case to assess the performance of the other more advanced techniques in this comparative study.

2.2.2 *Ad Hoc On-Demand Distance Vector*

AODV [1] has the same on-demand routing characteristics of DSR. Similar to DSR, when a node S needs to send data to a node D, node S first needs to broadcast a ROUTE REQUEST message including the last known sequence number for that destination node D. This ROUTE REQUEST message will flood the network until it reaches the node D or a node with information on a route to node D. This node then replies with a ROUTE REPLY message containing the number of hops needed to reach D and a sequence number for D. Similar to traditional routing tables, each node participating in the route creates an entry in the routing table, where each route entry is maintained using timer-based states. To maintain the state, each node periodically broadcasts a HELLO message. If a node does not receive the HELLO message from a neighbor node for three consecutive periods, then the link is considered broken. When a link breaks, any upstream node recently using this broken link will be notified using an UNSOLICITED ROUTE REPLY message with an infinite metric for the destination. When node *S* receives this UNSOLICITED ROUTE REPLY, it will start another route discovery as described previously.

Note that, it is unfair to compare AODV and DSR in this study because they do not require GPS. Nevertheless, including these two well-known routing techniques in this study provides a reference to assess the benefit of using location information from GPS.

2.2.3 Location-Aided Routing

The LAR [12] protocol leverages location information to limit the area of route request when searching for a new route. Using a previously known location and average speed of the destination node, this scheme estimates the "expected zone" of the destination node as a circular region that may contain the destination node. The request zone is then defined as the smallest bounding rectangle that includes the current location of the source node and the expected zone. Once the request zone has been defined, only nodes in the request zone need to forward the route request message ROUTE REQUEST.

A variant of the LAR scheme is LAR2 [12]. In LAR2, the distance ($DIST_s$) between the source node and the destination node, and the location information of the destination node are included in the route request message. When any node receives a routing request, it compares its distance ($DIST_i$) to the destination node and the distance ($DIST_s$) in the route request message. If $DIST_s + \delta \geq DIST_i$, this node will forward the route request and change the $DIST_s$ to $DIST_i$ in the request message; otherwise, it will discard the route request. The parameter δ determines the tolerant distance for a node to forward or not. Since this scheme still needs to maintain hop-by-hop routes, it does not ease the route maintaining problem.

2.2.4 GRID-Based Routing

GRID [10] is a cluster-based approach. This technique reduces the cost of route maintenance by dividing the network area into fixed-size grid cells with nodes within each cell forming a cluster. For each cluster, a node is selected as the gateway and only gateway nodes may rebroadcast messages. When gateway nodes leave their current cell, they call a handover procedure to select a new gateway. This strategy requires each gateway node to periodically broadcast its existence to other nodes within the cell.

In this environment, nodes establish connections by selecting a list of consecutive grid cells as a route. A gateway node in each of the selected grid cell participates in forwarding the data packets. Thus, the number of hops in a route is determined by the number of selected grid cells. Although the number of hops can be reduced by using a larger grid cell, this can result in a higher frequency of broken links because two gateway nodes of two adjacent cells might move out of the radio range of each other. In the worst scenario, the reestablishment of the connection might not be possible if the same situation is observed with all of the eight neighboring cells. To avoid this problem, we set the grid size to $x \times x$, where $x =$ (nominal radio range R)/$(2\sqrt{2})$, in our simulation study. This ensures that any node in one cell can communicate with any node in the neighboring cells within the same street or within in two intersected streets.

2.2.5 Greedy Perimeter Stateless Routing

To address the scalability issue, GPSR [9] makes a greedy decision at each forwarding node to minimize the total number of necessary hops. This is achieved by having each node broadcast its location information to its neighboring nodes. To send a data packet, the source node includes the location information of the destination node with the data packet. This information enables a forwarding node to greedily select a neighbor that is geographically closest to the packet's destination as the next hop. This process is repeated for each hop until the packet reaches its destination.

2.2.6 Terminode Routing

Terminode routing (TMNR) [16] is a hybrid of location-based routing and the table-driven technique. Each node maintains a routing table with the location information of its one-hop and two-hop neighbors. To establish a connection, a source node first attempts to determine the route to the destination using only information in its local routing table. If this fails, a direct path technique is used. This scheme determines the direct path as an approximation of the straight line from the source to the destination. Each data packet is forwarded by nodes along this direct path until the packet reaches its destination. A perimeter mode is used to circumvent any topology hole by planar graph traversal, similar to GPSR [9]. In the initial stage of this mode, the next forwarding node is generally farther from the destination than the last forwarding node. This mode is switched back to the direct path technique when a forwarding node is found to become closer to the destination.

Also, predetermined geographical anchor locations are used in Ref. [16] as waypoints toward the destination node. In this environment, a direct path is established between two adjacent waypoints such that a data packet is forwarded by intermediate nodes in the direction of the next waypoint in the list until the packet reaches a node close to that waypoint, at which point the next waypoint is attempted in the same manner until the data packet eventually reaches the destination node. Since any topology hole cannot be predetermined due to nodes moving randomly, the perimeter mode is not included in this chapter.

2.2.7 Contention-Based Forwarding for MANET and Vehicular Network

In CBF [5], a forwarding node transmits a data packet as a single-hop broadcast to all its neighbors. These neighbors compete with each other for the "right" to forward the packet. During this contention period, a node determines how well it is suited to be the next hop for the packet. The node that wins the contention *suppresses* the other nodes, thus establishing itself as the next forwarding node. This contention is based on the distances of the nodes to the destination. A drawback of this strategy in a high-density environment is that several neighboring nodes might have similar distances to the destination. Consequently, they will all establish themselves as the next hop and forward the data packet. This incurs unnecessary network traffic and wastes power in the mobile nodes. A solution, suggested in Ref. [5], is for each contestant node to report its qualification for forwarding the data packet and wait for the current forwarding node to select the winner for the next hop. Since this strategy is similar to GPSR, we consider its extension version for a street environment. In the extended CBF for a city environment (CBF-street) [20], a source needs to specify a forwarding path as a list of junctions. Between two junctions, a CBF approach is applied.

2.2.8 Connectionless Approach for MANET and Vehicular Network

In the CLA [4], the network area is divided into small "virtual cells." Instead of maintaining a hop-by-hop route between the source and destination nodes, the source selects a list of grid cells that form a connecting path between the source and destination. Nodes within each of these cells alternate in forwarding data toward the next cell using a delay function. This function computes a shorter delay for a node farther from the sender and closer to the destination. In this environment, a connecting path is considered broken if one of its cells becomes empty. This is addressed by replacing the empty cell with a neighboring cell. The fundamental advantages of CLA are twofold. First, a connection path (*i.e.*, a list of grid cells) is much less likely to become broken than a standard route used in conventional techniques. Second, unlike standard routes, the robustness of connection paths is not sensitive to the mobility inherent in the network. In the modified CLA for a street environment [21] (*i.e.*, CLA-S), streets are divided into small virtual cells. These cells are divided according to intersections and blocks. A source selects a list of cells as forwarding zones that form a connecting forwarding area between the source and destination. Nodes within each of these cells alternate in forwarding data toward the destination node.

2.3 Simulation Environment and Methodology

The overall goal of this chapter is to present the experiments that measure the ability of the eight routing protocols in handling network topology changes due to high mobility of nodes in a street environment. In this type of environment, the radio range and mobility are often constrained by buildings and streets. For example, if a mobile node is within a block, a mobile node's effective radio range is only within the street it is on (see Figure 2.2). Similarly, if a mobile node is at an intersection, the mobile node's effective radio range will be along the two intersecting streets (see Figure 2.3).

The simulators used for this study were implemented using the GloMoSim [33] library, a packet-level simulator specifically designed for ad hoc networks. It follows the Open Systems Interconnection (OSI) 7-layer network communication model. The field configuration is a

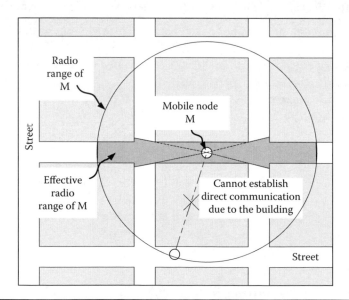

Figure 2.2   Effective radio range of a mobile node within a block.

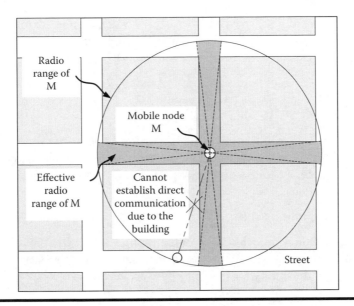

Figure 2.3 Effective radio range of a mobile node at an intersection.

1000 m × 1000 m field with a street width of 10 m and a building block size of 100 m × 100 m, unless it is specified otherwise by the network scenarios. The radio propagation range for each node is 375 m and channel capacity is 2 Mbits/sec. Initially, nodes are placed uniformly with two nodes per intersection and eight nodes per block; then the nodes move in the directions permitted in the streets. Upon arriving at an intersection, a node pauses for a period (*i.e.*, specify in the different simulation setups) of time, and then the node probabilistically changes its direction of movement (*e.g.*, turn left, turn right, or continue in the same direction). Traffic applications are constant bit rate (CBR) sessions involving 1/10 of all the nodes. Each data packet is 512 bytes and the senders are chosen randomly among the nodes. Multiple simulation runs (100 runs per setup on average) with different seed numbers were conducted for each scenario and collected data were averaged over those runs with a 95% confidence interval.

The routing protocols are compared according to the following four metrics.

1. **Fraction of packets delivered:** It measures the ratio of the data packets delivered to the destinations and the data packets generated by the CBR source. This number indicates the effectiveness of a protocol.
2. **End-to-end delay:** It is measured in milliseconds and includes processing, route discover latency, queuing delays, retransmission delay at the MAC, and propagation and transmission times. This number measures the total delay time from a sender to a destination.
3. **Normalized routing load:** It measures the number of routing packets transmitted per distinct data packet delivered to a destination. The routing overhead is an important metric for comparing these protocols as it measures the scalability of a protocol, and its efficiency in terms of throughput and power consumption.
4. **Packet duplication:** It measures the average number of duplicate packets per distinct data packet received by the destinations. A protocol with a high number of duplicate packets can congest the network and waste power in the mobile nodes.

The first three metrics were suggested by the IETF MANET working group for routing protocol evaluation [34], and were also used in Refs. [22,23]. The fourth metric is important to understand the trade-offs between network congestion, power consumption, and delivery ratio for the compared protocols.

2.4 Simulation Results

For this chapter, sensitivity analysis is performed in the simulation study to investigate the effect of various network parameters on the eight routing protocols—AODV, DSR, LAR, GRID, TMNR, GPSR, CBF (street version), and CLA (street version). We present our simulation results in this section.

2.4.1 Effect of Mobile Speed

This study is based on 200 nodes with 20 communication sessions. The simulation is setup with zero pause time to stress the mobility in the network. To understand the effect of mobile speed on performance, we varied the speed of the mobile nodes between 10 m/s (or 22 miles/h) and 25 m/s (or 56 miles/h).

The simulation results are presented in Figures 2.4 through 2.7. There are performance trade-offs in some techniques. Although DSR performs comparably to CLA (street version) and CBF (street version) in terms of end-to-end delay (Figure 2.5) and number of control packets transmitted per data packet (Figure 2.6), DSR does poorly in delivering data to their destination (Figure 2.4). This can be attributed to the fact that DSR needs to rediscover routes more frequently as node mobility increases. Similarly, GRID, TMNR, GPSR, LAR, and AODV have high end-to-end delay and control packet overhead because links break frequently due to high node mobility (Figures 2.5 and 2.6). Under this condition, they need to send more ROUTE DISCOVERY messages. In addition, LAR suffers from inaccurate prediction of the request zone (used to limit the flooding area), which makes flooding the entire network more common.

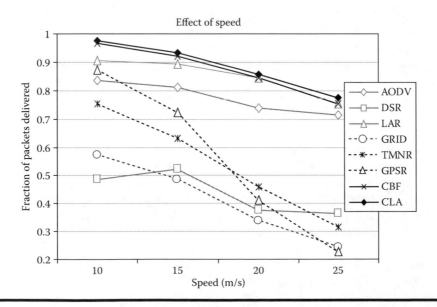

Figure 2.4   Effect of speed on fraction of packets delivered.

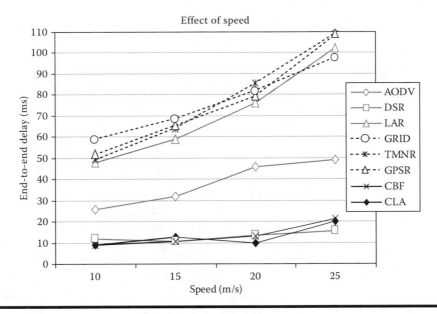

Figure 2.5   Effect of speed on end-to-end delay.

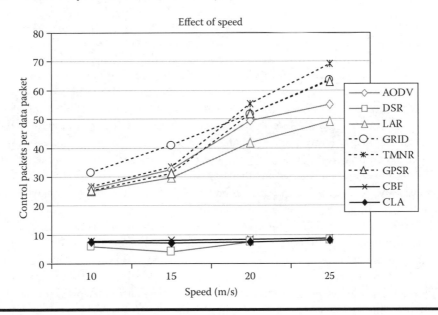

Figure 2.6   Effect of speed on normalized routing load.

In the cases of TMNR and GPSR, the high control overhead is caused by maintaining neighbor information (locations) and high end-to-end delay is caused by the inaccurate (outdated) neighbor information. The inaccurate neighbor information causes TMNR and GPSR to forward to non-existing neighboring nodes. In Figure 2.4, as the mobility increases, the performance of TMNR and GPSR degrades rapidly due to outdated information. Similarly, GRID also has high control overhead caused by maintaining information on the gateway node for each grid (see Figure 2.6). In Figure 2.5, the high end-to-end delay in GRID is not only caused by the inaccurate gateway

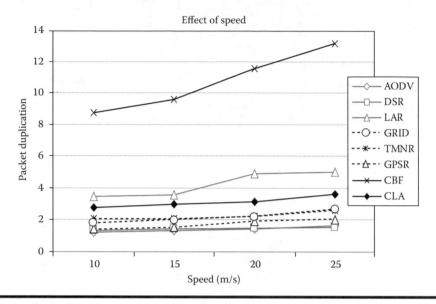

Figure 2.7   Effect of speed on packet duplication.

information but also the fact that only a few selected gateway nodes can forward data. The limited number of forwarding nodes (gateway nodes) causes the network throughput to decrease in Figure 2.4. This demonstrates that the connection-oriented approaches such as multihop based (*e.g.*, AODV, DSR, and LAR), cluster based (*e.g.*, GRID), and one-hop based (*e.g.*, TMNR and GPSR) are not suitable to be used in street environments. For example, it only takes half a second to traverse an intersection of 10 m wide at a speed of 20 m/s. This means that the connection-oriented approaches either drop a large amount of data packets (in the case of DSR in Figure 2.4) or require a large amount of control overhead to keep routes from the sources to the destinations up to date (in the case of AODV and LAR in Figure 2.6), neighboring nodes information (in the case of TMNR and GPSR in Figure 2.6), and cluster membership (in the case of GRID in Figure 2.6).

In contrast, since connection*less*-oriented approaches such as CLA and CBF have no connection to break and maintain, or neighbor information to update, they have low control overhead, short end-to-end delay, and high successful delivery ratio. Between these two techniques, Figure 2.7 shows the number of packet duplication for CBF is three times higher compared to that of CLA. This is due to the fact that CLA only allows nodes in selected grid paths to forward data packets. In addition, CBF has the "fan-out" effect that is similar to the broadcast storm problem [35,36] when forwarding data packets. We note that we did not study the effect of mobility beyond 25 m/s (or 56 miles/hour) because the performance comparisons can be extrapolated from the trends in the performance behavior.

2.4.2 Effect of Pause Time

In this study, we fixed the number of nodes at 200, their speed at 20 m/s, the number of communication sessions at 20, and varied the pause time between 0 and 600 seconds to investigate its effect on performance.

The simulation results are plotted in Figures 2.8 through 2.11. We note that as the pause time becomes very long, communication connections are less likely to break and most protocols display about the same performance. Nevertheless, we observe trade-off in the performance metrics

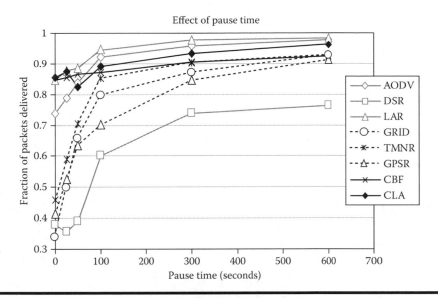

Figure 2.8   Effect of pause time on fraction of packets delivered.

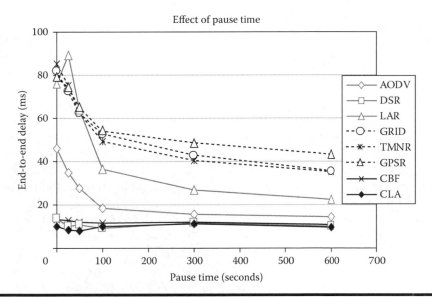

Figure 2.9   Effect of pause time on end-to-end delay.

among different protocols as the pause time is shorter, that is, higher mobility. We discuss these conditions as follows.

In Figure 2.8, AODV has very high fractions of packet delivered under short pause time. This is due to the fact that AODV periodically maintains a local routing table in each node, and a data package can be dynamically rerouted to a new next hop if the current next hop has moved away. This helps to reduce the number of lost packets. This strategy, however, incurs a high number of control packets per data packet due to the maintenance of the local routing tables, as seen in Figure 2.10.

Figure 2.9 indicates that DSR performs well in terms of end-to-end delay. This is, however, due to the fact that DSR takes a relatively longer time to establish a route. Longer routes take

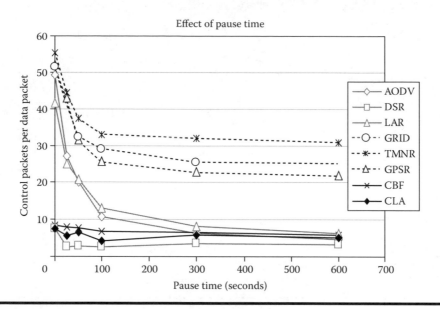

Figure 2.10   Effect of pause time on normalized routing load.

too long to connect and many of them become broken soon after they are established under high mobility (*i.e.*, short pause time). Consequently, we observed mostly short routes, two to three hops, in our simulation study with small end-to-end delay. This also explains the low fractions of packets delivered in DSR because many packets delivered over long routes are lost (see Figure 2.8). DSR also has low control overhead according to Figure 2.10. Nevertheless, this is due to the high percentage of packet loss (see Figure 2.8) and we do not take into account the control packets for these lost packets. For LAR, frequently flooding of the entire network is caused by inaccurate prediction of the request zone due to short pause time. Flooding the entire network will cause high end-to-end delay (see Figure 2.9) and a high number of packet duplication (see Figure 2.11). As the pause time increases, LAR performs better due to the more accurate predication of the request zone.

For GRID, the periodic update of a gateway node is required to notify its existence to the other nodes in its grid even if it stays at the same location. Therefore, the number of control packets stays the same after the pause time reaches 100 seconds in Figure 2.10. Since only gateway nodes can forward a data packet, the performance of end-to-end delay also stays the same after the pause time reaches 100 seconds in Figure 2.9. Similar to GRID, TMNR and GPSR have basic maintenance cost (*i.e.*, control overhead) associated with periodic update of neighbor information in Figure 2.10. Compared to GPSR, TMNR has higher control overhead caused by maintaining additional routing table information. This routing table is used when the destination is near.

From Figures 2.8 through 2.10, we see that the performance curves of CLA are essentially flat. CBF has similar performance in terms of fraction of packet delivered, end-to-end delay, and normalized routing load. This indicates that both CBF and CLA are unaffected by node mobility (*i.e.*, speed or pause time). This means that CBF and CLA are very robust and suitable for a wide range of mobile applications. In terms of the number of duplicate packets received by destinations, most of the routing protocols have on average two packets, except for CBF and CLA (Figure 2.11). Between these two, CBF has three times more packet duplication (10–12 duplicate packets) compared to CLA.

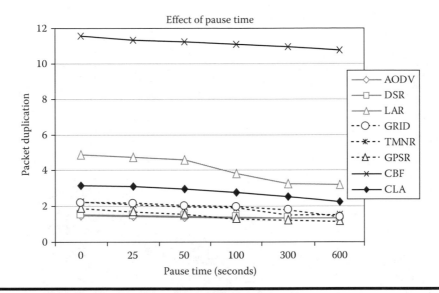

Figure 2.11   Effect of pause time on packet duplication.

2.4.3 Effect on Number of Communication Sessions

In this study, we performed sensitivity analysis with respect to the number of communication sessions. We ran our simulation with speed fixed at 20 m/s, pause time at 0, and number of nodes at 200. We varied the number of communication sessions between 5 and 40.

The simulation results are shown in Figures 2.12 through 2.15. Again, they show the trade-off among fractions of packets delivered, end-to-end delay, control overhead, and packet duplication. In Figures 2.12 and 2.14, AODV and LAR achieve high packet-delivered ratio with the cost of

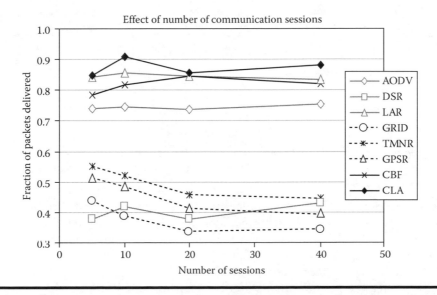

Figure 2.12   Effect of number of communication sessions on fraction of packets delivered.

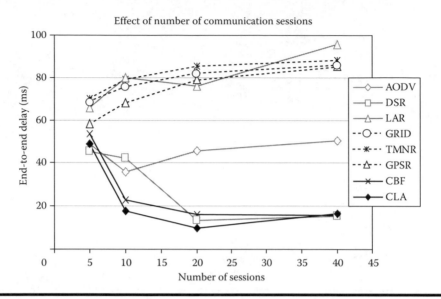

Figure 2.13   Effect of number of communication sessions on end-to-end delay.

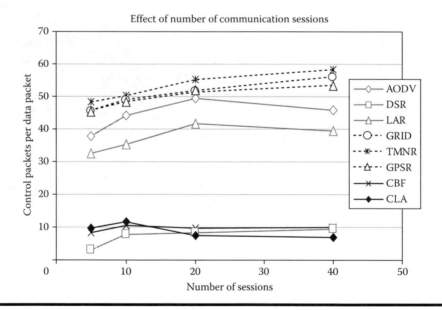

Figure 2.14   Effect of number of communication sessions on normalized routing load.

high control overhead. For AODV, each node periodically maintains the state of the routing table. This is the reason that AODV has high overhead in Figure 2.14. For LAR, the overhead cost comes from maintaining and updating the request zone and expected zone in Figure 2.14. An inaccurate request zone causes an LAR reissue route request by flooding the network. As the result, LAR has higher end-to-end delay and higher control overhead.

From the simulation, we notice that DSR does not perform well when mobility reaches above 20 m/s (see Figures 2.12 through 2.14). In DSR, only source nodes maintain the route. When a

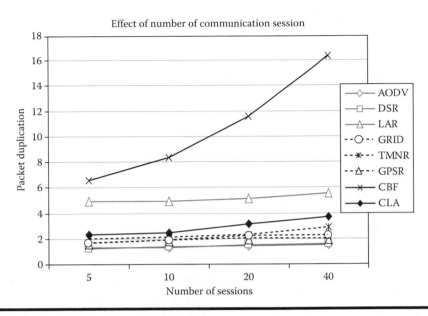

Figure 2.15   Effect of number of communication sessions on packet duplication.

route breaks, the source node will attempt to use any other route that it happens to know about or issues another route request to find a new route. However, with mobility that reaches above 20 m/s, a source node cannot robustly adapt to the changes of topology due to high mobility.

For GRID, TMNR, and GPSR, the low fractions of packet-delivered ratio and high-control overhead are the result of outdated information. For GRID, this out-of-date information is the gateway nodes in each of the grid cells. For TMNR and GPSR, this out-of-date information is the neighboring nodes of each node. And since only gateway nodes will forward the data packets, the gateway nodes become bottleneck due to unbalance workload in GRID.

Only CBF and the CLA can robustly adapt to the changes in the number of communication sessions to maintain good performance regardless of the network conditions. By robust, we mean that both CBF and CLA achieved high successful delivered rate (see Figure 2.12), low end-to-end delay (see Figure 2.13), and low control overhead (see Figure 2.14). Between these two, CLA has a significantly lower number of packet duplications compared to CBF. This can be attributed to the fact that CLA limits the forwarding area to a grid path.

2.4.4 Effect of Network Density

In this study, the nodes move constantly at 20 m/s without pausing and each maintains 20 concurrent communication sessions. To examine the effects of network density, the simulation is varied with different numbers of nodes: 50, 100, 150, 200, and 400 nodes.

The results of this study are plotted in Figures 2.16 through 2.19. In terms of fraction of packets delivered in Figure 2.16, AODV and LAR tend to perform the best out of eight routing protocols under a density of 100 nodes and 200 nodes, respectively. However, as the number of nodes increases, the number of control packets per data packet also increases for AODV and LAR (see Figure 2.18).

Again, DSR does not perform well, even if the number of nodes increases. For GRID, the number of gateway nodes is fixed as the number of grids is fixed (*i.e.*, one gateway node per grid).

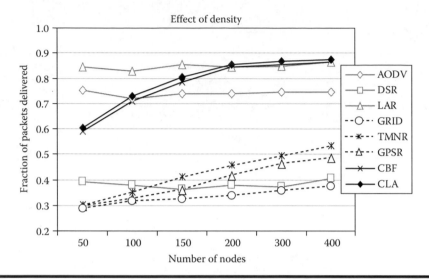

Figure 2.16   Effect of node density on fraction of packets delivered.

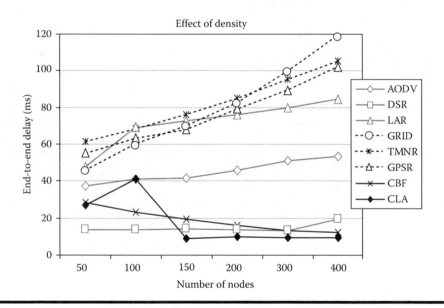

Figure 2.17   Effect of node density on end-to-end delay.

Therefore, since only gateway nodes allow to forward data packets, increasing the number of nodes in the network did not improve the performance of GRID.

For TMNR and GPSR, the increasing number of nodes in the network causes the two protocols to have more neighbor nodes to maintain. As a result, the number of control packets is also increased in Figure 2.18. As density increases, the time to determine which neighboring nodes are closer to the destination/next anchor to be the next hop also increases. This selection process becomes more time consuming as the number of nodes increases. Therefore, end-to-end delay also increases for TMNR and GPSR in Figure 2.17. Since this greedy forwarding approach will choose the next forwarding

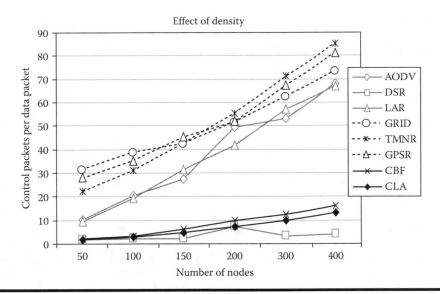

Figure 2.18   Effect of node density on normalized routing load.

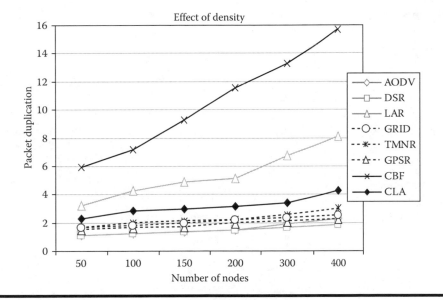

Figure 2.19   Effect of node density on normalized packet duplication.

hop closest to the destination and furthest from the current node, the connection between the current node and the selected next hop will be very weak (*i.e.*, faster out of radio range of each other). Therefore, this causes a low fraction of packets delivered as shown in Figure 2.16.

When the node density is sufficiently high (*i.e.*, 150 nodes or more), the CLA is the only scheme that consistently displays good performance under all four metrics. We note that 150 nodes in a 1000 m × 1000 m terrain or 100 grid cells is still a reasonably practical scenario. We observe that the end-to-end delay curve of CLA behaves irregularly when the node density is very low, that is, 50 and 100. This is due to the fact that data packets that fail to reach their final destination are not taken into account in the computation of the end-to-end delay. As a result, the average end-to-end delay is small because only packets relayed over a few hops make it to the destination. This measure increases when there are 100 nodes in the network because the node density now becomes sufficiently high to support longer hop-by-hop connections. In fact, when the density drops below 50 nodes in a 1000 m × 1000 m field (*i.e.*, 140 m × 140 m per node or three grids per node), TMNR, CBF, and CLA can no longer forward the data packets. If we continue to increase the number of nodes in the network, the CBF and CLA eventually have the option to select the shorter hop-by-hop connections for each packet transmission, resulting in very good end-to-end delay. From this point forward, the performance of the CBF and the connectionless technique (CLA) becomes flat given the fixed terrain dimensions. As the density increases, the number of packet duplications increases rapidly for CBF. Thus, CBF is not scalable for high-density environments.

2.4.5 Effect of Terrain Area (Scalability)

To study whether the techniques under consideration can scale up to facilitate large-area deployment, both the network area and the number of nodes are increased in order to maintain a constant node density (*i.e.*, averaging 70 m × 70 m per node or two nodes per grid cell):

■ 500 m × 500 m area and 50 nodes
■ 1000 m × 1000 m area and 200 nodes
■ 1500 m × 1500 m area and 450 nodes
■ 2000 m × 2000 m area and 800 nodes

In these four simulation runs, we fixed the node speed at 20 m/s, the pause time at 0, and the number of communication sessions at 20.

The results are presented in Figures 2.20 through 2.23. As we increase the network area, the performance of DSR, GRID, TMNR, and GPSR degrade very quickly in the case of the fraction

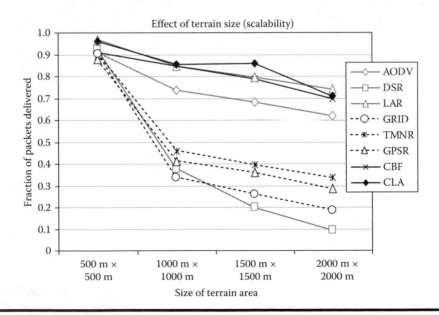

Figure 2.20　Effect of terrain area on fraction of packets delivered.

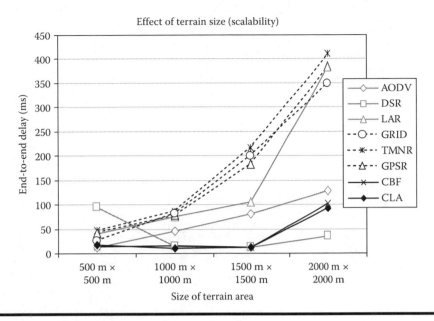

Figure 2.21   Effect of terrain area on end-to-end delay.

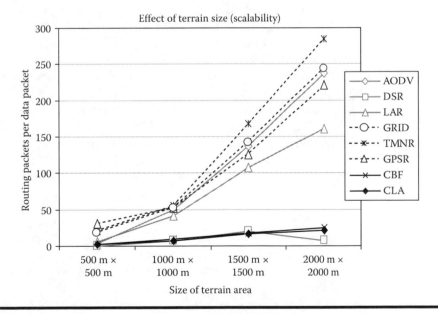

Figure 2.22   Effect of terrain area on normalized routing load.

of data packets delivered successfully. Similarly, LAR, AODV, GRID, TMNR, and GPSR degrade rapidly in the cases of end-to-end delay and control overhead. When scaling up the network in terms of area and number of nodes, the number of hops or the distance of a connection between source and destination becomes longer. Therefore, route maintenance becomes costlier in terms of control packets per data packet for most routing protocols. Also, a longer route has more chance to

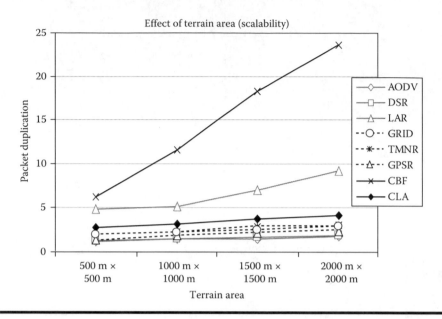

Figure 2.23   Effect of terrain area on packet duplication.

break due to any one of the nodes in a connection failing or being out of reach. Thus, the fraction of packets delivered is also lower as the scale of the network increases for DSR, GRID, TMNR, and GPSR. Although AODV and LAR can achieve a high fraction of packets delivered, both protocols do not perform well in the cases of control overhead and end-to-end delay.

Only the CBF and CLA perform well under three metrics: fraction of packets delivered, end-to-end delay, and normalized routing load. However, we observe that only CLA does not increase the number of packet duplications with the increases in the network area and the number of nodes. Thus, only CLA can scale up to support larger networks. In contrast, CBF degrades rapidly in the case of number of packet duplication when the network area and number of nodes increase.

2.5 Conclusion

This chapter presented an introduction and a comparative study of eight routing protocols for VANETs in street environments with high mobility. The detailed simulation studies, implemented using GloMoSim, allow us to perform fair and accurate comparisons of these techniques with a broad range of network parameters including mobility, pause time, the number of communication sessions, density, and size of terrain.

The performance characteristics of these techniques are summarized in Table 2.1. In this study, it is observed that AODV, DSR, LAR, GRID, TMNR, and GPSR have to make trade-offs between the fraction of packets delivered, the end-to-end delay, and the normalized routing load. Although both CBF and CLA perform well in terms of all three metrics, CBF has a much higher number of packet duplications compared to all of the other protocols. CLA, on the other hand, is not suitable to low-density environments (*i.e.*, below 50 nodes in a 1000 m × 1000 m field).

The performances of all protocols are simulated under high mobility and street environment.

Table 2.2 summarizes the characteristics of the environments suitable for each protocol. Multihop-based approaches, such as DSR, AODV, and LAR, are more suitable for applications

Table 2.1 Performance Trade-Off between Different Approaches

Approaches	Performance Metrics			
	Fraction of Packet Delivered[a]	End-to-End Delay[b]	Normalized Routing Load[c]	Packet Duplication[d]
AODV	Good	Average	**Fair**	Low
DSR	**Fair**	Good	Good	Low
LAR	Good	**Fair**	Fair	Low
GRID	**Fair**	**Fair**	Good	Low
TMNR	Average	**Fair**	Fair	Low
GPSR	Average	**Fair**	Fair	Low
CBF	Good	Good	Good	**High**
CLA	Good	Good	Good	Low

[a] For the fraction of packets delivered, *Good* means the protocol is able to achieve above 75% success rate, *Average* means the protocol is able to achieve an average of 50% success rate, and **Fair** means the protocol is only able to achieve an average of less than 50% success rate.

[b] For the end-to-end delay, *Good* means the protocol has 70% less (shorter) delay than the average, *Average* means the protocol has an average performance compared to other protocols, and **Fair** means the protocol has longer delays than the average.

[c] For the normalized routing load, *Good* means the protocol is able to achieve an average of less than ten control packets per data packet, *Fair* means the protocol has an average of more than 30 control packets per data packet.

[d] For the packet duplication, *Low* means the protocol has an average of less than three duplicated packets received by the destination node, and *High* means the protocol has average of greater than ten duplicated packets received by the destination node.

with low mobility, long pause time, light communication load, low density, and small terrain size. Notice that only AODV and DSR do not require any location information provided by GPS. For the cluster-based approach in general, longer pause time is needed to maintain up-to-date cluster membership. GRID can support moderate node speed because it only maintains the gateway nodes. However, short pause time can cause GRID to constantly reelect the gateway nodes. Thus, GRID adapts well in an environment with a moderate speed and long pause time. The simulation results showed the pause time can have great effect on GRID's performance. However, network throughput is limited by gateway nodes only allowed to forward for the cluster approach. For one-hop-based approaches such as TMNR and GPSR, the need to maintain neighboring information causes the performance to decrease under high mobility. Comparing the two protocols, GPSR can adapt to low density due to geographical anchor locations. Cluster based (*i.e.*, GRID) and one-hop based (*i.e.*, TMNR and GPSR) are suitable for a midrange scale of network such as disaster recovery or sensor network with location service available. For a CLA, both CBF and CLA perform well and are suitable for most environments. However, CBF tends to have a higher number of packet duplication. This leads to media congestion, waste of power, and lower network throughput. Not only can the connectionless-based approach (*i.e.*, CBF and CLA) perform well in a more static environment where nodes move in low speed and have long pause time,

Table 2.2 Applicable Environments of Different Approaches

Approaches	Environment Characteristics				
	Mobility	*Pause time*	*Comm load*	*Density*	*Terrain size*
AODV	*Low*	*Long*	*Low*	*Low*	*Small*
DSR	*Low*	*Long*	*Low*	*Low*	*Small*
LAR	*Low*	*Long*	*Low*	*Low*	*Small*
GRID	*Medium*	*Long*	*Medium*	*Medium*	*Medium*
TMNR	*Medium*	*Medium*	*Medium*	*Medium*	*Medium*
GPSR	*Medium*	*Medium*	*Medium*	*Low*	*Medium*
CBF	*High[a]*	*Short[b]*	*Medium*	*Medium*	*Large*
CLA	*High[a]*	*Short[b]*	*High*	*High*	*Large*

[a] High mobility means the protocols are able to adapt to high mobility as well as low mobility.

[b] Short pause time means the protocols are able to adapt to short pause times as well as long pause times.

Note that such a property (*i.e.*, ability to adapt to high mobility and also able to adapt to low mobility) is similar with other environment characteristics.

but it can also adapt to large terrain, high node density, and nodes moving at a high speed with short pause time, such as a vehicle ad hoc network. Between CBF and CLA, CLA tends to have better communication load efficiency and lower end-to-end delay which is ideal for voice and video applications. While the results of this chapter can provide guidelines, the final selection of a routing protocol should also take into account considerations specific to a given application.

References

1. C.E. Perkins and E.M. Royer, Ad Hoc On Demand Distance Vector (AODV) Routing, *IETF Inter Draft, draft-ietf-manet-aodv-09.txt*, November 2001.
2. F. Kuhn, R. Wattenhofer, and A. Zollinger. Asymptotically Optimal Geometric Mobile Ad-Hoc Routing, *Proceedings Dial-M '02*, pp. 24–33, Atlanta, GA, September 2002.
3. C.C. Chiang, H.K. Wu, W. Liu, and M. Gerla, Routing in Clustered Multihop, Mobile Wireless Networks with Fading Channel, *Proceedings of IEEE SICON'97*, pp. 197–211, Singapore, April 14–17, 1997.
4. Y.H. Ho, A.H. Ho, and K.A. Hua, Routing Protocols for Inter-Vehicular Networks: A Comparative Study in High-Mobility and Large Obstacles Environments. *Computer Communications Journal (ComCom)*, Elsevier, Vol. 31, no. 12, pp. 2767–2780, April 17–18, 2008.
5. H. Fubler, J. Widmer, M. Kasemann, M. Mauve, and H. Hartenstein. Contention-Based Forwarding for Mobile Ad-Hoc Networks. *Elsevier's Ad-Hoc Networks*. Vol. 1, no. 4, pp. 351–369, 2003.
6. S. Basagni, I. Chlamtac, V.R. Syrotiuk, and B.A. Woodward. A Distanced Routing Effect Algorithm for Mobility (DREAM), *Proceedings of the Fourth Annual ACM/IEEE International Conference on Mobile Computing and Networking*. pp. 76–84, Dallas, TX, October 25–30, 1998.
7. C.E. Perkins and E.M. Royer, Highly Dynamic Destination-Sequenced Distance-Vector Routing (DSDV) for Mobile Computer, *Computer Communications Review*, Vol. 24, no. 4, pp. 234–244, October 1994.

8. D.B. Johnson and D.A. Maltz. Dynamic source routing in ad hoc wireless networks. In Mobile Computing, T. Imielinski and H. Korth, Eds. Kluwer Academic Publishers, 1996, ch. 5, pp. 153–181.

9. B. Karp and H. T. Kung. GPSR: Greedy Perimeter Stateless Routing for Wireless Networks, *Proceedings of MOBICOM '00*, pp. 243–254, Boston, MA, August 2000.

10. W.-H. Liao, Y.-C. Tseng, and J.-P. Sheu, GRID: A Fully Location-Aware Routing Protocol for Mobile Ad Hoc Network, *Telecommunication System*, Vol. 18, pp. 37–60, 2001.

11. T. Chan and M. Gerla, Global State Routing: A New Routing Scheme for Ad Hoc Wireless Networks. *IEEE International Conference on Communications (ICC)*. pp. 171–175, Atlanta, GA, June 7–11, 1998.

12. Y.B. Ko and N.H. Vaidy, Location-Aided Routing (LAR) in Mobile Ad Hoc Network, *Proceedings of ACM/IEEE MOBICOM '98*, pp. 66–75, Dallas, TX, October 1998.

13. S.J. Lee, W. Su, and M. Gerla, On-Demand Multicast Routing Protocol in Multihop Wireless Mobile Network, Mobile Networks and Application '02, pp. 441–453, December 2002.

14. D. Niculescu and B. Nath, Trajectory Based Forwarding and Its Applications, *Proceedings of the ACM/ IEEE MOBICOM '03*, pp. 260–272, San Diego, CA, September 2003.

15. S. De, S.K. Das, H. Wu, and C. Qiao, Trigger-Based Distributed QoS Routing in Mobile Ad Hoc Networks, *ACM SIGMOBILE Mobile Computing and Communications Review*, Vol. 6, Issue 3, pp. 22–35, June 2002.

16. L. Blazevic, J. Le Boudec, and S. Giordano. A Location-Based Routing Method for Mobile Ad Hoc Networks. *IEEE Transactions on Mobile Computing*. Vol. 4, No. 1. pp. 97–110, March 2005.

17. S. Murthy and J.J. Garcia-Luna-Aceves, An Efficient Routing Protocol for Wireless Network, *Mobile Networks and Applications*. Vol. 1, pp. 183–197, October 1996.

18. Z. Hass and M. Pearlman, Performance of Query Control Schemes for the Zone Routing Protocol, *ACM SIGCOMM Computer Communication Review*, Vol. 28, no. 4, pp. 167–177, 1998.

19. M. Heissenbuttel, T. Braun, T. Bernoulli and M. Wälchli, BLR: Beacon-Less Routing Algorithm for Mobile Ad-Hoc Networks, *Computer Communications Journal*, vol. 27, no. 11, 1, pp. 1076–1086, 2004.

20. C. Lochert, H. Hartenstein, J. Tian, D. Herrmann, H. Füßler, and M. Mauve, Routing Strategy for Vehicular Ad Hoc Networks in City Environments, *Proceedings of IEEE Intelligent Vehicles Symposium (IV2003)*, pp. 156–161, Columbus, OH, June 2003.

21. A.H. Ho, Y.H. Ho, and K.A. Hua, A Connectionless Approach to Mobile Ad Hoc Network in Street Environments, *Proceedings of IEEE Intelligent Vehicles Symposium* (IV 2005), Nevada, June 2005.

22. J. Broch, D.A. Maltz, D.B. Johnson, Y.-C. Hu, and J. Jetcheva. A Performance Comparison of Multi-Hop Wireless Ad Hoc Network Routing Protocols, *Proceedings of the Fourth Annual ACM/IEEE International Conference on Mobile Computing and Network (MOBICOM)*, pp. 85–97, Dallas, TX, October 1998.

23. S.R. Das, C.E. Perkins, and E.M. Royer, Performance Comparison of Two On-Demand Routing Protocols for Ad Hoc Network, *Proceedings of Nineteenth Annual Joint Conference of IEEE Computer and Communications Societies* (INFOCOM), Vol. 1, pp. 3–12, Tel-Aviv, Israel, March 2000.

24. P. Johansson, T. Learsson, N. Nedman, B. Mielczarek, and M. Degermark. Scenario-Based Performance Analysis of Routing Protocols for Mobile Ad Hoc Networks, *Proceedings of the ACM/ IEEE International Conference on Mobile Computing and Networking (MOBICOM)*, pp. 195–206, August 1999.

25. ISI. The Network Simulator—ns-2. http://www.isi.edu/nsnam/ns/

26. J. Hsu, S. Bhatia, K. Tang, R. Bagrodia, and M.J. Acriche, Performance of Mobile Ad Hoc Network Routing Protocols in Large Scale Scenarios, *Proceedings of the IEEE Military Communications Conference (MILCOM)*, pp. 21–27, Monterey, CA, October 31–November 03, 2004.

27. C.A. Santivanez, B. McDonald, I. Stavrakakis, and R. Ramanthan, On the Scalability of Ad Hoc Routing Protocols, *Proceedings of the IEEE INFOCO*, 2002.

28. S.-C.M. Woo, and S. Singh, Scalable Routing Protocol for Ad Hoc Network, *Wireless Network*, Vol. 7, no. 5, pp. 513–529, 2001.

29. X. Zhang and G.F. Riley. Scalability of Ad Hoc On-Demand Routing Protocol in Very Large-Scale Mobile Wireless Networks, *Simulation: Transactions of the Society of Modeling and Simulation International*, 2006.

30. G. Dommety and R. Jain, Potential Networking Applications of Global Positioning System (GPS), Tech. Rep. TR-24, CS Dept., The Ohio State University, April 1996.
31. J. Li, J. Jannotti, D.S.J. De Couto, D.R. Karger, and R. Morris. A Scalable Location Service for Geographic Ad Hoc Routing, *Proceedings of the ACM/IEEE MOBICOM*, pp. 120–130, Boston, MA, August 2000.
32. A. Savvides, C.-C. Han, and M. Srivastava, Dynamic Fine-Grained Localization in Ad-Hoc Networks of Sensors, *MOBICOM '01*, pp. 166–179, July 2001.
33. X. Zeng, R. Bagrodia, and M. Gerla, GloMoSim: A Library for Parallel Simulation of Large-Scale Wireless Network, *Proceedings of the Twelfth Workshop on Parallel and Distributed Simulation*, pp. 154–161, Banff, Alberta, Canada, May 1998.
34. M.S. Corson and J. Macker, Mobile Ad Hoc Networking (MANET): Routing Protocol Performance Issues and Evaluation Consideration. *Request for Comments 2501*, Internet Engineering Task Force, January 1999.
35. S. Ni, Y. Tseng, Y. Chen, and J. Sheu. The Broadcast Storm Problem in a Mobile Ad Hoc Network, *Proceedings of the ACM/IEEE International Conference on Mobile Computing and Networking (MOBICOM)*, pp. 151–162, Seattle, WA, 1999.
36. B. Williams, and T. Camp, Comparison of Broadcasting Techniques for Mobile Ad Hoc Networks, *Proceedings of the ACM International Symposium on Mobile Ad Hoc Networking and Computing (MOBIHOC)*, pp. 194–205, Lausanne, Switzerland, 2002.
37. Y.H. Ho, A.H. Ho, K.A. Hua, and G.L. Hamza-Lup, A Connectionless Approach to Mobile Ad hoc Networks, *Proceedings of Ninth International Symposium on Computers and Communications (ISCC)*, Vol 1, pp. 188–195, Alexandria, Egypt, 2004.

Chapter 3

Cooperative Data Scheduling via V2V/V2I Communications in Software-Defined Vehicular Networks

Kai Liu
Chongqing University
Chongqing, China

Victor C. S. Lee
City University of Hong Kong
Kowloon Tong, Hong Kong, China

Sang H. Son
Daegu Gyeongbuk Institute of Science and Technology (DGIST)
Daegu, Korea

Joseph K. Y. Ng
Hong Kong Baptist University
Kowloon Tong, Hong Kong, China

Jiannong Cao
Hong Kong Polytechnic University
Hung Hom, Hong Kong, China

Hongwei Zhang
Department of Electrical and Computer Engineering
Iowa State University
Ames, Iowa

Contents

3.1 Introduction

With recent advances in computing, communication, control, and sensing, vehicular networks are envisioned as a promising paradigm to achieve breakthroughs in transportation safety, efficiency, and sustainability. In particular, efficient data dissemination in vehicular networks is a key enabler for implementing future innovative intelligent transportation systems (ITS), such as intersection control [1], platooning [2], and vehicular localization [3]. However, intrinsic characteristics of vehicular networks including highly dynamic network topology, sparse distribution of infrastructures, high mobility of vehicles, real-time traffic information, and heterogeneous wireless communication environments, make it challenging to optimize the utilization of network resources to improve service quality and enhance system scalability.

The dedicated short-range communication (DSRC) is being standardized as a *de facto* protocol in vehicular networks to support both vehicle-to-infrastructure (V2I) and vehicle-to-vehicle (V2V) communications [4]. A great number of studies have investigated data dissemination via V2I communication in the coverage of roadside units (RSUs) [5–9]. However, the data service in such an architecture suffers from intermittent connections due to the short V2I communication range and the sparse deployment of RSUs. On the contrary, since vehicles with onboard units (OBUs) can also communicate with each other, many studies have focused on distributed data dissemination via V2V communications [10–14] to enable large-scale information services in vehicular networks. Nevertheless, existing solutions can hardly best exploit the joint effect of V2V/V2I communications by maximizing the efficiency of limited wireless bandwidth in the heterogeneous vehicular communication environment.

In this chapter, we present a service architecture that exercises a software-defined network (SDN) concept [15] to enable efficient data services in vehicular networks by exploiting the synergy between V2V and V2I communications. SDN is an emergent paradigm, which brings flexibility and programmability to networks. The core idea of SDN is the separation of the control plane and the data plane. The SDN controller, which has a global network interconnection knowledge, formulates data broadcasting/forwarding rules for network nodes such as RSUs, base stations (BSs), and vehicles via the control plane. As a result, the SDN controller can define the behavior of individual vehicles, RSUs, and BSs to make scheduling decisions based on the global view, targeted at maximizing the overall system performance. Recently, the concept of software-defined vehicular networks (SDVN) has been proposed and discussed in a number of studies [16–19]. Nevertheless, most of them focused on the design of system architectures, while how to fulfill efficient data services based on specific application requirements on top of SDVN is still an open problem.

With the above motivations, in this chapter we present a data service architecture in SDVN. On this basis, we consider two typical application scenarios and investigate efficient data scheduling policies by exploring the cooperation of V2V/V2I communications.

- We investigate a real-time data service scenario in SDVN, where the requested information has to be retrieved by vehicles via the hybrid of V2V/V2I communications within certain time constraints. Specifically, we present a service architecture to enable collaborative V2V/V2I data dissemination. Then, we formulate the problem of cooperative data scheduling (CDS) and prove that it is NP-hard. Further, we propose a heuristic online scheduling algorithm for realizing efficient data dissemination in the concerned scenario. Finally, we give a performance evaluation to show the superiority of the algorithm.

- We investigate a delay-tolerant data service scenario in SDVN, where the information can be jointly provided by heterogeneous wireless communication interfaces such as DSRC and cellular networks. Specifically, we present a service architecture by incorporating vehicular caching and network coding into the scheduler at the control plane. Then, we formulate the problem of coding-assisted broadcast scheduling (CBS) to maximize the bandwidth efficiency and prove that it is NP-hard. Further, inspired by the superiority of memetic computing [20] in solving complex problems, we propose a memetic algorithm for efficiently tackling the formulated data dissemination problem in SDVN. Finally, we give a performance evaluation to show the superiority of the algorithm.

The remainder of this chapter is organized as follows. In Section 3.2, we present a data service architecture in SDVNs. In Section 3.3, we explore a real-time service scenario in SDVN by incorporating V2V/V2I communications. We formulate a CDS problem and prove that it is NP-hard. Then, a heuristic online scheduling algorithm is presented and evaluated. In Section 3.4, we explore a delay-tolerant service scenario in SDVN by exploiting vehicular caching and network coding for data dissemination in heterogeneous vehicular communication environments. We formulate a CBS problem and prove that it is NP-hard. Then, an online scheduling algorithm via memetic computing is presented and evaluated. Finally, Section 3.5 summarizes this chapter.

3.2 Software-Defined Vehicular Network

Figure 3.1 shows a general framework of the SDVN [17]. Typically, RSUs, BSs, vehicles, and other wireless devices are abstracted as switches in conventional SDN, representing the data plane, while the scheduling decisions such as the bandwidth allocation, routing protocol, and data scheduling are exercised by the control plane. Based on the received control messages, switches such as RSUs and BSs will operate accordingly. On the contrary, considering special characteristics in vehicular networks, such as limited wireless bandwidth, highly dynamic topology, high mobility of vehicles, heterogeneous communication interfaces, and large-scale services, distributed infrastructures such as RSUs may also exercise certain intelligence and make part of scheduling decisions individually in such an architecture to enhance system scalability.

In SDVN, vehicles ask for a variety of information of interest such as traffic conditions, parking slots, or gas stations, which can be jointly provided by the heterogeneous wireless interfaces such as DSRC, 4G cellular networks, or Wi-Fi APs. Different infrastructures can maintain their local databases to provide location-based services. Alternatively, they can also retrieve information from

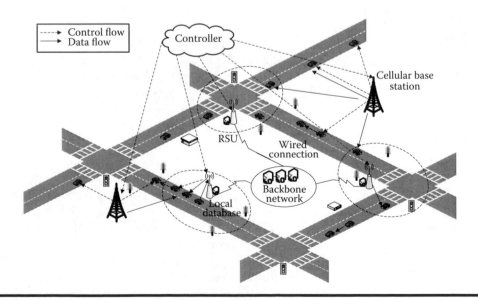

Figure 3.1   Software-defined vehicular networks.

the backbone network via wired connections to provide broader information services. Vehicles in the service range of corresponding infrastructures can retrieve their requested information via V2I communication. On the contrary, due to the limited communication range and sparse deployment of infrastructures such as RSUs, it is desirable to explore the heterogeneous wireless communication characteristic as well as the V2V communication capacity to improve the service efficiency and enhance the system scalability.

With the above motivation, the presented SDVN is expected to better support the cooperation of heterogeneous wireless communications and better explore the potential of cooperative V2V/V2I communications. Specifically, the utilization of heterogeneous network resources is expected to be enhanced via logical centralized scheduling by the control plane. For example, considering the distinguishing characteristics of different wireless interfaces, including cost, data rate, and communication range, the control plane is able to make scheduling decisions based on global information of traffic conditions, resource status, and application requirements, so as to maximize overall system performance. On the contrary, with the presented SDVN, it is promising to enhance both the temporal and spatial reusability of the wireless bandwidth for cooperative V2V/V2I communications by considering vehicular communication characteristics such as interference of concurrent V2V communication, switching between V2V and V2I modes, and half-duplex transmission of OBUs. Two detailed data dissemination problems based on such an architecture and corresponding solutions will be presented in the following.

3.3 Real-Time Services via Cooperative V2V/V2I Communications

3.3.1 *System Architecture*

In this section, we present a system architecture for real-time services in SDVN via a hybrid of V2V/V2I communications. As shown in Figure 3.2, we consider real-time services provided by the RSU installed along the road. Specifically, we define the *service region* of the system, which is the V2I communication coverage of the RSU plus a V2V communication distance (allowing vehicles which have left the RSU's coverage to have the chance for retrieving data items via V2V communication by their neighbors which can still receive instructions from the RSU). Note that only one-hop V2V communication is considered in the current framework for assisting V2I-based data dissemination, while the routing problems via multihop V2V communications will be considered in our future work. The requested data items have to be retrieved by vehicles within the service region. Otherwise, the service is failed. Application scenarios include intersection control, speed advisory, and location-based services, where information has to be disseminated to vehicles around the RSU.

We consider one control channel and two service channels in the system. The control channel is used for disseminating management information, service advertisements, and control messages, by which the system is able to monitor real-time locations and velocities of vehicles which are in the coverage of the RSU. One of the service channels is used for disseminating data items via V2I communication, while the other one is used for disseminating data items via V2V communication. The single radio OBUs are considered as they are commonly adopted in vehicular networks due to both deployment and economic concerns. Therefore, vehicles can tune in to only one of the channels at a time. The time unit adopted in this section refers to a *scheduling period*, which consists of three phases as introduced below.

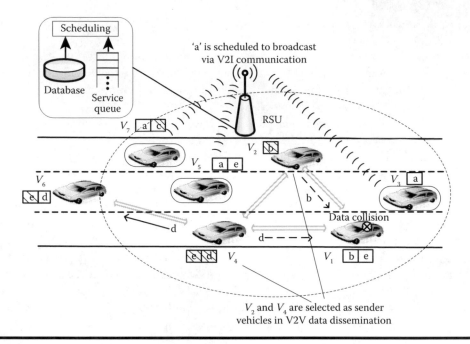

Figure 3.2   Real-time services via cooperative V2V/V2I communications in SDVN.

In the first phase, all the vehicles are set to the V2V mode and broadcast their heartbeat messages (*i.e.*, the *basic safety message* as defined in SAE J2735 [4]), so that each vehicle is able to identify a list of its neighboring vehicles. For instance, by measuring the signal-to-noise ratio through the heartbeat messages received from other vehicles, a vehicle can recognize a set of vehicles, to which it can transmit and receive data items. Meanwhile, service status (*e.g.*, cached information and outstanding requests) and kinetic information (*e.g.*, velocities and accelerations) of neighboring vehicles are shared in this phase.

In the second phase, all the vehicles switch to the V2I mode and communicate with the RSU. Specifically, each vehicle informs the RSU with its updated information, including the list of its current neighbors, the identifiers of the retrieved and newly requested data items, and real-time kinetic information. Note that the vehicles in the RSU's coverage also upload the service status and kinetic information of their neighbors, so that the system can still monitor the status of those vehicles which are outside the RSU's coverage but still in the service region. This information is piggybacked into the probe vehicle message as defined in SAE J2735 [4]. Each request is made only for one data item, and the request is satisfied as long as the corresponding data item is retrieved via either V2I or V2V communication. Outstanding requests are pended in the service queue. According to a certain algorithm, the scheduling decisions are announced via the control channel.

In the third phase, each vehicle participates in either V2I or V2V communication based on the scheduling decisions. Multiple instances of data dissemination may take place simultaneously in this phase. Specifically, some vehicles will be instructed to tune in to the V2I mode and retrieve the data item transmitted from the RSU, while some others will be instructed to tune in to the V2V mode for data transmission or retrieval.

To enable a collaborative V2I and V2V data dissemination, the algorithm is expected to make the following scheduling decisions. First, the algorithm divides the vehicles into two groups. One group is for V2I communication, while the other is for V2V communication. Second, the algorithm selects one data item to be transmitted from the RSU, so that vehicles in the V2I group can retrieve this data item via the V2I service channel. Third, for vehicles in the V2V group, the algorithm determines a set of sender vehicles and the corresponding data items disseminated by each sender vehicle, so that the neighbors of each sender vehicle may have chance to retrieve their requested data items via the V2V service channel. The vehicles are assumed to stay in the same neighborhood for a short period of time (*i.e.*, during a scheduling period) [14].

Figure 3.2 illustrates a simple example: Vehicles in the V2I group can retrieve the data item only when they are in the RSU's coverage, which is represented by the ellipse. In this example, three vehicles are designated into the V2I group (*i.e.*, V_3, V_5, and V_7). The data block without shadow represents that the corresponding data item has been requested by the vehicle but has not yet been retrieved. In contrast, the shadowed data block means that the corresponding data item has been retrieved and cached. Accordingly, when a is broadcast from the RSU, V_3, V_5, and V_7 can retrieve it via the V2I service channel. In V2V communication, the double-edge arrow represents that the two vehicles are neighbors. The data item to be disseminated by a sender vehicle has to be cached in advance. For instance, V_4 has cached e and d, and it could be selected as a sender for disseminating d with the target receiver of V_6. Similarly, V_2 has cached b, and it could be selected as another sender for disseminating b to serve V_1. However, due to the nature of broadcast in wireless communications, simultaneously disseminating data items from the vehicles which are in the immediate or adjacent neighborhoods will lead to data collision [14]. In this example, V_1 is in the neighborhood of both V_2 and V_4. Therefore, data collision happens at V_1 when V_2 and V_4 are disseminating data items at the same time. Thus, V_1 cannot receive b from V_2 because of the interference caused by V_4. Note that for V_3, V_5, and V_7, as they are tuned in to the V2I service channel, they will not be influenced by V2V communication.

3.3.2 Cooperative Data Scheduling Problem

3.3.2.1 Problem Formulation

In this section, we formally describe the aforementioned CDS problem. The database $D = \{d_1, d_2, ..., d_{|D|}\}$ consists of $|D|$ data items. The set of vehicles is denoted by $V(t) = \{V_1, V_2,...,V_{|V(t)|}\}$, where $|V(t)|$ is the total number of vehicles at time t. Note that the unit of time t is the scheduling period defined in Section 3.3.1. Depending on the communication mode of vehicles, $V(t)$ is divided into two sets: $V_I(t)$ and $V_V(t)$, where $V_I(t)$ represents the set of vehicles in the V2I mode and $V_V(t)$ represents the set of vehicles in the V2V mode. Each vehicle stays in either V2I or V2V mode at a time, namely, $V_I(t) \cap V_V(t) = \phi$ and $V_I(t) \cup V_V(t) = V(t)$.

Each V_i ($1 \le i \le |V(t)|$) has a set of requests, which is denoted by $Q_{V_i}(t) = \{q_{V_i}^1, q_{V_i}^2,...q_{V_i}^{|Q_{V_i}(t)|}\}$, where $|Q_{V_i}(t)|$ is the total number of requests submitted by V_i at time t. Each $q_{V_i}^j$ ($1 \le j \le |Q_{V_i}(t)|$) corresponds to a data item in the database, and it is satisfied once this data item is retrieved by V_i. According to the service status of requests (*i.e.*, satisfied or not), $Q_{V_i}(t)$ is divided into two sets: $SQ_{V_i}(t)$ and $PQ_{V_i}(t)$, where $SQ_{V_i}(t)$ represents the set of satisfied requests, while $PQ_{V_i}(t)$ represents the set of pending requests. Then, we have $SQ_{V_i}(t) \cap PQ_{V_i}(t) = \phi$ and $SQ_{V_i}(t) \cup PQ_{V_i}(t) = Q_{V_i}(t)$. Since each request $q_{V_i}^j$ corresponds to a data item d_k, without causing ambiguity, the expression $d_k \in SQ_{V_i}(t)$ is adopted to represent that d_k is requested by V_i and it has been retrieved (*i.e.*, $q_{V_i}^j$ has been satisfied).

For each V_i in the V2V mode, the set of its neighboring vehicles (*i.e.*, the set of vehicles in the V2V mode and within the V2V communication range of V_i) is denoted by $N_{V_i}(t)$, where $N_{V_i}(t) \subset V_V(t)$. The RSU maintains an entry in the service queue for each V_i, which is characterized by a three-tuple: $\langle V_i, Q_{V_i}(t), N_{V_i}(t) \rangle$. The values of $Q_{V_i}(t)$ and $N_{V_i}(t)$ are updated in every scheduling period. To facilitate the formulation of CDS, relevant concepts are defined as follows.

In V2I communication, the RSU broadcasts one data item in each scheduling period, which is denoted by $d_I(t)$, where $d_I(t) \in D$. Denote $V_{RSU}(t)$ as the set of vehicles within the RSU's coverage. Only when $V_i \in V_{RSU}(t)$, can it retrieve $d_I(t)$ via the V2I service channel. Specifically, the receiver vehicle set in V2I communication is defined as follows.

Definition 1: Receiver vehicle set in V2I communication.

Given the data item $d_I(t)$ transmitted from the RSU, the set of receiver vehicles for $d_I(t)$, denoted by $RV(d_I(t))$, consists of any vehicle V_i, which satisfies the following conditions: (a) V_i is in the RSU's coverage, (b) V_i is in the V2I mode, and (c) $d_I(t)$ is requested by V_i and it has not yet been retrieved. That is:

$$RV(d_I(t)) = \{V_i \mid V_i \in V_{RSU}(t) \wedge V_i \in V_I(t) \wedge d_I(t) \in PQ_{V_i}(t)\} \tag{3.1}$$

In V2V communication, a set of sender vehicles is designated to disseminate data items, which is denoted by $SV(t) = \{SV_1, SV_2, ..., SV_{|SV(t)|}\}$, where $|SV(t)|$ is the number of designated sender vehicles. All sender vehicles are in the V2V mode. That is, $SV(t) \subseteq V_V(t)$. The set of data items to be disseminated is denoted by $D(SV(t)) = \{d(SV_1), d(SV_2), ..., d(SV_{|SV(t)|})\}$, where $d(SV_i)$ $(1 \leq i \leq |SV(t)|)$ is the data item disseminated by SV_i. Note that $d(SV_i)$ has to be retrieved by SV_i in advance, namely, $d(SV_i) \in SQ_{SV_i}(t)$. Due to the broadcast effect, simultaneous data dissemination of multiple sender vehicles may cause data collision at receivers. Specifically, the set of receiver vehicles suffering from data collision is defined as follows.

Definition 2: Receiver vehicle set suffering from data collision.

Given the set of sender vehicles $SV(t)$, for any V_k in the V2V mode, if V_k is in the neighborhood of both SV_i and SV_j ($SV_i, SV_j \in SV(t)$), then data collision happens at V_k. Accordingly, the receiver vehicle set suffering from data collision is represented by $\{V_k \mid V_k \in V_V(t) \wedge V_k \in N_{SV_i}(t) \wedge V_k \in N_{SV_j}(t)\}$ ($\forall SV_i, SV_j \in SV(t)$).

Considering data collision, given a sender vehicle SV_i with the transmitted data item $d(SV_i)$, the set of receiver vehicles for $d(SV_i)$ is defined as follows.

Definition 3: Receiver vehicle set for d(SV$_i$).

The receiver vehicle set for $d(SV_i)$, denoted by $RV(d(SV_i))$, consists of any vehicle V_j, which satisfies the following four conditions: (1) V_j is in the neighborhood of SV_i, (2) $d(SV_i)$ is requested by V_j but it has not yet been retrieved, (3) V_j is not in the sender vehicle set, and (4) V_j is not in the neighborhood of any other sender vehicles excepting for SV_i. That is:

$$RV(d(SV_i)) = \{V_j \mid V_j \in N_{SV_i}(t) \wedge d(SV_i) \in PQ_{V_j}(t) \wedge$$

$$V_j \notin SV(t) \wedge V_j \notin N_{SV_k}(t), \forall SV_k \notin \{SV(t) - SV_i\}\} \tag{3.2}$$

The first two conditions are straightforward. The third condition means that a vehicle cannot be the sender and the receiver at the same time. The fourth condition guarantees that no data collision happens at the receiver. On this basis, given the set of sender vehicles $SV(t)$ with the corresponding data items $D(SV(t))$, the receiver vehicle set in V2V communication is defined as follows.

Definition 4: Receiver vehicle set in V2V communication.

Given $D(SV(t))$, the receiver vehicle set in V2V communication, denoted by $RV(D(SV(t)))$, is the union of receiver vehicle sets for each $d(SVi) \in D(SV(t))$. That is:

$$RV(D(SV(t))) = \cup_{d(SV_i) \in D(SV(t))} RV(d(SV_i)) \tag{3.3}$$

In view of the dynamic traffic workload and the heavy data service demand, it is imperative to enhance the system scalability via cooperative data dissemination (CDD). Therefore, one of the primary objectives is to maximize the total number of vehicles which can be served via either V2I or V2V communication in each scheduling period. To this end, the gain of scalability is defined as follows.

Definition 5: Gain of scalability.

Given the data item $d_I(t)$ transmitted from the RSU, the set of sender vehicles $SV(t)$ and the corresponding set of data items $D(SV(t))$ in V2V communication, the gain of scalability, denoted by $G(t)$, is the total number of vehicles which can be served via either V2I or V2V communication in a scheduling period, which is computed by:

$$G(t) = |RV(d_I(t))| + |RV(D(SV(t)))| \tag{3.4}$$

where $|RV(d_I(t))|$ is the number of receiver vehicles in V2I communication, and $|RV(D(SV(t)))|$ is the number of receiver vehicles in V2V communication.

To be elaborated in the algorithm design, serving different vehicles may have different impacts on overall system performance due to the real-time service requirement. To capture this feature, we define the weighted gain as follows.

Definition 6: Weighted gain.

Denote $W_{Vi}(t)$ as the weight of serving V_i at t. The weighted gain, denoted by $G^w(t)$, is the summation of the weight for each served vehicle in a scheduling period, which is computed by:

$$G^w(t) = \sum_{V_i \in RV(d_I(t)) \cup RV(D(SV(t)))} W_{V_i}(t) \tag{3.5}$$

With the above knowledge, CDS is formulated as follows. Given the database $D = \{d_1, d_2, ..., d_{|D|}\}$, the set of vehicles $V(t) = \{V_1, V_2, ..., V_{|V(t)|}\}$, the set of requests $Q(t) = \{Q_{V_1}, Q_{V_2}, ..., Q_{V_{|V(t)|}}\}$, and the set of weights for serving each vehicle $W(t) = \{W_{V_1}, W_{V_2}, ..., W_{V_{|V(t)|}}\}$, the algorithm makes the following scheduling decisions. First, it divides the vehicles into V2I and V2V sets, namely, $V_I(t)$ and $V_V(t)$. Then, a data item $d_I(t)$ is selected to broadcast via V2I communication. Meanwhile, a set of sender vehicles $SV(t)$ together with the corresponding set of data items $D(SV(t))$ are selected in V2V

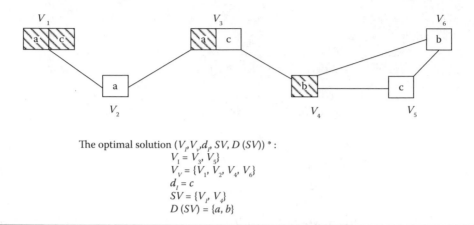

The optimal solution $(V_I, V_V, d_I, SV, D(SV))^*$:
$V_I = \{V_3, V_5\}$
$V_V = \{V_1, V_2, V_4, V_6\}$
$d_I = c$
$SV = \{V_1, V_4\}$
$D(SV) = \{a, b\}$

Figure 3.3 An example of CDS problem.

communication. Given D, V, Q, and W, let Λ (D, V, Q, W) be the set of scheduling decisions for V_I, V_V, d_I, SV, and $D(SV)$. CDS is to find an optimal scheduling decision, denoted by $(V_I, V_V, d_I, SV, D(SV))^*$, such that the weighted gain $G^w(t)$ is maximized. That is:

$$(V_I, V_V, d_I, SV, D(SV))^* = \arg \max_{(V_I, V_V, d_I, SV, D(SV)) \in \Lambda(D, V, Q, W)} G^w(t) \tag{3.6}$$

Figure 3.3 shows an example of CDS. The vehicle set $V = \{V_1, V_2, ..., V_6\}$ and requests of each vehicle are represented by data blocks. Specifically, the shadowed data block represents that the corresponding data item has been retrieved. In contrast, the data block without shadow represents the outstanding request. The edge represents the neighborhood relationship of vehicles. Assuming all the vehicles are within the RSU's coverage (*i.e.*, $V = V_{RSU}$), and assuming the gain of serving each vehicle is 1 (*i.e.*, $W_{V_i} = 1$, $i = 1, 2, ...6$), then it is not difficult to observe the following optimal solution. (a): c is scheduled to be transmitted from the RSU (*i.e.*, $d_I = c$). (b): V_3 and V_5 are set to the V2I mode (*i.e.*, $V_I = \{V_3, V_5\}$), while the other four vehicles are set to the V2V mode (*i.e.*, $V_V = \{V_1, V_2, V_4, V_6\}$). (c): V_1 and V_4 are designated as sender vehicles (*i.e.*, $SV = \{V_1, V_4\}$) for disseminating a and b, respectively. Accordingly, the set of data items in V2V communication is $D(SV) = \{a, b\}$. Given such a schedule, the receiver vehicle sets in V2I and V2V communications are $RV(d_I) = \{V_3, V_5\}$ and $RV(D(SV)) = \{V_2, V_6\}$, respectively. As shown, all the outstanding requests can be served in this scheduling period, and $G^w = |RV(d_I)| + |RV(D(SV))| = 4$.

3.3.2.2 NP-Hardness

We prove that CDS is NP-hard by constructing a polynomial time reduction from a well-known NP-hard problem called maximum weighted independent set (MWIS) [21]. Before presenting the formal proof, a sketch of the idea is outlined as follows. First, we introduce a viable scheduling operation (to be defined as the "tentative schedule" formally), which forms the basis of finding an optimal solution of CDS. Second, based on certain constraints on CDD in vehicular ad hoc networks (VANETs), we establish a set of rules to identify conflicting operations such that any pair of conflicting operations cannot coexist in an optimal solution of CDS. Third, we construct an undirected graph G by creating a vertex for each operation and adding an edge between any two conflicting operations. The weight of each vertex is set as the weight of the corresponding operation.

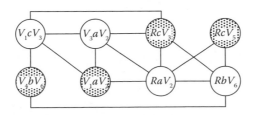

Figure 3.4   An example of reduction from MWIS to CDS.

With the above mapping, we demonstrate that the optimal schedule of CDS is derived if and only if the MWIS of G is computed. Therefore, CDS is NP-hard. To have clearer exposition, we further illustrate the idea with an example.

As shown in Figure 3.4, an undirected graph G is constructed based on the example shown in Figure 3.3. The identifier of each vertex represents a viable operation. For instance, the vertex V_1cV_3 represents that V_1 transmits c to V_3. Referring to Figure 3.3, it is viable because V_1 has cached c, while its neighbor V_3 is requesting. Therefore, this operation has the potential to serve V_3. So, there is a one-to-one mapping between each operation and each vertex. An edge between two vertices represents that the two corresponding operations are in conflict with each other. For instance, the edge between V_1cV_3 and V_3aV_2 means that the two operations (*i.e.*, V_1 transmits c to V_3 and V_3 transmits a to V_2) cannot be scheduled at the same time. This is due to the constraint that V_3 cannot be the sender and the receiver simultaneously. We will define a set of rules to capture all the constraints on CDD so that very pair of conflicting operations can be identified. In accordance with the assumption in Figure 3.3 (*i.e.*, $W_{V_i} = 1$), the weight of each vertex is set to 1. Given the constructed G, we can check that the four shadowed vertices shown in Figure 3.4, namely RcV_3, RcV_5, V_1aV_2, and V_4bV_6 are the MWIS of G. Accordingly, the total weight is 4, which is consistent with the maximum weighted gain derived from Figure 3.3. The formal description of the above idea is presented as follows.

3.3.2.3 *Tentative Schedules*

A TS refers to an operation which has the potential to serve one pending request via either V2I or V2V communication. In this regard, we classify the TS into two sets, $TS_{V2I}(t)$ and $TS_{V2V}(t)$, where $TS_{V2I}(t)$ is the set of TSs serving requests via V2I communication, and $TS_{V2V}(t)$ is the set of TSs serving requests via V2V communication. To facilitate the analysis, a TS in $TS_{V2I}(t)$ is parsed by $R\hat{d}V_r$, where R represents the RSU, $\hat{d}$ represents the data item transmitted from the RSU, and V_r represents the receiver vehicle for $\hat{d}$. Note that $\hat{d}$ has to be in the pending request set of V_r (*i.e.*, $\hat{d} \in PQ_{V_r}(t)$). Similarly, a TS in $TS_{V2V}(t)$ is parsed by $V_s\hat{d}V_r$, where V_s represents the sender vehicle, $\hat{d}$ represents the data item to be disseminated by V_s, and V_r represents the receiver vehicle for $\hat{d}$. Note that $\hat{d}$ has to be in the satisfied request set of V_s. In the meantime, it has to be in the pending request set of V_r (*i.e.*, $\hat{d} \in SQ_{V_s}(t) \wedge \hat{d} \in PQ_{V_r}(t)$). As specified, each TS has the potential to serve an outstanding request via either V2I or V2V communication. For instance, as shown in Figure 3.3, V_1aV_2 is a TS, which can be interpreted as the potential service by assigning V_1 as the sender and V_2 as the receiver with respect to the data item a. In contrast, V_1aV_3 is not a TS, because V_3 has already received a, and this schedule cannot serve any outstanding request.

3.3.2.4 Conflicting TSs

Different TSs may be in conflict with each other due to the following constraints on CDD: (a) the RSU can only broadcast one data item at a time, (b) each sender vehicle can only disseminate one data item at a time, (c) a vehicle cannot be both the sender and the receiver at the same time, (d) data collision cannot happen at receivers, and (e) each vehicle can only be in one of the modes (*i.e.*, V2I or V2V) at a time. The corresponding five rules for identifying conflicting TSs are introduced as follows.

1. If the two TSs are both for V2I communication (*i.e.*, $R\hat{d}V_r \in TS_{V2I}(t)$ and $R\hat{d}'V_r' \in TS_{V2I}(t)$), but they specify different data items to broadcast (*i.e.*, $\hat{d} \neq \hat{d}'$), then $R\hat{d}V_r$ is in conflict with $R\hat{d}'V_r'$, because the RSU can only broadcast one data item at a time. For example, RcV_3 and RaV_2 are in conflict with each other.

2. If the two TSs are both for V2V communication (*i.e.*, $V_s\hat{d}V_r \in TS_{V2V}(t)$ and $V_s'\hat{d}'V_r' \in TS_{V2V}(t)$), but they designate the same sender vehicle to disseminate different data items (*i.e.*, $V_s = V_s'$ and $\hat{d} \neq \hat{d}'$), then $V_s\hat{d}V_r$ is in conflict with $V_s'\hat{d}'V_r'$, because each sender vehicle can only disseminate one data item at a time. For example, V_1cV_3 and V_1aV_2 are in conflict with each other.

3. If the two TSs are both for V2V communication (*i.e.*, $V_s\hat{d}V_r \in TS_{V2V}(t)$ and $V_s'\hat{d}'V_r' \in TS_{V2V}(t)$), where one TS designates a vehicle as the sender, while the other TS designates the same vehicle as the receiver (*i.e.*, $V_s = V_r'$ or $V_r = V_s'$), then $V_s\hat{d}V_r$ is in conflict with $V_s'\hat{d}'V_r'$, because a vehicle cannot be both the sender and the receiver at the same time. For example, V_1cV_3 and V_3aV_2 are in conflict with each other.

4. If the two TSs are both for V2V communication (*i.e.*, $V_s\hat{d}V_r \in TS_{V2V}(t)$ and $V_s'\hat{d}'V_r' \in TS_{V2V}(t)$), but a receiver is the neighbor of both the senders (*i.e.*, $V_r \in N_{V_s'}(t)$ or $V_r' \in N_{V_s}(t)$), then $V_s\hat{d}V_r$ is in conflict with $V_s'\hat{d}'V_r'$, because data collision happens at one of the receivers. For example, V_1cV_3 and V_4bV_6 are in conflict with each other.

5. If one TS is for V2I communication (*i.e.*, $R\hat{d}V_r \in TS_{V2I}(t)$) and the other TS is for V2V communication (*i.e.*, $V_s'\hat{d}'V_r' \in TS_{V2V}(t)$), but the receiver in V2I communication is the same with either the sender or the receiver in V2V communication (*i.e.*, $V_r = V_s'$ or $V_r = V_r'$), then $R\hat{d}V_r$ is in conflict with $V_s'\hat{d}'V_r'$, because a vehicle can be only in one of the modes (*i.e.*, V2I or V2V) at a time. For example, RcV_3 and V_1cV_3 are in conflict with each other.

3.3.2.5 Constructing the Graph

First, we find the set of TSs by traversing both the retrieved and outstanding data items for each vehicle. Then, the undirected graph G can be constructed by the following procedures: (a) create a vertex for each TS, (b) set the weight of each vertex to the weight of the receiver vehicle in the corresponding TS, and (c) for any two conflicting TSs, add an edge between the two corresponding vertices. Apparently, G can be constructed in polynomial time. There is a one-to-one mapping between each vertex and each TS. Any two non-adjacent vertices in G represent that the two corresponding TSs are not in conflict with each other, and hence their weighted gain can be accumulated, which is equivalent to the summation of the weight of the two corresponding vertices.

Overall, the maximum weighted gain of CDS is achieved if and only if the MWIS of G is computed. The above proves that CDS is NP-hard.

3.3.3 A Heuristic Algorithm

In this section, we present a heuristic algorithm called CDD to solve the CDS problem. Note that simply maximizing the gain of scalability as defined in Section 3.3.2 in each scheduling period cannot guarantee global optimal scheduling performance, as it cannot distinguish the urgency of different vehicles in a particular scheduling period. Therefore, we capture the property of service urgency by considering the remaining time units of vehicles ($Slack_{V_i}(t)$) in the service region, which is computed by $Slack_{V_i}(t) = \dfrac{Dis_{V_i}(t)}{Vel_{V_i}(t)}$, where $Dis_{V_i}(t)$ is the current distance to the exit of the service region and $Vel_{V_i}(t)$ is the current velocity of V_i. Note that there could be a feasible schedule only when $Slack_{V_i}(t) \geq 1$. To give higher priority on serving more urgent vehicles, the weight is inversely proportional to its slack time, which is defined as $W_{V_i}(t) = \dfrac{1}{Slack_{V_i}(t)^\alpha}$, where α ($\alpha > 0$) is the tuning parameter to weigh the urgency factor. To give an overview, CDD schedules with the following three steps.

- **Step 1:** CDD examines all the TSs in both $TS_{V2I}(t)$ and $TS_{V2V}(t)$ to find every pair of conflicting TSs and compute the weight of each TS.
- **Step 2:** CDD constructs the graph G and transforms CDS to MWIS. Then, it selects a subset of TSs based on a greedy method.
- **Step 3:** CDD generates the following outputs: (a) the data item $d_l(t)$ to be transmitted from the RSU, (b) the set of receiver vehicles $RV(d_l(t))$ for $d_l(t)$, (c) the set of sender vehicles $SV(t)$ in V2V communication, (d) the set of data items $D(SV(t))$ for each sender vehicle, and (e) the set of receiver vehicles $RV(D(SV(t)))$ in V2V communication.

Details of each step are presented as follows.

3.3.3.1 Identify Conflicting TSs and Compute the Weight

This step consists of four operations. First, CDD determines the set of vehicles $V_{RSU}(t)$, which are in the RSU's coverage. Second, it finds all the TSs in both V2I and V2V communications and constructs $TS_{V2I}(t)$ and $TS_{V2V}(t)$. Third, it identifies any pair of conflicting TSs. Finally, it computes the weight for each TS. The implementation is elaborated as follows.

In order to determine $V_{RSU}(t)$, CDD checks each entry $< V_i, Q_{V_i}(t), N_{V_i}(t) >$ maintained in the service queue. Since every vehicle within the RSU's coverage shall update its information in each scheduling period, in case there is no update received for an entry $< V_i, Q_{V_i}(t), N_{V_i}(t) >$, it implies that V_i has left the RSU's coverage. In order to construct $TS_{V2I}(t)$ and $TS_{V2V}(t)$, CDD examines the entry $< V_j, Q_{V_j}(t), N_{V_j}(t) >$ for each $V_j \in V_{RSU}(t)$. Specifically, for each pending request of V_j (i.e., $\forall q_{V_j}^m \in PQ_{V_j}(t)$), there is a TS: $\{Rq_{V_j}^m, V_j\} \in TS_{V2I}(t)$, which represents that the RSU disseminates $q_{V_j}^m$ to V_j via V2I communication. On the contrary, for each satisfied request of V_j (i.e., $\forall q_{V_j}^m \in SQ_{V_j}(t)$), if $q_{V_j}^m$ is a pending request of V_k (i.e., $q_{V_j}^m \in PQ_{V_k}(t)$) and V_k is the neighbor of V_j (i.e., $V_k \in N_{V_j}(t)$), then there is a TS $\{V_j q_{V_j}^m V_k\} \in TS_{V2V}(t)$, which represents that V_j disseminates $q_{V_j}^m$ to V_k via V2V communication. Note that V_k may not necessarily be in the

RSU's coverage. In order to identify each pair of conflicting TSs, CDD follows the five rules as specified in Section 3.3.2. Finally, for each vehicle in the service queue, CDD updates its current velocity $Vel_{V_i}(t)$ and its current distance to the exit of the service region $Dis_{V_i}(t)$, so that the remaining dwell time is estimated by $Slack_{V_i}(t) = \dfrac{Dis_{V_i}(t)}{Vel_{V_i}(t)}$. Given any TS with a receiver vehicle V_r, its weight is computed by $\dfrac{1}{Slack_{V_r}(t)^\alpha}$. Note that vehicles which are out of RSU's coverage but within the service region are still maintained in the service queue (to be elaborated in Step 3). According to the description in Section 3.3.1, the status of these vehicles can be reported by their neighbors which are still in the RSU's coverage.

3.3.3.2 Construct G and Select TSs

This step consists of three operations. First, CDD constructs the graph G based on the description in Section 3.3.2. Second, it approximately solves the weighted independent set problem using a greedy method. Third, it constructs the set of selected TSs. The implementation is elaborated as follows.

In order to construct G, CDD creates a vertex v for each TS. The weight of v is denoted by $w(v)$, and it is set to the weight of the receiver vehicle of the corresponding TS. For each pair of the identified conflicting TSs, they are mapped to the corresponding vertices (*e.g.*, v_i and v_j), and then an edge e_{ij} is added between v_i and v_j. In order to select independent vertices, CDD adopts the greedy method presented in Ref. [22]. The general idea is recapitulated below. First, it computes the value of $w(v)/(d(v) + 1)$ for each vertex v in G, where $w(v)$ and $d(v)$ represent the weight and the degree of v, respectively. Second, it selects the vertex $v_{selected}$ with the maximum value of $w(v)/(d(v) + 1)$. Third, it updates G by removing the set of vertices $N^+(v_{selected})$, where $N^+(v_{selected})$ contains $v_{selected}$ and all of its adjacent vertices. Fourth, it repeats the above operations until there is no vertex remaining in G (*i.e.*, $V(G) = \emptyset$). Denote $W(I)^*$ as the weight of the maximum independent set of G and denote $W(I)$ as the weight of the independent set obtained by this greedy method. It has been proven that the performance ratio satisfies $\dfrac{W(I)}{W(I)^*} \geq \dfrac{1}{\Delta_G}$, where Δ_G is the maximum degree of any vertex in G. Based on the above described operations, a set of TSs are selected, which is denoted by $TS_{selected}(t)$.

3.3.3.3 Generate Outputs and Update Service Queue

This step consists of two operations. First, CDD parses each selected TS to make the scheduling decisions, including the determination of $d_i(t)$, $RV(d_i(t))$, $SV(t)$, $D(SV(t))$, and $RV(D(SV(t)))$. Second, it updates the service queue by adding the entries for newly arrived vehicles and removing the entries for vehicles which are out of the service region. The implementation is elaborated as follows.

In order to generate the outputs, CDD parses each TS in $TS_{selected}(t)$. Specifically, for any selected TS belonging to V2I communication (*i.e.*, $\forall R\hat{d}V_r \in TS_{selected}(t)$), the data item $\hat{d}$ will be the same, because all the selected $R\hat{d}V_r$ are not in conflict with each other. Accordingly, $\hat{d}$ is scheduled to be transmitted from the RSU, which determines $d_i(t)$. Then, the union of V_r from each $R\hat{d}V_r$ is the set of receiver vehicles in V2I communication, which determines $RV(d_i(t))$. On the contrary, for any selected TS belonging to V2V communication (*i.e.*, $\forall V_s\hat{d}V_r$ in $TS_{selected}(t)$), the union of V_s from each $V_s \hat{d} V_r$ forms the set of sender vehicles so that $SV(t)$ is determined. Meanwhile, the union of $\hat{d}$ from each $V_s \hat{d} V_r$ forms the set of data items to be disseminated via V2V communication, and thus $D(SV(t))$ is determined. Last, the union of V_r from each $V_s \hat{d} V_r$ forms the set of receiver vehicles so that $RV(D(SV(t)))$ is determined. In order to maintain the service queue, the system needs to add an entry for each newly arrived vehicle and remove the entry for each vehicle that has left. Note that for a left vehicle V_i, its entry is removed only when V_i is out of the RSU's coverage (*i.e.*, $V_i \notin V_{RSU}(t)$) and V_i is not in the neighborhood of any vehicle within the RSU's coverage (*i.e.*, $V_i \notin N_{V_k}(t)$, $\forall V_k \in V_{RSU}(t)$). This is because if $V_k \in V_{RSU}(t)$ and $V_i \in N_{V_k}(t)$, then V_i may still have chance to retrieve the data item from V_k via V2V communication, even though $V_i \notin V_{RSU}(t)$.

3.3.4 Performance Evaluation

3.3.4.1 Setup

The simulation model is built based on the system architecture described in Section 3.3.1, and it is implemented by CSIM19 [23]. The traffic characteristics are simulated according to the Greenshield's model [24], which is widely adopted in simulating macroscopic traffic scenarios [25]. Specifically, the relationship between the vehicle velocity (v) and the traffic density (k) is represented by $v = V^f - \dfrac{V^f}{K^j} \cdot k$, where V^f is the free-flow speed (*i.e.*, the maximum speed limit) and K^j is the jam density (*i.e.*, the density which causes the traffic jam). Three lanes are simulated, and the free-flow speeds of the three lanes are set to $V_1^f = 120$ *km/h*, $V_2^f = 100$ *km/h*, and $V_3^f = 80$ *km/h*, respectively. The same jam density K^j is set for each lane, which is 100 vehicles/km. Consider that all the vehicles drive in the same direction and the arrival of vehicles in each lane follows the Poisson process. In order to evaluate the system performance under different traffic workloads, a wide range of vehicle arrival rates is simulated. Given a specific vehicle arrival rate on each lane, the corresponding vehicle velocities and vehicle densities are also collected. Detailed traffic statistics are summarized in Table 3.1, where the traffic workload is getting heavier from Scenario 1 to Scenario 5.

The communication characteristics are simulated based on DSRC. In particular, the radius of RSU's coverage is set to 300 m, and the V2V communication range is set to 150 m. We do not specify absolute values of the data size and the wireless bandwidth, but set the scheduling period as 1 s. The database size is set to 100. Each vehicle may submit a request at a random time when passing through the RSU. The number of requests submitted by each vehicle is uniformly distributed from one to seven. The data access pattern follows the Zipf distribution [26] with the parameter $\theta = 0.7$. Specifically, the access probability of a data item d_i is computed by $\dfrac{(1/i)^\theta}{\sum\limits_{j=1}^{|D|} (1/j)^\theta}$, where $|D|$ is the size of the database.

We implement two well-known algorithms for performance comparison. One is first come first served (FCFS) [27], which broadcasts data items according to the arrival order of requests. The other is most requested first (MRF) [28], which broadcasts the data item with the maximum number of pending requests. The tuning parameter α of CDD is set to 4%, which gives it the best

Table 3.1 Simulation Statistics under Different Traffic Scenarios

Traffic Scenarios (Workload)	Mean Arrival Rate (Vehicles/h)			Mean Velocity (km/h)			Mean Density (Vehicles/km)		
	Lane 1	Lane 2	Lane 3	Lane 1	Lane 2	Lane 3	Lane 1	Lane 2	Lane 3
1 (very light)	1200	1000	800	104.32	86.83	70.59	13.06	13.17	11.79
2 (light)	1600	1400	1200	98.75	81.17	65.03	17.69	18.82	18.71
3 (medium)	2000	1800	1600	91.31	74.64	56.03	23.89	25.34	29.95
4 (heavy)	2400	2200	2000	83.43	60.99	40.37	30.44	38.96	49.50
5 (very heavy)	2800	2600	2400	64.77	39.30	28.14	45.96	60.66	64.80

performance in the default setting. The following metrics are considered to quantitatively analyze the algorithm performance.

■ **Distribution of gains:** This metric partitions the served requests into two sets. One set contains those requests served by the RSU, and the other set contains the requests served by neighboring vehicles. The proportion of each set reflects the contribution of V2I and V2V communications to the overall performance.

■ **Service ratio:** Given the total number of served requests (n_s) and the total number of submitted requests (n) by all vehicles, the service ratio is computed by n_s/n.

■ **Service delay:** It measures the waiting time of served requests, which is the duration from the instance when the request is submitted to the time when the corresponding data item is retrieved.

3.3.4.2 Simulation Results

Figure 3.5 examines the distribution of gains contributed by V2I and V2V communications for CDD. As observed, when the traffic workload is light, most vehicles are served via V2I data dissemination. This is because the vehicle density on each lane is low (*i.e.*, as shown in Table 3.1, only around 13 vehicles/km on each lane in Scenario 1). Therefore, with few neighbors of each vehicle, the chance to retrieve a data item of interest via V2V communication is slim. With an increase of the traffic workload, the contribution of V2V data dissemination increases notably. This makes sense due to the following reasons. First, the vehicle density is getting higher in a heavier traffic workload environment, resulting more neighbors of each vehicle. Accordingly, the chance to have common requests among neighboring vehicles is higher. Note that these common requests of different vehicles may be submitted at different times. Therefore, it is not likely to serve all of them via a single V2I broadcast, leaving the remaining requests to have a higher possibility to be served via V2V communication. Second, a higher traffic workload also causes longer dwell time of vehicles in the service region. This gives a higher chance for each vehicle to retrieve more requested data items, which gives a better opportunity for V2V data sharing. Last, there are more requests submitted by vehicles asking for different data items when the traffic workload is getting heavier, but only one data item can be broadcast from the RSU in each scheduling period. This limits the proportion

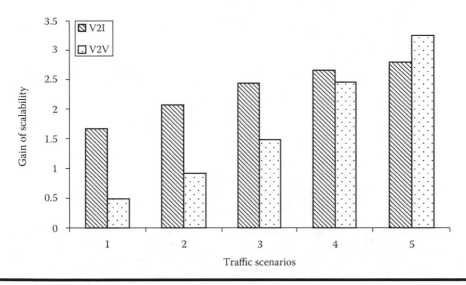

Figure 3.5 Distribution of gains under different traffic scenarios.

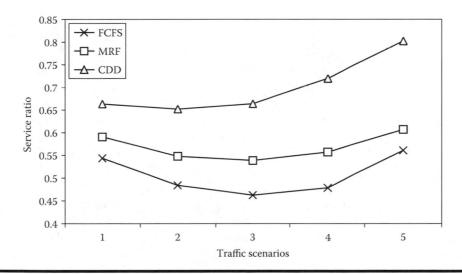

Figure 3.6 Service ratio under different traffic scenarios.

of contribution via V2I communication. In contrast, by appropriately exploiting the spatial reusability, multiple data items can be disseminated via V2V communication simultaneously without conflicting, which enhances the portion of V2V contribution in a heavy traffic scenario. To sum up, CDD is able to strike a balance between V2I and V2V communications on data services.

Figure 3.6 shows the service ratio of algorithms under different traffic scenarios. As observed, the service ratio of all the algorithms decline to a certain extent when the traffic workload starts to get heavier. When the traffic workload keeps increasing, the service ratio of all the algorithms is getting higher. The reasons are explained as follows. At the beginning (*i.e.*, in Scenario 1), although vehicles pass through the service region with pretty high velocities due to low traffic density, the system can still achieve reasonably good performance due to the small number

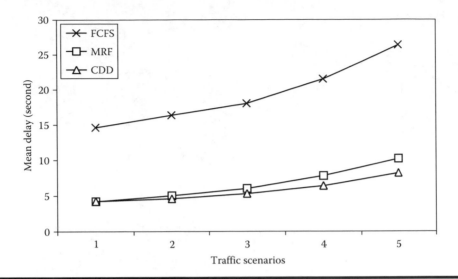

Figure 3.7 Service delay under different traffic scenarios.

of total submitted requests. When the vehicle arrival rate starts to increase (*i.e.*, in Scenario 2), although the velocity drops accordingly, the increased number of requests dominates the performance, which results in the decline of the service ratio. When the vehicle velocity keeps dropping in a heavier traffic workload environment, the long dwell time of vehicles gradually dominates the performance. Accordingly, the service ratio is getting higher. As shown, CDD outperforms other algorithms significantly in all scenarios. Note that although this work only focuses on data dissemination from a single RSU, it is straightforward to extend the solution and further enhance system performance when multiple RSUs cooperate to provide data services.

Figure 3.7 shows the service delay of algorithms under different traffic scenarios. Although CDD performs closely to MRF in light traffic scenarios, it gradually achieves shorter service delay when the traffic workload is getting heavier. This is because as analyzed in Figure 3.5, the benefit of V2V communication achieved by CDD becomes significant in a heavy traffic environment. Furthermore, note that the mean service delay is derived from all the served requests. As demonstrated in Figure 3.6, CDD serves more requests than both MRF and FCFS in all scenarios, which further demonstrates the superiority of CDD as it is not trivial to achieve both shorter service delay and higher service ratio.

3.4 Delay-Tolerant Services via Coding-Assisted Scheduling

3.4.1 System Architecture

In this section, we present an SDVN-based system architecture for delay-tolerant services in a heterogenous vehicular communication environment. As shown in Figure 3.8, we consider the service scenario where vehicles ask for a set of common interest information such as traffic conditions, which are jointly provided by RSUs via the DSRC interface and BSs via the cellular network interface (*e.g.*, 3G and 4G). RSUs are installed sparsely along the roads. Vehicles in the RSU's coverage (denoted by the dotted circle in Figure 3.8) can retrieve information via DSRC. Nevertheless, due to the limited communication range and intermittent connections of RSUs, for large data volume

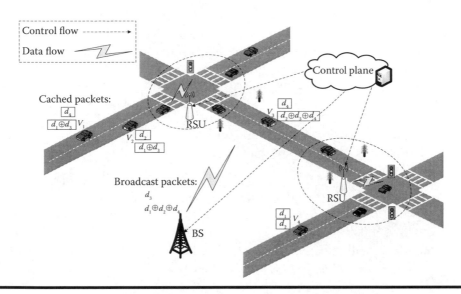

Figure 3.8 Coding-assisted data broadcast system in SDN-based vehicular networks.

and delay-tolerant services, it is desirable to explore the characteristics of SDVN to better support the cooperation of heterogeneous wireless communications in vehicular networks.

The considered application scenario is described as follows. Vehicles periodically update their status to the controller via the cellular network interface, including positions, cached contents, and kinematic information. Vehicles in the RSU's coverage will be instructed to turn into the DSRC interface and communicate with the RSU, so that the RSU is aware of the set of vehicles in its coverage as well as their service status. RSUs can either follow the control instructions from the logically centralized control plane, or alternatively, they may exercise certain intelligence and make part of scheduling decisions individually. When vehicles leave the RSU's coverage, they will be instructed by the controller to turn to the cellular network interface. With the global information of vehicle service status, the control plane makes scheduling decisions and informs the BS via the control message, so that the BS will provide data services via the cellular network interface accordingly.

With the above description, the target of the system is to coordinate data services between BSs and RSUs by implementing an efficient scheduling policy at the control plane. In particular, the bitwise exclusive-or ($\oplus$) coding operation is adopted at the BS for data broadcast due to its trivial implementation overhead. For example, given an encoded packet $p = d_1 \oplus d_2$, to decode a data item (say d_1) from p, it requires the remaining data items in p (say d_2) by computing $d_1 = p \oplus d_2$. Meanwhile, different from traditional wireless sensor nodes which have very limited cache size, vehicles can support large storage, and the cache size will not be a hurdle of system performance. In view of this, we consider that vehicles are allowed to cache those encoded packets even if these packets cannot be decoded out immediately.

Figure 3.8 shows a toy example to illustrate the benefit of caching encoded packets, and meanwhile, it reveals the challenges in designing an efficient scheduling policy. Considering there are four vehicles V_1, V_2, V_3, and V_4, which are out of the RSUs' coverage. The set of commonly interested data items is d_1, d_2, d_3, and d_4. The current cached contents are d_4 and $d_1 \oplus d_3$ for V_1; d_2 and $d_1 \oplus d_3$ for V_2; d_4 and $d_2 \oplus d_3 \oplus d_4$ for V_3; and d_1 and d_2 for V_4. To maximize the bandwidth efficiency of the BS, it is expected to complete the service to all the vehicles with the minimum number of broadcast transmissions. Consider the duration for broadcasting a data item (or an encoded packet) as a time unit. In this example, it is observed that at least two time units are required to complete the service by broadcasting d_3 and $d_1 \oplus d_2 \oplus d_4$, respectively. Taking V_1 as an example, when d_3 is broadcast, it can decode out d_1 from $d_1 \oplus d_3$. Then, with d_1 and d_4, it can further decode out d_2 from $d_1 \oplus d_2 \oplus d_4$, and hence the service is completed for V_1. Similar decoding procedures can be applied to other vehicles to complete the service. In contrast, if it does not allow vehicles to cache encoded packets, namely, V_1, V_2, V_3 only cache d_4, d_2, d_4, respectively, and V_4 caches d_1 and d_2. Then, it can be checked that one of the best solutions is to broadcast $d_2 \oplus d_4$, d_3, and d_1 in sequence to serve all the vehicles, which requires three time units.

3.4.2 Coding-Assisted Broadcast Scheduling Problem

3.4.2.1 Problem Formulation

In this section, we formally describe the concerned CBS problem. Denote the set of vehicles as $V = \{V_1, V_2, \ldots V_{|V|}\}$, where $|V|$ is the total number of vehicles. The common interested data set is denoted by $D = \{d_1, d_2, \ldots d_{|D|}\}$, where $|D|$ is the total number of data items. An encoded packet is denoted by p, which corresponds to a D-dimension coefficient vector $\mathbf{a}(p) = [a(p)_1 \ a(p)_2 \ \ldots a(p)_{|D|}]$, where $a(p)_i$ in $\{0,1\}$, $(1 \le i \le |D|)$. Specifically, if d_i is encoded in p, then $a(p)_i$ is set to 1. Otherwise, $a(p)_i$ is set to 0. Note that a non-encoded data item (*e.g.*, d_j) can be represented by a special case of an encoded packet, where $a(p)_j = 1$ and $a(p)_k = 0$ for $k = 1, 2, \ldots |D|$, $k \ne j$. In the rest of this paper, the term *packet* is used to refer to either encoded or non-encoded data items without causing ambiguity. Given a vehicle V_m $(1 \le m \le |V|)$, the set of its cached packets is denoted by $C(V_m) = \{p_1^m, p_2^m, \ldots, p_{|C(V_m)|}^m\}$. With the above knowledge, the CBS problem is described as follows. Given the set of vehicles $V = \{V_1, V_2, \ldots V_{|V|}\}$, the set of data items $D = \{d_1, d_2, \ldots d_{|D|}\}$, and the set of cached packets of each vehicle $C(V_m)$ $(V_m \in V)$, the control plane targets at maximizing the bandwidth efficiency of the BS by determining the set of encoded packets P, which can complete the service to all the vehicles with the minimum number of time units.

3.4.2.2 NP-Hardness

We prove the NP-hardness of CBS by constructing a polynomial time reduction from the *simultaneous matrix completion problem*, which is a well-known NP-hard problem [29]. The general concept of the simultaneous matrix completion problem is described as follows. Given a set of matrices, and each matrix contains a mixture of numbers and variables, each particular variable can only appear once per matrix but may appear in several matrices. The objective is to find values for these variables such that all resulting matrices simultaneously have full rank. To have a clear exposition, before giving the formal proof, we illustrate a sketch of the idea with an example.

Suppose $D = \{d_1, d_2, d_3, d_4\}$, and there are two vehicles V_1 and V_2 with cache sets $C(V_1) = \{d_4, d_1 \oplus d_3, d_1 \oplus d_3 \oplus d_4\}$ and $C(V_2) = \{d_2, d_1 \oplus d_3\}$. For V_i, the number of cached packets is denoted by $|C(V_i)|$, and each cached packet is represented by a $|D|$-dimension coefficient vector. Then, we can construct a $|C(V_i)| \times |D|$ matrix for V_i, representing its cached contents. For example, the constructed matrix for V_1 is $\begin{bmatrix} 0 & 0 & 0 & 1 \\ 1 & 0 & 1 & 0 \\ 1 & 0 & 1 & 1 \end{bmatrix}$.

In this example, we observe that the packet p_3^1 (*i.e.*, $d_1 \oplus d_3 \oplus d_4$) in $C(V_1)$ can be derived from the other two packets p_1^1 and p_2^1 in $C(V_1)$ by computing $p_1^1 \oplus p_2^1$. It indicates that given p_1^1 and p_2^1, p_3^1 actually has no further contribution to decode out any new data item for V_1. Therefore, it is expected to find the set of cached packets which are independent with each other. To this end, we further transform the matrix to its reduced row echelon form by Gaussian elimination. In this example, the reduced row echelon form of the constructed matrix is $\begin{bmatrix} 1 & 0 & 1 & 0 \\ 0 & 0 & 0 & 1 \end{bmatrix}$.

The transformed matrix for V_i is denoted by $A(V_i)$, which is a $|C'(V_i)| \times |D|$ matrix, represented by $\begin{bmatrix} \mathbf{a}'(p_1^i) & \mathbf{a}'(p_2^i) & ... & \mathbf{a}'(p_{|C'(V_i)|}^i) \end{bmatrix}^T$, where $|C'V_i|$ is the number of independent vectors and $|C'(V_i)| \leq |C(V_i)|$. In addition, since the rows of $A(V_i)$ are independent, we have $|C'V_i| \leq |D|$. In particular, if $|C'(V_i)|==|D|$, $A(V_i)$ is a full rank matrix (it is an identity matrix in the reduced row echelon form). Therefore, all the data items can be decoded out from $C(V_i)$. Otherwise (*e.g.*, $|C'(V_i)|<|D|$), there are at least $|D|-|C'(V_i)|$ packets are required to serve V_i completely.

By obtaining $A(V_1) = \begin{bmatrix} 1 & 0 & 1 & 0 \\ 0 & 0 & 0 & 1 \end{bmatrix}$ and $A(V_2) = \begin{bmatrix} 1 & 0 & 1 & 0 \\ 0 & 1 & 0 & 0 \end{bmatrix}$, we may observe that there are at least two packets (*e.g.*, p_1^* and p_2^*) which are required to serve both V_1 and V_2. Suppose we could find such two packets, by pending the two corresponding coefficient vectors $\mathbf{a}(p_1^*)$ and $\mathbf{a}(p_2^*)$ to both $A(V_1)$ and $A(V_2)$, the two matrices $\begin{bmatrix} \mathbf{a}'(p_1^1) & \mathbf{a}'(p_2^1) & \mathbf{a}(p_1^*) & \mathbf{a}(p_2^*) \end{bmatrix}^T$ and $\begin{bmatrix} \mathbf{a}'(p_1^2) & \mathbf{a}'(p_2^2) & \mathbf{a}(p_1^*) & \mathbf{a}(p_2^*) \end{bmatrix}^T$ should be full rank. In this example, we can check that by setting $\mathbf{a}(p_1^*) = [0\,0\,1\,0]$ and $\mathbf{a}(p_2^*) = [0\,1\,0\,1]$, the two matrices will become full rank. Therefore, both V_1 and V_2 will be completely served.

Theorem 1 *The coding-assisted broadcast scheduling problem is NP-hard.*

Proof. Given V, D and $C(V_i)$ ($V_i \in V$), we construct a set of coefficient matrices $A = \{A(V_i), A(V_i), ..., A(V_{|V|})\}$, where $A(V_i)$ is represented by $\begin{bmatrix} \mathbf{a}'(p_1^i) & \mathbf{a}'(p_2^i) & ... & \mathbf{a}'(p_{|C'(V_i)|}^i) \end{bmatrix}^T$, which is the reduced row echelon form of the matrix $\begin{bmatrix} \mathbf{a}(p_1^i) & \mathbf{a}(p_2^i) & ... & \mathbf{a}(p_{|C(V_i)|}^i) \end{bmatrix}^T$. As a vehicle can decode out at most one data item by receiving a packet, it requires at least another $|D| - |C'(V_i)|$ packets to serve V_i (*i.e.* to decode out $|D|$ data items). Denote $L = \min(|C'(V_i)|)$ ($\forall V_i \in V$). To complete the service to all the vehicles, it requires at least $|D|-L$ packets. Accordingly, to maximize the bandwidth efficiency, the CBS problem is to schedule and broadcast $|D|-L$ packets so that all the vehicles can be served.

Denote the set of scheduled packets as $P^* = \left\{ p_1^*, p_2^*, ... p_{|D|-L}^* \right\}$. The coefficient vector of p_j^* is denoted by $\mathbf{a}(p_j^*)$ ($p_j^* \in P^*$). For each $A(V_i)$ in A, we append the row vectors $\mathbf{a}(p_j^*)(\forall p_j^* \in P^*)$ to the matrix and the new matrix is denoted by $A^*(V_i)$, namely, $A^*(V_i) = \begin{bmatrix} \mathbf{a}'(p_1^i) & ... & \mathbf{a}'(p_{|C'(V_i)|}^i) & \mathbf{a}(p_1^*) & ... & \mathbf{a}(p_{|D|-L}^*) \end{bmatrix}^T$. Since V_i can be served if and only if $A^*(V_i)$ can be transformed to an identity matrix by Gaussian elimination (*i.e.* $A^*(V_i)$ is full rank), the service to all the vehicles is equivalent to the simultaneous matrix completion for $A^*(V_i)$, $\forall V_i$ in V, with the variables $\mathbf{a}\left(p_1^* \right), \mathbf{a}\left(p_2^* \right), ... \mathbf{a}\left(p_{|D|-L}^* \right)$. The above proves that the CBS problem is NP-hard.

3.4.3 A Memetic Algorithm

With the fast growth of learning and optimization techniques, it is promising to have dedicated intelligent algorithms for solving complex data dissemination problems in vehicular networks. Memetic computing is a new paradigm proposed in the literature, which has been successfully applied to solve many real-world problems such as permutation flow shop scheduling, quadratic assignment problem, and feature selection. [20]. In this section, we present a memetic algorithm called MA to solve the CBS problem. Before elaborating the detailed implementation, we describe the general idea of the MA as follows.

First, the algorithm generates a population of solutions. Each solution represents one possible selection of the broadcast packets. Subsequently, parent selection is performed to identify the solutions to undergo crossover and mutation for offspring solution generation. Further, a local search process kicks in to refine the solution. Afterward, the fitter solutions among both parent solutions and the generated offspring solutions will survive for the next generation through the population replacement process. The above procedures will repeat until a predefined stopping criterion is satisfied. The detailed procedure of MA is presented as follows.

3.4.3.1 Initialization

Denote a solution as χ, and it is represented by a binary vector with N-dimension, where N is the total number of the randomly generated coefficient vectors with dimension $|D|$. Specifically, we set $\chi[i] = 1$ if the vector $\mathbf{a}_i$ ($1 \leq i \leq N$) is selected. Otherwise, $\chi[i] = 0$. Given the maximum scheduling period T, as most T packets can be scheduled at a time. Accordingly, a feasible solution should satisfy $0 < \sum_{i=1}^{N} \chi_m[i] \leq T$. Figure 3.9 shows the representation of a feasible solution. With this binary representation, the population initialization is then completed by randomly generating M feasible solutions, which is denoted by $\Lambda_0 = \{\chi_1, \chi_2, \ldots \chi_M\}$.

3.4.3.2 Offspring Generation

Parent selection, crossover, and mutation are the reproduction operators conducted for generating offspring solutions. In particular, the tournament selection [30] is employed to identify the two parents in the population for reproduction. The basic idea of tournament selection is to select the best two solutions χ_l and χ_n in terms of the fitness value f (to be defined in Equation 3.7) from k individuals in Λ. In addition, the uniform crossover and bit mutation [31] are adopted. Specifically, each bit of the offspring is randomly selected either from χ_l or χ_n, and hence each

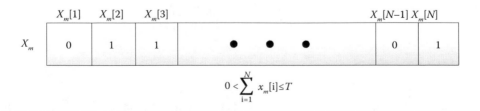

Figure 3.9   Representation of a feasible solution.

offspring has approximately half of the genes from each parent. Further, to increase the diversity of the search process, the bit mutation operator is applied to all dimensions in each offspring solution with a predefined mutation probability ρ.

3.4.3.3 Local Search

The local search is implemented by a sequential flipping operations conducted on each solution. In particular, given a solution χ_m, the local search will be executed from the first to the last dimension with value flipping, that is, 0 to 1 or 1 to 0. Given the fitness function f (to be defined in Equation 3.7), only if the flipping with fitness improvement would be accepted and stored in the solution. For example, suppose χ'_m is the newly obtained solution by flipping the ith bit of χ_m. Then, we have $\chi'_m[i] \leftarrow 1 - \chi_m[i]$ and $\chi'_m[j] \leftarrow \chi_m[j], \forall j \neq i$. Afterward, we compute and compare the fitness values of χ_m and χ'_m, denoted by f_{χ_m} and $f_{\chi'_m}$, respectively. If $f_{\chi'_m} > f_{\chi_m}$, then χ_m is replaced by χ'_m in this round of local search.

Note that the operations of crossover, mutation, and local search may induce infeasible solutions where $\sum_{i=1}^{N} \chi_m[i] > T$. To address this issue, we design the following repair operator. First, it finds the set of '1' bit in the solution χ_m, which is $S = \{i | \chi_m[i] = 1\}$. Second, it randomly selects $|S| - T$ elements from S and sets them to 0. Clearly, we have $\sum_{i=1}^{N} \chi_m[i] = T$ by applying this repair operator.

3.4.3.4 Population Replacement

To determine the solutions for survival in the next generation and keep the population size unchanged, MA compares the fitness value of solutions in the fusion of parent and offspring. Then, the top-M solutions are kept in the population for further evolution. Specifically, the fitness function is derived as follows.

Recall that the objective of the CBS problem is to maximize the bandwidth efficiency by finding the minimum set of packets to complete the service. With this in mind, we design the following function to evaluate the fitness of a solution. Given a solution χ_m ($1 \leq m \leq M$), if $\chi_m[i] = 1$ ($1 \leq i \leq N$), then we attach the vector $\mathbf{a}_i$ to the coefficient matrix of each vehicle. The total number of attached vectors is denoted by k_m, where $k_m = \sum_{i=1}^{N} \chi_m[i]$. For vehicle V_j, denote its original and newly obtained matrices as $A(V_j)$ and $A(V_j)'$, respectively, and their corresponding ranks are denoted by r_j and r'_j. Then, the fitness function f of a solution is defined as the average matrix rank improvement contributed by each packet in this solution, which is computed by:

$$f = \left(\sum_{j=1}^{|V|} (r'_j - r_j) / |V| \right) / k_m \tag{3.7}$$

As defined, if the two solutions contain the same number of coefficient vectors (*i.e.*, the same number of packets is scheduled), then the one improving more of the average matrix rank is preferred, because more data items can be retrieved by vehicles with the same bandwidth consumption. On the contrary, if the two solutions improve the same amount of average matrix rank (*e.g.*, the same number of data items can be retrieved by vehicles), then the one containing fewer coefficient vectors is preferred, because it can achieve the same portion of services with less bandwidth consumption.

3.4.3.5 Termination

For simplicity, the stopping criterion in this study is defined as the condition that the predefined maximum number of generations or the global optimum is reached. Note that it is also straightforward to adopt other criterion considered in the literature, such as maximal number of fitness evaluations.

3.4.4 Performance Evaluation

3.4.4.1 Setup

The simulation model is built based on the system architecture presented in Section 3.4.1 for performance evaluation and the MA is implemented by MATLAB®. By default, the population size is set to 100 for the proposed MA. The maximum number of generations is set to 100 and the maximum scheduling period is set to five time units. Two algorithms are implemented for performance comparison. One is the MRF [28], which broadcasts the data item with the maximum number of pending requests. The other is Round-Robin [32], which broadcasts all the data items periodically. Note that although both of the algorithms do not encode data in scheduling, vehicles may still cache encoded data packets via the service from other wireless interfaces such as RSUs. In the default setting, the number of requested data items is set to 50 and 200 vehicles are simulated in the service region. The number of cached packets of each vehicle is uniformly distributed from 15 to 25. Unless stated otherwise, all the experiments are conducted with the default setting. The following metrics are considered to quantitatively analyze the algorithm performance.

- **Number of broadcast packets (NBP):** It is defined as the NBP to serve all the requests. According to the analysis in Section 3.4.2, the lower bound of NBP is computed by $|D|-L$, where $|D|$ is the set of requested data items and $L = \min_{\forall V_i \in V} (|C'(V_i)|)$.
- **Average service delay (ASD):** It is designed for evaluating the algorithm performance on satisfying delay-tolerant services. Specifically, denote the service delay for V_j as l_j. That is, it takes l_j time units to achieve a full rank of the coefficient matrix for V_j. Then, ASD is defined as the summation of each vehicle's service delay over the number of vehicles, which is computed by:

$$ASD = \left(\sum_{j=1}^{|V|} l_j \right) / |V| \tag{3.8}$$

- A lower value of ASD indicates a better performance of the algorithm on satisfying delay-tolerant services.

■ **Broadcast productivity (BP):** It is designed for evaluating the algorithm performance on enhancing the bandwidth efficiency. Specifically, denote the ranks of the original (before service) and the newly obtained (after service) matrices of V_j as r_j and r'_j, respectively, and the service duration is denoted by t_j. Then, BP is defined as the summation of each vehicle's improved rank per time unit over the number of vehicles, which is computed by:

$$BP = \left(\sum_{j=1}^{|V|} \frac{\left(r'_j - r_j \right)}{t_j} \right) / |V| \qquad (3.9)$$

■ A higher value of BP indicates a better performance of the algorithm on enhancing the efficiency of broadcast bandwidth. The upper bound of BP is 1 because at most the average rank improves by one in a time unit when all the vehicles receive an outstanding data item.

3.4.4.2 Simulation Results

The first set of experiments evaluates the algorithm performance under different system workloads (Figures 3.10 through 3.12). The more data items requested by vehicles represents a higher system workload. Specifically, Figure 3.10 compares the NBP of algorithms. As analyzed, the lower bound of the NBP can be computed by $|D| - \min_{\forall V_i \in V} (|C'(V_i)|)$, while the upper bound of the NBP equals $|D|$. Accordingly, we observe from Figure 3.10 that MA can achieve the optimal performance in terms of minimizing NBP. In contrast, either MRF or Round-Robin almost reaches

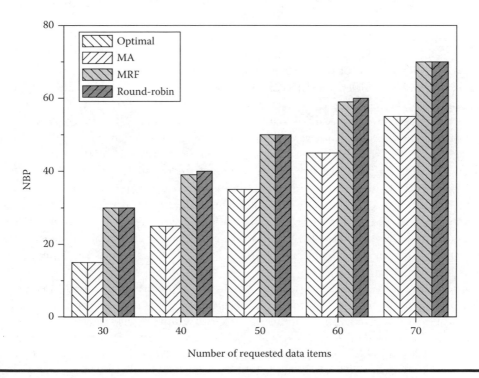

Figure 3.10 Number of broadcast packets (NBP) under different system workloads.

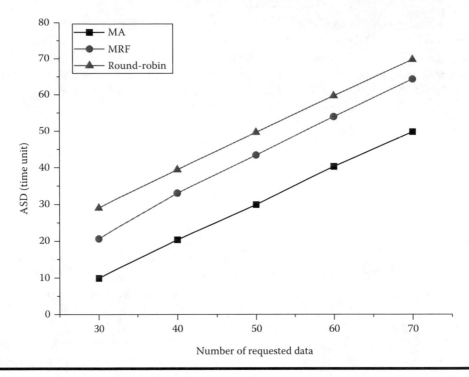

Figure 3.11   Average service delay (ASD) under different system workloads.

the upper bound in all cases. This is because neither MRF nor Round-Robin adopted network coding by considering the cached packets of vehicles, which indicates that a data item has to be broadcast even if only one of the vehicles is waiting for it. Due to the diversity of outstanding requests by different vehicles, most likely, the algorithm has to schedule all the data items for completing the service.

Figure 3.11 compares the ASD of algorithms under different system workloads. As shown, the ASD of all the algorithms increases with the increasing of the system workload. Nevertheless, MA always achieves the lowest ASD among all the algorithms. To better comprehend and verify such a superiority of MA, we further examine the result shown in Figure 3.12, where the BP of algorithms under different system workloads are compared. As analyzed, the higher value of BP demonstrates the better capability of the algorithm on improving the bandwidth efficiency. Evidently, MA achieves near optimal BP in all scenarios, which is much higher than other algorithms. Also, note that MRF performs better than Round-Robin, because MRF considers data productivity in scheduling by broadcasting the data item with the most pending requests.

The second set of experiments evaluates the algorithm performance under different number of cached packets (Figures. 3.13 through 3.15). Specifically, Figure 3.13 compares the NBP of algorithms under different numbers of cached packets. Intuitively, with the increasing number of cached packets, a fewer broadcast packets are required to complete the service. Nevertheless, we note that the NBP of MRF and Round-Robin is almost unchanged when there are more cached packets. This is because MRF and Round-Robin can barely take the advantage of those cached packets as most of them are in the encoded form. In contrast, MA can always achieve the optimal result by best exploiting the benefit of network coding.

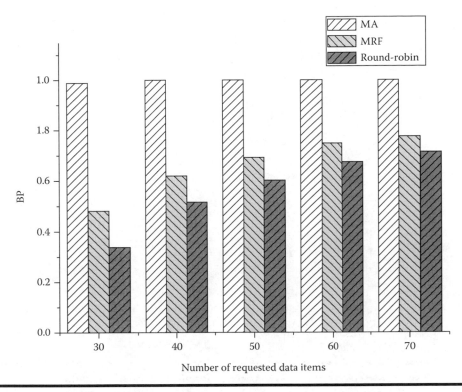

Figure 3.12   Broadcast productivity (BP) under different system workloads.

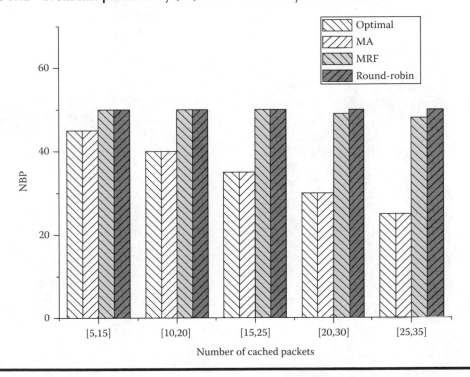

Figure 3.13   Number of broadcast packets (NBP) under different number of cached packets.

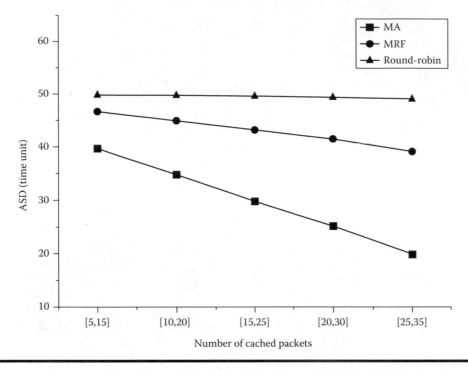

Figure 3.14   Average service delay (ASD) under different number of cached packets.

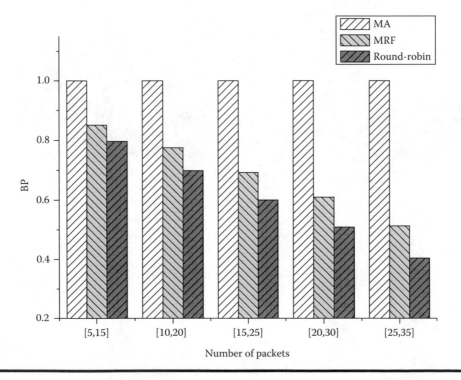

Figure 3.15   Broadcast productivity (BP) under different number of cached packets.

Figure 3.14 compares the ASD of algorithms under different numbers of cached packets. As shown, MA constantly outperforms other algorithms in all the scenarios, especially when there are more cached packets, which further demonstrates the superiority of MA on making coding decisions. We give further analysis by evaluating the BP of algorithms under different numbers of cached packets, which is shown in Figure 3.15. As observed, MA always manages to achieve the upper bound of BP in different scenarios, while the performance of MRF and Round-Robin is decreasing. This is because when the number of cached packets is increasing, it is more likely that a broadcast data item by MRF and Round-Robin can only serve a few number of vehicles, which results in the decreasing of BP.

3.5 Summary

In this chapter, we present a data service architecture in SDVN and discussed the potential benefit as well as challenges on data scheduling in such an environment. On this basis, we investigate two data service scenarios in SDVN, including real-time services via cooperative V2V/V2I data dissemination and delay-tolerant services in heterogeneous vehicular communication environments. Accordingly, two problems, namely the CDS problem and the CBS problem, are formulated. For CDS, we prove that it is NP-hard from the reduction of MWIS. A heuristic online scheduling algorithm called CDD is proposed to enhance the data dissemination performance by best exploiting the synergy between V2I and V2V communications. For CBS, we prove that it is NP-hard from the reduction of the simultaneous matrix completion problem. A memetic algorithm called MA is proposed to solve the CBS problem, which consists of a binary vector representation for encoding CBS solutions, a local search method for solution enhancement, and a fitness function for solution evaluation. We evaluate the performance of the two algorithms in their respective service scenarios and show the effectiveness of the presented solutions.

Acknowledgments

This work was supported in part by the National Natural Science Foundation of China (Grant No. 61572088); a grant from CityU (7004412), ICT R&D program of MSIP/IITP (14-824-09-013, Resilient Cyber-Physical Systems Research), GRL Program (2013K1A1A2A02078326) through NRF, and DGIST Research and Development Program (CPS Global Center) funded by MSIP.

References

1. P. Dai, K. Liu, Q. Zhuge, E. H.-M. Sha, V. C. S. Lee, and S. H. Son, Quality-of-experience-oriented autonomous intersection control in vehicular networks, IEEE *Transactions on Intelligent Transportation Systems*, vol. 17, no. 7, pp. 1956–1967, 2016.
2. S. Santini, A. Salvi, A. Valente, A. Pescape, M. Segata, and R. L. Cigno, A consensus-based approach for platooning with inter-vehicular communications, in *IEEE Conference on Computer Communications (INFOCOM'15)*, Kowloon, Hong Kong, pp. 1158–1166, April 26–May 01, 2015.
3. K. Liu, H. B. Lim, E. Frazzoli, H. Ji, and V. C. Lee, Improving positioning accuracy using GPS pseudorange measurements for cooperative vehicular localization, *IEEE Transactions on Vehicular Technology*, vol. 63, no. 6, pp. 2544–2556, 2014.

4. Y. L. Morgan, Notes on DSCR & WAVE standards suite: Its architecture, design, and characteristics, *IEEE Communications Surveys & Tutorials*, vol. 12, no. 4, pp. 504–518, 2010.

5. K. Liu, J. K. Ng, V. C. Lee, S. H. Son, and I. Stojmenovic, Cooperative data scheduling in hybrid vehicular ad hoc networks: VANET as a software defined network, *IEEE/ACM transactions on networking*, vol. 24, no. 3, pp. 1759–1773, 2016.

6. K. Liu, J. K.-Y. Ng, J. Wang, V. C. Lee, W. Wu, and S. H. Son, Network-coding-assisted data dissemination via cooperative vehicle-to-vehicle/-infrastructure communications, *IEEE Transactions on Intelligent Transportation Systems*, vol. 17, no. 6, pp. 1509–1520, 2016.

7. P. Dai, K. Liu, L. Feng, Q. Zhuge, V. C. Lee, and S. H. Son, Adaptive scheduling for real-time and temporal information services in vehicular networks, *Transportation Research Part C: Emerging Technologies*, vol. 71, pp. 313–332, 2016.

8. K. Liu, V. C. S. Lee, J. K.-Y. Ng, J. Chen, and S. H. Son, Temporal data dissemination in vehicular cyber-physical systems, *IEEE Transactions on Intelligent Transportation Systems*, vol. 15, no. 6, pp. 2419–2431, 2014.

9. X. Shen, X. Cheng, L. Yang, R. Zhang, and B. Jiao, Data dissemination in VANETs: A scheduling approach, *IEEE Trans. on Intelligent Transportation Systems*, vol. 15, no. 5, pp. 2213–2223, 2014.

10. Q. Xiang, X. Chen, L. Kong, L. Rao, and X. Liu, Data preference matters: A new perspective of safety data dissemination in vehicular ad hoc networks, in *IEEE Conference on Computer Communications (INFOCOM'15)*, Kowloon, Hong Kong, pp. 1149–1157, April 26–May 01, 2015.

11. H. A. Omar, W. Zhuang, and L. Li, On multihop communications for in-vehicle Internet access based on a TDMA MAC protocol, in *IEEE Conference on Computer Communications (INFOCOM'14)*, Toronto, Canada, pp. 1770–1778, April 27–May 02, 2014.

12. H. Zhu, M. Dong, S. Chang, Y. Zhu, M. Li, and X. S. Shen, Zoom: Scaling the mobility for fast opportunistic forwarding in vehicular networks, in *IEEE Conference on Computer Communications (INFOCOM'13)*. Turin, Italy: IEEE, pp. 2832–2840, April 14–19, 2013.

13. Y. Bi, H. Shan, X. S. Shen, N. Wang, and H. Zhao, A multi-hop broadcast protocol for emergency message dissemination in urban vehicular ad hoc networks, *IEEE Transactions on Intelligent Transportation Systems*, vol. 17, no. 3, pp. 736–750, 2016.

14. J. Zhang, Q. Zhang, and W. Jia, VC-MAC: A cooperative MAC protocol in vehicular networks, *IEEE Transactions on Vehicular Technology*, vol. 58, no. 3, pp. 1561–1571, 2009.

15. O. N. Foundation, Software-defined networking: The new norm for networks, *ONF White Paper*, 2012.

16. X. Wang, C. Wang, J. Zhang, M. Zhou, and C. Jiang, Improved rule installation for real-time query service in software-defined Internet of vehicles, *IEEE Transactions on Intelligent Transportation Systems*, vol. 18, no. 2, pp. 225–235, 2017.

17. Z. He, J. Cao, and X. Liu, SDVN: Enabling rapid network innovation for heterogeneous vehicular communication, *IEEE Network*, vol. 20, no. 4, pp. 2–7, 2016.

18. K. Zheng, L. Hou, H. Meng, Q. Zheng, N. Lu, and L. Lei, Soft-defined heterogeneous vehicular network: Architecture and challenges, *IEEE Network*, vol. 30, no. 4, pp. 72–80, 2016.

19. C. Jiacheng, Z. Haibo, Z. Ning, Y. Peng, G. Lin, and S. Xuemin, Software defined Internet of vehicles: Architecture, challenges and solutions, *Journal of Communications and Information Networks*, vol. 1, no. 1, pp. 14–26, 2016.

20. X. Chen, Y.-S. Ong, M.-H. Lim, and K. C. Tan, A multi-facet survey on memetic computation, *IEEE Transactions on Evolutionary Computation*, vol. 15, no. 5, p. 591, 2011.

21. D. S. Hochba, Approximation algorithms for NP-hard problems, *ACM SIGACT News*, vol. 28, no. 2, pp. 40–52, 1997.

22. S. Sakai, M. Togasaki, and K. Yamazaki, A note on greedy algorithms for the maximum weighted independent set problem, *Discrete Applied Mathematics*, vol. 126, no. 2, pp. 313–322, 2003.

23. H. Schwetman, CSIM19: A powerful tool for building system models, in *Proceedings of the 33nd Conference on Winter Simulation (WSC'01)*. Arlington, VA: IEEE, pp. 250–255, December 09–12, 2001.

24. C. F. Daganzo, *Fundamentals of transportation and traffic operations*, Pergamon Press: Oxford, UK 1997.

25. P. Edara and D. Teodorović, Model of an advance-booking system for highway trips, *Transportation Research Part C: Emerging Technologies*, vol. 16, no. 1, pp. 36–53, 2008.
26. G. Zipf, *Human Behavior and the Principle of Least Effort: An Introduction to Human Ecology*, Cambridge, MA: Addison-Wesley Press, Cambridge, 1949.
27. J. Wong and M. H. Ammar, Analysis of broadcast delivery in a videotex system, *IEEE Transactions on Computers*, vol. 100, no. 9, pp. 863–866, 1985.
28. J. W. Wong, Broadcast delivery, *Proceedings of the IEEE*, vol. 76, no. 12, pp. 1566–1577, 1988.
29. N. J. Harvey, D. R. Karger, and S. Yekhanin, The complexity of matrix completion, in *Proceedings of the 7th Annual ACM-SIAM Symposium on Discrete Algorithm*, Miami, Florida, pp. 1103–1111, January 22–26, 2006.
30. B. L. Miller and D. E. Goldberg, Genetic algorithms, tournament selection, and the effects of noise, *Complex Systems*, vol. 9, no. 3, pp. 193–212, 1995.
31. H. Mühlenbein, M. Gorges-Schleuter, and O. Krämer, Evolution algorithms in combinatorial optimization, *Parallel Computing*, vol. 7, no. 1, pp. 65–85, 1988.
32. S. Khanna and S. Zhou, On indexed data broadcast, in *Proceedings of the 13th Annual ACM symposium on Theory of Computing*. Dallas, TX: ACM, pp. 463–472, May 24–26, 1998.

V2X NETWORK TYPES

Chapter 4

Application and Performance of CEN-DSRC to the Smart Tachograph

Sithamparanathan Kandeepan

RMIT University
Melbourne, Australia

Gianmarco Baldini

European Commission, Joint Research Center (JRC)
Ispra, Italy

Contents

4.1 Introduction

In Europe, the current digital tachograph (DT) system to monitor the driving time in commercial vehicles above 3.5 tonnes (*e.g.*, freight transport) is governed by Council Regulation (EEC) No 3821/85 of December 20, 1985, which was modified on several occasions through amendments. In July 2011, the European Commission proposed a new set of features including the capability for targeted early detection of fraud, through wireless communication technology. The final version of the new approved regulation was published in February 2014 [1] as Regulation (EU) No 165/2014.

The vehicular wireless technology developed under the standards of Comité Européen de Normalisation (CEN)-DSRC [2] is considered to be used for implementing the targeted early detection feature to support law enforcers and transmit specific DT data over short distances. The choice of the CEN-DSRC technology at 5.8 GHz has been made for various reasons. First, its communications parameters suit the objective well, including the short-range distance between the enforcer and the vehicle. Second, the CEN-DSRC technology is widely used in Europe for tolling purposes and it is a reasonable choice to expand the enforcement application to data other than tolling. Third, the same CEN-DSRC onboard equipment, very cost-effective, might be able to support the remote communication functions for current or future regulated applications in commercial vehicles, simplifying the vehicle electronics architecture. An example of future application is Directive (EU) 2015/719, on the maximal weights and dimensions for heavy goods vehicles, which will also be based on CEN-DSRC [2]. Finally, the Regulation (EU) No 165/2014 itself recommended the CEN-DSRC suite of standards in clause (23).

For the new revision of the DT application (called Smart Tachograph in Ref. [1]), we must ensure that the CEN-DSRC for DT can support the targeted early detection feature for the amount of data which must be transmitted, and in the operational scenarios. Therefore, the authors have performed a theoretical and a simulation study at the physical (PHY) and medium access (MAC) layers for transmitting DT data over CEN-DSRC. We consider two use cases for using CEN-DSRC to support the exchange of the DT data: targeted checks from (a) a patrol (police) car unit (PCU) and (b) a roadside unit (RSU), for a commercial vehicle on the road as shown in Figure 4.1.

The performance study presented here is (a) based on the general knowledge with known theories from wireless communications [3] considering the pathloss, random fading and shadowing effects of the wireless channel with multipath effects and Doppler variations, and (b) based on the technological limitations as per the standards specified for the CEN-DSRC technology [4] such as the maximum transmission power, minimum required BER, signal bandwidth, signal modulations, data frame structure, and the MAC protocol.

The constraints in using CEN-DSRC for DT considering the points in (a) and (b) above give a limitation on (i) the maximum achievable range for communication between the transmitter and the receiver, (ii) the true achievable BER, and (iii) the delay in completing a transaction between the interrogator and the vehicle considering the MAC layer protocol. It is also important to realize here that (i), (ii), and (iii) are related to each other, and we present a performance study

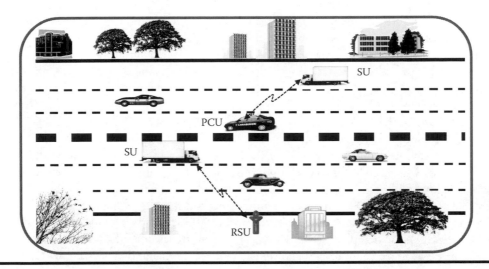

Figure 4.1   Digital tachograph use cases with the roadside unit (RSU)/patrol car (PC) for interrogating the subject unit (SU).

considering (i), (ii), and (iii). Moreover, we also make recommendations (where possible) during the course of this study for further improving the usage of the CEN-DSRC technology for the new DT application.

The structure of the paper is the following: Section 4.2 describes the DT application and how the provision of wireless communications for targeted checks by law enforcers on the road can greatly benefit the security and safety on the road. Section 4.3 describes the overall system model. Sections 4.4 and 4.5 provide, respectively, the analysis for communication range and BER. Section 4.6 shows the feasibility of completing the MAC transaction on the basis of the previous results. Finally, Section 4.7 concludes the paper and describes future developments.

4.2 CEN-DSRC for the Digital Tachograph

4.2.1 Benefits of CEN-DSRC for the New Version of the Digital Tachograph

The DT is a recording function intended to capture the driver's activities including distance, speed, driving times, and rest periods of the driver. It is used in the European Union to check the driving times of drivers and enforce the legislation on social rules (*e.g.*, the rest period or driving time limits) with the goal to improve road safety or to reduce the number of accidents and fatalities in the EU. The DT is a device installed in a truck composed of a computing device (*i.e.*, the vehicle unit or VU) connected, and cryptographically bound, to a gearbox-mounted motion sensor, which is used to measure the traveled distance through the movement of a toothed ring inside the gear box. The VU holds data on drivers of the vehicle, periods of driving, and duty for a 12-month period. In the new version of the DT, a global navigation satellite system (GNSS) receiver is also used to record the start and end points of the trip in addition to periodic locations (every 3 hours of cumulative driving).

Enforcing driving time is a very important goal. Tired drivers have a looser control of the vehicles and have a higher risk of accidents. Excessive driving hours also distorts competition between companies in a domain where salaries and labor taxes are one of the two main costs together with fuel.

Without a targeted early detection of fraud through wireless communication technology, law enforcers can only inspect a limited number of commercial vehicles for conformance to regulations. In the current conditions, law enforcers do not have any equipment providing data from the moving vehicles and they select the vehicles on the basis of subjective criteria. The output of the DT data for roadside inspections is currently executed by printing a hardcopy from the VU, or by downloading the data in the VU having the access granted by the physical insertion of a law enforcer card in the VU. The consequence is that (a) only a limited percentage of commercial vehicles is checked (b) check of conformant vehicle causes lost time for both driver and law enforcer. Note that it is not the intention of the legislator that the data transmitted through the DSRC interface will be used for direct fining, but they remain only a selection tool to stop the vehicles possibly offending the regulation for a complete manual check.

The data for targeted early detection are only a small subset of the complete set of data, which can be downloaded by a law enforcer from the VU, but it has been defined to specifically indicate nonconformances. The complete set of data to be transmitted is identified in the regulation [1]. Examples of fields in the set of data are (a) vehicle motion conflict and (b) driving without a valid card. In the first case, a vehicle motion conflict between the different sensors (*i.e.*, motion sensor and the GNSS receiver) can indicate a fraud because different recorded motions may indicate tampering of one or both sensors. The complete set of data can be transmitted in a maximum of two frames (*i.e.*, around 200 bytes) in the CEN-DSRC standard and the analysis described in the rest of the paper is based on this assumption.

Security requirements are also addressed in the transmission of data. The data are cryptographically signed and it contains the identifier of the commercial vehicle and the processing platform providing the data to be transmitted (*e.g.*, the vehicle registration number of the vehicle, which is also called license plate and the serial number of the processing platform). The distribution of the keys to support the cryptographic signature will be an evolution of the existing trust model of the DT (see Ref. [5]), whose detailed description is out of scope of this paper, which is focused on the wireless propagation and performance of the CEN-DSRC for this specific scenario.

While CEN-DSRC is widely used for electronic tolling (ET), its application to law enforcement is quite new and the operational scenarios shown in Figure 4.1 are different from ET because the law enforcer CEN-DSRC reader is located on the roadside or in a law enforcer's vehicle rather than mounted on a gantry above the road as in ET. The different configurations require an analysis of the performance of CEN-DSRC for the DT, to ensure that the different scenario still supports the successful transmission of the two frames. This is the main objective of this study, which has not been previously performed in literature to the knowledge of the authors.

4.2.2 CEN-DSRC Protocol and Parameters

From the CEN-DSRC point of view, the DT (including the VU) will be called the subject unit (SU) in the rest of the paper. The protocol architecture for using DT over CEN-DSRC is briefly shown in Figure 4.2, in which the DT application (DT App) running at the RSU/PCU side and the SU side respectively communicate with each other over the CEN-DSRC protocol stack. The CEN-DSRC protocol includes the physical layer (PHY), the data link layer (DLL) which is the MAC, and the application layer core (ALC). The DT App directly interacts with the ALC for communications.

According to the CEN-DSRC PHY standards [4] the downlink and the uplink are specified as summarized in Table 4.1. CEN-DSRC technology uses a backscattering technique and therefore the link budget for the uplink depends on the downlink link budget. Moreover, unity gain

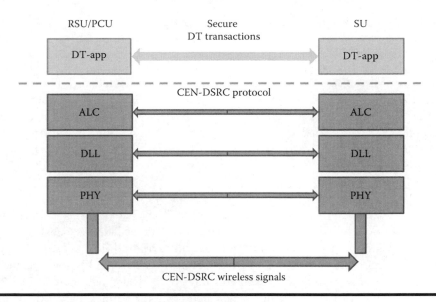

Figure 4.2   The DT over CEN-DSRC protocol architecture.

Table 4.1   Key Transmission Parameters for the CEN-DSRC-PHY Up- and Downlinks

	Uplink	*Downlink*
Data rate	250 kbps	500 kbps
Modulation	BPSK	2 Level AM
Max EIRP	–17 dBm	33 dBm
Frequency	5.8 GHz	5.8 GHz
BER	10^{-6} or less	10^{-6} or less
P_{min}	N/A	–43 dBm

antennas are considered together with 0 dB loss for the windscreen attenuation considering that the SU device is expected to be mounted on the windscreen of the vehicle. In fact, a minimal windscreen attenuation is one of the specific requirements of the technical specifications of the new version of the DT.

4.3 The Overall System Model for the Study

We consider a highway as an example where the technology is to be tested initially. A maximum of three lanes in a given direction of the traffic is assumed. In total, there will be a maximum of six lanes in the highway in both directions as depicted in Figure 4.1. The readers requesting the information from the SU are the RSU/PCU, where RSU/PCU are assumed to be on the same side of the highway as the SU.

4.3.1 The Highway Use Cases and Scenarios

Two use cases with three scenarios each are considered in our study. The first use case includes the SU and the RSU only. The RSU is placed on the curb side of the highway as shown in Figure 4.1, and the three corresponding scenarios that cover this use case are when the SU travels in Lane 1, Lane 2, and Lane 3, respectively, of the highway on the same side as the RSU (where Lane 1 is the closest to the RSU). The second use case includes the SU and the PCU only. The three scenarios that correspond to the second use case are (a) when the PCU is traveling in the same lane, (b) the first adjacent lane, and (c) the second adjacent lane, respectively.

4.3.2 The Vehicular Wireless Channel

The vehicular wireless channel plays a crucial part of this study that affects the performance of the digital communications overall [9]. The time varying nature of the fading wireless channel including the multipath effects make the communications vulnerable for the considered DT application. Using the results reported in literature and our own (previous) studies [10], the vehicular wireless channel is modeled using a log-normal shadowing model [6]. The log-normal model has two components: (a) the mean pathloss component calculated using the pathloss exponent α and (b) the random shadowing component modeled using a zero-mean Gaussian process with a shadow variance of σ_s^2. Based on this model, the mean signal loss becomes proportional to d^α with an additive term χ (dB) that is modeled by a zero-mean Gaussian process with a variance of σ_s^2. Note that this model is well-known in the field of wireless communications and is extensively used. However, the identification of the unique values for α and σ_s is the key to the model given our specific application, as these parameters range over different values for different applications depending on the operating environments. From reported experiments in the literature for vehicular communications [6] typical values for α range from 1.8 to 2.5 for the line-of-sight (LOS) conditions and from 2.4 to 3.5 for the non-line-of-sight (NLOS) conditions, and the shadow parameter σ_s ranges from 1.8 dB to 3.8 dB. In the study presented here, we use these values as references.

Moreover, in order to limit the study here, we assume that only a single SU is present and also there is no external wireless interference; though these are very important aspects related to the success of the technology we make these assumptions in order understand the range that could be achieved for a single SU. The study related to access and interference is outside the scope of this work.

4.4 Communication Range Analysis

The communication range for the application of interest very much depends on the link budget and more importantly, the mean pathloss the signal undergoes between the RSU/PCU and the SU. The wireless channel, therefore, plays a crucial role defined by the parameters α and σ_s. In this section, we present the achievable range for communication for the described scenarios considering the wireless channel model presented in Section 4.3, together with the CEN-DSRC-PHY standards that regulate the transmit power levels in the uplink and downlink.

We begin the range analysis by considering the antenna bore sight directions of the SU, RSU, and the PCU since the effective isotropically radiated power (EIRP) depends on the bore sight direction as per the standards.

- ■ The bore sight of the RSU antenna is best directed at $\phi = 70°$ horizontally as shown in Figure 4.3 to maximize the coverage region based on the antenna main lobe.

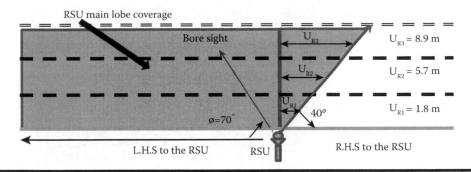

Figure 4.3 RSU bore sight direction for maximal coverage based on the CEN-DSRC-PHY EIRP regulation.

- The bore sight direction of the PCU is directed horizontally backward facing the vehicles behind the PCU where the antenna is mounted on the roof top of the PCU.
- The antenna (transceiver) for the SU is expected to be placed on the front windscreen of the vehicle.

For the scenarios with the RSU, we further observe that when the SU crosses the RSU and enters the R.H.S region, NLOS conditions are experienced since the SU antenna is placed on the front windscreen. Given the antenna placements as discussed above, the expected transmission range is mainly dependent on the maximum EIRP and the minimum acceptable power P_{min} at the receiver. In the remainder of this section we present the range analysis based on the maximum allowable EIRP.

4.4.1 Expected Downlink Range

Based on the EIRP within the main lobe of the RSU/PCU antenna and P_{min} values set in the CEN-DSRC standards, the maximum range for guaranteed communication in the downlink can be determined analytically considering a link budget analysis. Since the range predominantly depends on the pathloss between the transmitter and the receiver, we analyze the range for various values of α. Using the pathloss model described in Section 4.3 and 0 dBi antennas, the maximum range for the transmissions to achieve the minimum required power level P_{min} is computed for different values of α and the results are presented in Figure 4.4.

Note that for the first use case scenarios with the RSU, the figure here only depicts the L.H.S range from the RSU, the total range for the first use case therefore becomes the summation of the distance specified in Figure 4.4 and the R.H.S range as shown in Figure 4.3. From Figure 4.4, we observe that the downlink transmission range is around 38 m for good channel conditions with LOS conditions and is around 6 m for bad channel conditions with NLOS conditions. It is worthwhile again reminding here that this range is based on the reported values for α in the literature as discussed in Section 4.3.

The overall ranges for the communication between RSU and SU considering the L.H.S and the R.H.S ranges for the first use-case scenarios are given in Table 4.2a, and the overall ranges for the second use-case scenarios are given in Table 4.2b. The distances shown in Table 4.2b correspond to the displacement between the PCU and the SU along the direction of the highway but not the distance between the PCU and the SU. Later in this chapter, we will use these range values to understand whether a complete transaction between the RSU/PCU and the SU is possible after analyzing the expected delay per transaction.

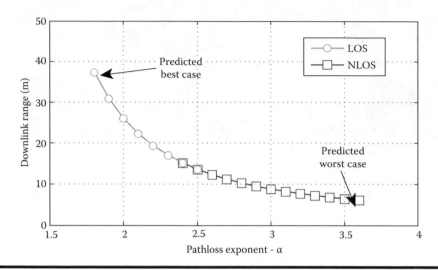

Figure 4.4   Maximum achievable range for the downlink for various values of α for P_{min} at the receiver.

Table 4.2   Summary of Achievable Ranges for the Six Scenarios, (a) First Use Case and (b) Second Use Case

(a)	L.H.S Worst Case ($\alpha = 3.6$)	L.H.S Best Case ($\alpha = 1.8$)	R.H.S Worst Case ($\alpha = 3.6$)	R.H.S Best Case ($\alpha = 2.4$)	Total Range (Worst Case)	Total Range (Best Case)
Lane 1	5.91 m	37.27 m	$U_{R1} = 1.8$ m	$U_{R1} = 1.8$ m	7.71 m	39.7 m
Lane 2	4.12 m	37.03 m	$U_{R2} = 5.7$ m	$U_{R2} = 5.7$ m	9.82 m	42.73 m
Lane 3	Unreachable	36.54 m	Unreachable	$U_{R2} = 8.9$	Unreachable	45.44 m
(b)	PCU Ahead of SU Worst Case ($\alpha = 3.6$)	PCU Ahead of SU Best Case ($\alpha = 1.8$)				
Same lane	6.1 m	37.30 m				
First adjacent	5.3 m	37.18 m				
Second adjacent	1.1 m	36.81 m				

4.4.2 Maintaining the Uplink Range

As mentioned previously, the uplink range depends on the received signal strength in the downlink since the backscattering technique is used. Here we present the key link design details that are crucial for the uplink in order to maintain the same range as the downlink. The link design is based on the required BER of 10^{-6} as set by the CEN-DSRC-PHY standard for the uplink.

Considering the Binary phase-shift keying (BPSK) modulation and a signal bandwidth of 500 kHz in the uplink, in order to achieve the given BER a minimum signal to noise power ratio of 7.52 dB is required.

Therefore, in order to achieve a transmission distance of 37.3 m at $\alpha = 1.8$, the design criteria of $G_c - P_n > 127.3$ dBm has to be met in the uplink, where G_c is the conversion gain at the SU for converting the received signal strength to the uplink signal and P_n is the noise power level at the RSU/PCU. According to the CEN-DSRC-PHY standards [4], G_c is limited to a value of 10 dB. The noise power level of commercial base stations are typically around −110 dBm to −120 dBm. These constraints on G_c and P_n set a limitation in the link design.

For example, a noise level of $P_n = -117$ dBm is required for SUs with a conversion gain of $G_c = 10$ dB to achieve the same range as the downlink in the uplink. Note here that 0 dBi antenna gains were considered in the design study and any additional gains from the antennas will trade off P_n. The antenna gains and polarization are according to the CEN-DSRC standards. As we observe from the range study, the overall link design (both for uplink and downlink) becomes much constrained by the uplink transmissions.

4.4.3 *The Outage of Range*

The ranges provided in Table 4.2 are the expected values based on the value of α. This range will change randomly depending on the shadowing variance σ_s^2 in both up- and downlinks. The expected ranges presented in Table 4.2, therefore, need to be considered together with the outage probability for achieving the particular range depending on the shadowing component. The outage probability for the expected range is given by the probability that the received signal power P_r falls below the minimum acceptable power level P_{min}, denoted as $P_o = \mathbf{Pr}[P_r < P_{min}]$ for a given range d, α and σ_s. Figure 4.5 depicts the outage probability for achieving a particular range for various values of σ_s^2 for the worst and best cases of α. As we observe from the figure, in the worst-case situation ($\alpha = 3.6$) almost one in every a hundred SU experiences an outage of 3 m for the worst case of shadowing, or in other words 99% of the vehicles will have a range of at least 3 m. Note that the case of an intervening vehicle in the propagation path and the related shadowing effect is not considered in this study, because a law enforcer would visually notice the presence of the vehicle

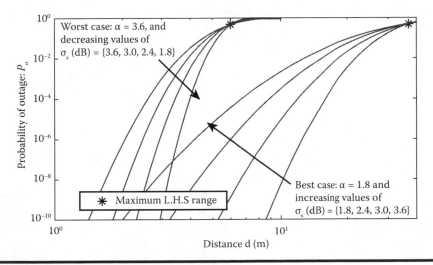

Figure 4.5   Probability of outage for the L.H.S range.

and justify the degraded performance of the communication. Section 4.6 shows that this range is quite sufficient to complete a single transaction between the RSU/PCU and the SU. Furthermore, it is also notable that the outage is 50% at the maximum achievable ranges corresponding to the worst and the best cases, which is obtained due to the symmetry of the Gaussian distribution about the mean value, and the mean value (mean receiver signal strength) at the maximum achievable ranges become P_{min} for both the worst and best cases of α. As the vehicle approaches closer to the RSU/PCU from the maximum range, the outage probability reduces as observed in the figure.

4.5 BER Analysis with Small-Scale Fading

The small-scale fading model contributed by the factors of multipath transmission and Doppler effects in the vehicular environment play an important role for the BER analysis. By analyzing the Doppler frequency and the multipath spread, we can identify whether the channel experiences slow, fast, flat, or frequency selecting fading correspondingly.

4.5.1 The Doppler Effect and Coherence Time

The Doppler shift for the second use case with the PCU can be ignored since the PCU is expected to travel almost at the same speed as the SU. The maximum Doppler shift f_m for the second use case is experienced when the SU is directly approaching (*i.e.*, with an angle of arrival 0⁰) the RSU at a maximum considered speed of 150 kmph, which is given by f_m = 805 Hz. This maximum Doppler shift is much less than the standardized tolerance of carrier frequency level as stated in the CEN-DSRC-PHY standard which is ±5 ppm (approximately ±29 kHz). Therefore, in our study the Doppler frequency errors can be ignored considering that the receiver should be able to tolerate/correct any frequency errors up to ±29 kHz.

The channel coherence time, T_c, defines whether the channel changes in time over a symbol period or not and hence classifying the channel as either a fast- or slow-fading channel [3]. The 50% correlation coherence time is given by T_c = 9/16πf_m= 222 µs. Note that the largest Doppler will have the worst-case situation for a fast-channel variation. The maximum bit duration for CEN-DSRC is given by T_b = 1/250e3 = 4 µsec in the uplink, and since $T_c \gg T_b$, according to theory the corresponding wireless channel is classified as a slow-fading channel that does not change over a symbol period.

4.5.2 The Multipath Effect and Coherence Bandwidth

The multipath delay quantified by the RMS delay spread σ_τ indicates whether the vehicular wireless channel will experience frequency-selective fading or flat fading. The channel coherence bandwidth B_c (Hz) is calculated using σ_τ and is compared with the signal bandwidth B_w to classify whether the channel will be frequency-selective fading or flat fading. The reported values for σ_τ in the literature [7] are in the range of 40 ns to 400 ns.

For the worst case, multipath delay with σ_τ = 400 ns, the coherence bandwidth with 90% frequency correlation is given by $B_c^{90\%}$ = 1/50σ_τ = 50 kHz and the coherence bandwidth with 50% frequency correlation is given by $B_c^{50\%}$ = 1/50σ_τ = 500 kHz. From these values we observe that, for the worst-case scenario, the coherence bandwidth is in the same order of magnitude as the channel bandwidth which makes it hard to decide whether multipath will be present or not according to theory. However, since the transmission symbol rates (2 µs and 4 µs) are much greater than the RMS delay spread, we could rule out multipath fading here. Therefore, the vehicular wireless channel here is modeled as a flat fading channel.

In summary, based on the analyses presented above, the small-scale fading channel here can be modeled as a slow flat fading channel. In our BER study presented subsequently, we consider the well-known models for slow flat fading channels with a Rayleigh fading model for NLOS conditions and a Rician fading channel for LOS conditions [3].

4.5.3 The Expected BER Performance

The CEN-DSRC-PHY standard requires a BER of 10^{-6} for both uplink and downlink. Due to the random variation of the received bit energy to noise spectral density ratio $\gamma = E_b/N_0$ and due to the random gain in the wireless channel, the BER will also be randomly varying. In such situations, we analyze the BER in two ways according to common practice: (a) analyzing the expected BER for a random channel assuming a distribution for γ given by $f_\gamma(\gamma)$ and (b) analyzing the outage probability for the given channel $\Pr[\gamma < \gamma_{\min}]$ with the BER calculated using only the additive Gaussian noise where $\gamma_{\min}$ is the minimum required γ to achieve the required BER. Note that we analyze the BER at a specific distance of interest, that is at the edge of the coverage area over a short vehicular movement of several cm (*i.e.*, a snapshot at the edge of the cell in our analysis); the shadowing correlation therefore can be assumed to be one because the shadowing variation is only over large distances in space.

Downlink BER Analysis: The BER in the downlink is not at all the issue due to the high transmit power from the RSU/PCU. Calculations show that the expected BER even for the worst-case small-scale fading condition with NLOS (Rayleigh fading) conditions an expected BER of 10^{-6} can be achieved at the maximum downlink range. We, therefore, present no further details on the BER for downlink here.

Uplink BER Analysis: Due to the low transmit power from the SU to the RSU/PCU, the BER in the uplink becomes the bottleneck for the overall link. Figure 4.6 depicts the received mean γ for the worst and best cases of α for two different receiver noise levels P_n and $G_c = 10$. The figure also depicts the minimum required $\gamma = (E_b/N_0)$ to achieve a BER of 10^{-6}. The BER graph for the uplink is given in Figure 4.7, calculated for BPSK modulation. From the figure, we observe that for the AWGN-only case, a BER of 10^{-6} is achievable over the range of guaranteed communication. However, for the Rayleigh fading situation the required mean BER is not achievable at

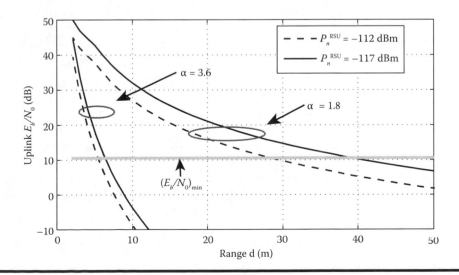

Figure 4.6 Expected E_b/N_0 at the receiver in the uplink.

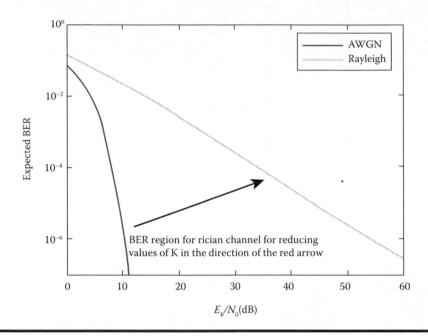

Figure 4.7   BER curves for the uplink: The dark curve (AWGN) approaches the light (Rayleigh) curve when the Rice factor decreases.

all (note that this is only the mean BER under the Rayleigh fading conditions). The result that this minimum mean BER is not achieved does not necessarily mean that this BER will never be achieved; it only means that the expected BER of 10^{-6} will not be achieved at any time (*i.e.*, when you average out the BER statistically). There are instances where the required BER can be achieved as further described by the outage probability which we present below.

It is also important to note that Rayleigh fading occurs for the entire range of the coverage only when the SU is completely shadowed from the RSU all the time by a large vehicle in the adjacent lane, appearing between the SU and the RSU/PCU. This specific case can be waived from an operational point of view, because the law enforcer will acknowledge the reason why the required BER cannot be met. As mentioned, a better way to analyze the BER problem in fading is by considering the outage probability. The outage probability under a Rayleigh fading channel can be directly computed using the Rayleigh distribution for a minimum required $\gamma = (E_b/N_0)_{min}$ to achieve a BER of 10^{-6}, and is given in Figure 4.8. From the figure, we observe that at the edge of the coverage area for NLOS (Rayleigh) condition the required BER can only be achieved for 36% of the time, however, at a distance of 10 m for $\alpha = 1.8$, $P_n = -112$ dBm, which gives us $E_b/N_0 = 26$ dB, the required BER can be achieved for around 95% of the time. Therefore, even though the overall communication is limited by the uplink transmissions, the results here show that the required BER can be achieved with a good success probability. Moreover, we iterate on this point that these outage results shown in Figure 4.8 are only for the extreme worst-case scenario that experiences Rayleigh fading.

4.6 MAC Layer Delay analysis for Completing a Transaction

Some preliminary analysis on the MAC layer delay is presented here to supplement the range analyses presented in the previous sections. Considering the CEN-DSRC MAC protocol [8] for

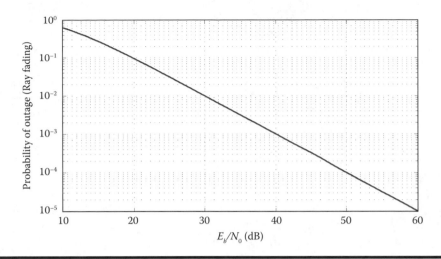

Figure 4.8 Outage of BER for Rayleigh fading channel.

link access and data transmissions between the RSU/PCU and the SU, there will be a minimum delay for completing a transaction with any given SU. This minimum delay corresponds to the distance traveled by the SU depending on the speed. It is important to understand whether the range achieved by the CEN-DSRC-PHY as reported in the previous sections is sufficient to complete a transaction with the data flow described by CEN-DSRC-MAC. Moreover, this delay is expected to increase when data frames are lost due to noise or channel effects. In the remainder of this section, we analyze these delays at the MAC layer. Note that the delay calculated here does not consider the scenario where multiple SUs are present. The delay analysis associated with multiple SUs and the CEN-DSRC MAC process is outside the scope of this chapter.

4.6.1 Minimum Delay

Based on the DT application requirement, two data frames (in the worst case) from the SU are expected to be transmitted each with a size of 128 kbytes to the RSU/PCU. The transmissions of these two data frames are coordinated by the control frames between the RSU/PCU and the SU such as the downlink public window allocation frame, the uplink private window request frame, and a couple of downlink private window allocation frames as per the CEN-DSRC MAC standards. The message structures and the minimum response times for the SU and RSU/PCU to respond to each other to any received control messages are set by the standards [8]. Based on these values, we calculate a total of T_{min} = 16.12 msec required to complete a transaction with the SU. This minimum required time corresponds to a distance of 0.67 m at an SU speed of 150 kmph and a distance of 0.49 m at a speed of 110 kmph. Considering the overall range achieved by the CEN-DSRC-PHY (as in Table 4.2), we can clearly see that the transaction can be successfully completed even in worst-case channel conditions.

4.6.2 Delay with Frame Errors

Based on the single frame loss scenarios identified in the CEN-DSRC MAC standards [8], we study the corresponding delays in completing the transaction between the RSU/PCU and the SU. We only consider a single frame loss situation here by assuming that the likelihood of losing more than one frame is low. Based on the MAC message exchange process, we identify that the maximum delay due to a frame loss in a given transaction occurs when the uplink private window request frame is lost, in which case the total time required to complete a single transaction increases to 19.65 msec. This amount of delay corresponds to a distance of 0.82 m at a speed of 150 kmph and a distance of 0.60 m at a speed of 110 kmph. Again, considering the overall range for the guaranteed communication (as in Table 4.2), we can clearly see that the transaction can be successfully completed even in the worst-case channel conditions.

4.7 Conclusions

In this chapter, we have presented an analysis for using the CEN-DSRC based wireless standards for transmitting DT information between a vehicle and a RSU or a PCU. The PHY layer aspects of the CEN-DSRC technology were treated for understanding the feasibility of using this technology for the application.

Based on the range analysis, approximately 5 m to 37 m of distance (*e.g.*, depending on which lane the vehicle is traveling) is available for the communication between the vehicle and the RSU/PCU. Within this range, it is possible to achieve the required BER as per the CEN-DSRC standard. It is worthwhile noting that the critical part of the entire communication is the uplink where the vehicle captures the signal strength from the downlink signal and uses that energy to send its uplink signal using back-scatter technology. Only one vehicle was considered in this analysis and the presence of other interferences in the frequency band was also assumed to be negligible. These assumptions were made primarily to understand the feasibility of using CEN-DSRC for the DT application.

Based on the delay analysis at the MAC layer, we identified that the distance required to complete a transaction between the vehicle and the RSU/PCU at a velocity of 150 kmph is less than a meter depending on whether frame loss/errors occur or not, at the same time this distance becomes closer to 0.5 m as the velocity is reduced to less than 100 kpmh. This means only a meter or a fraction of a meter out of the total range (5 m–37 m) is required for the transaction to be successful. If the transaction cannot take place in the first meter within the range (at the edge of the coverage area) then there will multiples of opportunities to complete the transaction in the subsequent meters. Considering that the outage probability is low when the vehicle gets closer to the RSU/PCU the chances, therefore, for completing the transaction become high. This gives us a positive view in terms of the feasibility of using CEN-DSRC for the new version of the DT or other future applications. Future developments will validate the study presented in this chapter with measurement campaigns.

References

1. Regulation (EU) No 165/2014 of the European Parliament and of the Council of 4 February 2014 on tachographs in road transport, 2014.
2. DIRECTIVE (EU) 2015/719 on the maximal weights and dimensions for heavy goods vehicles, 2015.
3. Rappaport T., *Wireless Communications: Principles and Practice*, 2nd edition, New York: Pearson/Prentice Hall, 2002.
4. UNI-EN-12253-2005, Road Transport and Traffic Telematics—Dedicated Short Range Communication (DSRC)—Physical Layer Using Microwave at 5.8GHz, 2005.

5. Marc S., and Karaklajic D., *Internet of Trucks and Digital Tachograph—Security and Privacy Threats. ISSE 2014 Securing Electronic Business Processes*, Wiesbaden: Springer Fachmedien, pp. 230–238, 2014.

6. Karedal J., Tufvesson F., Czink N., Paier A., Dumard C., Zemen T., Mecklenbrauker C.F., and Molisch A.F., A geometry-based stochastic MIMO model for vehicle-to-vehicle communications in *Wireless Communications, IEEE Transactions on,* vol.8, no.7, pp. 3646–3657, July 2009.

7. He R., Molisch A.F., Tufvesson F., Zhong Z., Ai B., and Zhang T., Vehicle-to-vehicle propagation models with large vehicle obstructions, in *Intelligent Transportation Systems, IEEE Transactions on*, vol.15, no.5, pp. 2237–2248, Vienna, Austria, Oct. 2014.

8. UNI-EN-12795-2005, Road Transport and Traffic Telematics—Dedicated Short Range Communication (DSRC)—DSRC Data Link Layer—Medium Access and Logical Control, 2005.

9. Mecklenbrauker C.F., Molisch A.F., Karedal J., Tufvesson F., Paier A., Bernado L., Zemen T., Klemp O., and Czink N., Vehicular channel characterization and its implications for wireless system design and performance, *Proceedings of the IEEE*, vol.99, no.7, pp. 1189–1212, 2011.

10. Hourani A., Chandrasekharan S., Baldini G., and Kandeepan S., Propagation measurements in 5.8GHz and pathloss study for CEN-DSRC, in *Connected Vehicles and Expo (ICCVE), 2014 International Conference on*, pp. 1086–1091, New York, NY, Nov. 3–7, 2014.

Chapter 5

Wi-Fi on the Wheels Modeling and Performance Evaluation

Xi Chen

McGill University
Montreal, Canada

Contents

Automobile manufacturers are delivering a new generation of connected vehicles with in-cabin Wi-Fi devices, enabling advances in a wide range of in-vehicle communications and infotainment capabilities. The communication performance is a key factor enabling instant multimedia streaming. A fundamental question remains to be explored: as every running vehicle is equipped with an in-cabin Wi-Fi, how will the communication performance be affected by the varied number of surrounding vehicles, the transmission power, and the data rate? In order to answer this question, we establish analytical models that embody the novel in-cabin Wi-Fi features. We set up a new evaluation platform for in-cabin Wi-Fi communications based on extensive simulations. We report how the performance of in-cabin Wi-Fi communications is affected by traffic conditions and Wi-Fi hotspot settings. Through extensive analysis and evaluations, we show how to configure the in-cabin Wi-Fi to guarantee the quality of service.

5.1 Background

The automotive industry is currently undergoing a technology revolution. A newly developed Wi-Fi system* provides a wireless network that allows a driver and passengers to connect to the Internet using a smartphone, or a car-embedded infotainment system. The built-in Wi-Fi hotspot (we call it in-cabin Wi-Fi in the rest of the chapter) provides a car with the ability to become an Internet hotspot, powering Wi-Fi devices throughout the vehicle. Hence, cars become portable offices and homes, providing a shared mobile experience for all passengers in a vehicle.

The in-cabin Wi-Fi technology enables passengers to instantly stream music and movies and makes the driving experience easier for everyone. However, along with the popularity of this emerging technology, a critical research question arises: *as every running vehicle on a heavy-traffic road is equipped with an in-cabin Wi-Fi, will the performance (e.g., goodput) of in-cabin Wi-Fi communications degrade much due to the interference from Wi-Fi hotspots in surrounding vehicles?* This is a difficult yet practical problem that is faced by car manufacturers at this initial deployment stage. It is very expensive and almost impossible to run a real testbed with 30 to 1200 vehicles to obtain the answer

---

* Autonet mobile, Buckle up because this is a whole new way to experience the Internet! URL http://www. autonetmobile.com

to this question. However, the performance of in-cabin Wi-Fi communications is the key factor impacting passenger experiences. This requires fundamental study of in-cabin Wi-Fi communications to help researchers and engineers understand the issues and prepare solutions to tackle them.

To address the above challenge, we conduct comprehensive studies of in-cabin Wi-Fi communications in this chapter. Wi-Fi communication performances have been widely studied through analytical models as well as simulation and experimental evaluations in existing literature [1,2]. The existing work mainly considers Wi-Fi hotspots built in static structures and does not take the features of in-cabin Wi-Fi communications into consideration. Due to the mobility of vehicles, the density of Wi-Fi hotspots can vary considerably and it can be very large on a congested road. Though the communication distance of an in-cabin Wi-Fi can be much smaller than those of conventional Wi-Fi hotspots, it still remains an open problem what the in-cabin Wi-Fi communications performance looks like on a densely deployed road. To take these features into account and provide answers for the research question, we build a new evaluation platform for in-cabin Wi-Fi communications based on analytical models and evaluations.

Concretely, we first analyze the features of in-cabin Wi-Fi communications. To understand the impacts of these features on in-cabin Wi-Fi performance, we develop new analytical models based on a two-dimension Markov chain. We conduct extensive simulations to validate proposed analytical models. With theoretical analysis and simulation study, we are able to show how in-cabin Wi-Fi performance is affected by traffic density, transmission power, and data rate. We further recommend some default settings for in-cabin Wi-Fi, so as to guarantee the communication performance under different traffic conditions.

In summary, the main contribution of this chapter is threefold.

1. We investigate an important and novel problem of evaluating in-cabin Wi-Fi communications. Our studies can help point out issues with in-cabin Wi-Fi communications at this early deployment stage and provide valuable recommendations of how to address these issues.
2. We establish a set of analytical models which incorporate unique features of in-cabin Wi-Fi systems. The established models are shown to match simulation results based on real-world settings.
3. With a new evaluation platform, we conduct extensive simulations to investigate the impact of different factors including traffic conditions, transmission power, and data rate. We also discuss some seemingly counterintuitive findings.

The rest of this chapter is organized as follows. Section 5.2 describes the in-cabin Wi-Fi unique features and their impacts. Section 5.3 presents the design of the analytical models. Section 5.4 validates the models in a practical scenario with real-world settings. Section 5.5 conducts extensive simulations and explains important observations with the analytical models. Section 5.6 discusses the related work. Section 5.7 concludes the chapter.

5.2 In-Cabin Wi-Fi Communication Features

In this section, we first describe how the in-cabin Wi-Fi service is provided for each passenger. We then point out the unique features of in-cabin Wi-Fi communications, as well as their impacts on system performance. We further define the average goodput and the regional goodput for in-cabin Wi-Fi communications.

5.2.1 In-Cabin Wi-Fi Services

As shown in Figure 5.1, the in-cabin Wi-Fi service is provided through two wireless links. The long-range cellular link carries the data from base stations to a built-in cellular interface, while the short-range Wi-Fi link delivers the data to the in-vehicle clients directly. There is an Ethernet connection relaying the data from the cellular interface to the in-cabin Wi-Fi. When a client wants to access the Internet, it first sends a request to the in-cabin Wi-Fi. This request is then relayed to the cellular interface via the Ethernet connection. Based on the request, the cellular interface communicates with the base station and receives the corresponding data from the Internet. These data are then delivered back to the in-cabin Wi-Fi and finally the client. By doing the above, the in-cabin Wi-Fi is able to provide passengers services including streaming video for entertainment as well as services like real-time traffic updates and navigation driving directions. The in-cabin Wi-Fi system framework not only allows consumers to bring in and connect to personal mobile devices, but also lets the vehicle act as its own mobile device, enabling embedded vehicle capabilities.

Comparing with long-term evolution (LTE) connected smart devices, the in-cabin Wi-Fi service offers distinct advantages including a more powerful antenna array to improve signal quality, a constant energy source to power this service, and an integrated design that is optimized for in-vehicle use. Meanwhile, compared to other vehicular communication technologies (*e.g.*, dedicated short-range communications [DSRC] [3]), the in-cabin Wi-Fi technology provides a much higher service. The long-range cellular link utilizes 4G LTE technology, and provides a

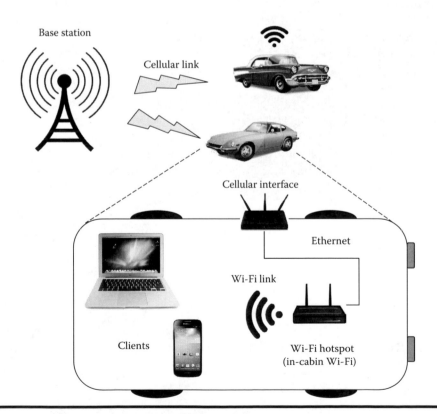

Figure 5.1   An example of the in-cabin Wi-Fi system.

peak download rate of up to 299.6 Mbps. The short-range Wi-Fi link adopts IEEE 802.11n or the newly deployed IEEE 802.11ac, and offers a maximum downlink speed of 433 Mbps for single antenna devices.

In this chapter, we mainly focus on the in-cabin Wi-Fi links. Compared to the long-range cellular links, the Wi-Fi links are more likely to become the bottleneck of the whole system. While the long-range cellular links are controlled and scheduled by the base stations, the Wi-Fi links contend for the channel based on the carrier sense multiple access with collision avoidance (CSMA-CA) scheme. Further, we do not consider the interference of cellular signals on in-cabin Wi-Fi links, since LTE and Wi-Fi use different carrier frequencies.

5.2.2 Differences from Other Communication Techniques

The in-cabin Wi-Fi system is specially designed to meet the requirements of high-speed vehicular communications. As a new communication technique, in-cabin the Wi-Fi communication technique has a number of differences from existing communication techniques.

5.2.2.1 Differences between In-Cabin Wi-Fi Communications and Traditional Cellular Communications

Although the in-cabin Wi-Fi system employs LTE communications, the in-cabin Wi-Fi communications are essentially different from traditional cellular communications: (1) Unlike traditional cellular communications, the in-cabin Wi-Fi communications do not require any cellular interface in the client side. Such a cellular interface is missing in most of potential in-cabin Wi-Fi client devices (*e.g.*, laptops and tablets). Without the in-cabin Wi-Fi communications, these client devices can hardly access the high-speed cellular networks in a mobile scenario. (2) Compared to client devices directly connected to a cellular network, client devices served by an in-cabin Wi-Fi system consume less battery energy. The reason is that, instead of connecting to a distant cellular tower, in-cabin Wi-Fi client devices connects to a nearby onboard Wi-Fi hotspot, which greatly reduces the energy consumption [4].

5.2.2.2 Differences between In-Cabin Wi-Fi Communications and Traditional VANET Communications

(1) The primary objective of traditional vehicular ad hoc network (VANET) communications is to enhance the safety, while the in-cabin Wi-Fi system aims to equip vehicles with the capability to provide infotainment services to passengers. To achieve this goal, compared to traditional VANET systems (*e.g.*, IEEE 802.11p-based systems), the in-cabin Wi-Fi system has a larger bandwidth in both the Wi-Fi and LTE links. (2) At the same time, the in-cabin Wi-Fi network has a unique layout. An in-cabin Wi-Fi serves only the passengers inside the vehicle. Therefore, the communication distance between an in-cabin Wi-Fi and its clients (typically 0.5 m–2 m) is much smaller than the traditional communication distance. (3) Moreover, the vehicle cabin introduces a path loss to the interference signals from hotspots on other vehicles.

5.2.3 Unique Features and Their Impacts

In this section, we discuss some unique features and their impacts of in-cabin Wi-Fi communications, comparing with conventional Wi-Fi communications.

Feature 1: *(Space Layout) Each in-cabin Wi-Fi device is very close to its users.*

The ultimate goal of in-cabin Wi-Fi communications is to provide high-quality Internet access for passengers in the vehicle. The distance between the in-cabin Wi-Fi and passengers ranges from 0.5 m to 2 m, typically.

Impact of Feature 1: Due to Feature 1, the desired coverage of in-cabin Wi-Fi is much less than that in conventional scenarios. To cover a very small area such as a vehicle, each Wi-Fi device only requires a small transmission power to provide a high-quality wireless service. Any signal beyond the vehicle cabin becomes interference to passengers in other vehicles.

Moreover, due to the unique space layout, the number of hidden terminals and exposed terminals is very limited in an in-cabin Wi-Fi system. Take Figure 5.2 as an example. One client C1 is 1 m away from its in-cabin Wi-Fi, A1. The sensing range is set to 400 m, which is a typical value for Wi-Fi communications. The transmission range is smaller than the sensing range, which follows the practice in the industry. If a node is outside the sensing range of C1, then the signal from that node to C1 is very weak and is deemed as noise. If a node is inside the sensing ranges of both A1 and C1, then that node will not collide with A1. Therefore, any other node can be a hidden terminal to A1, only if that node is inside the sensing range of C1 and at the same time outside the sensing range of A1 [5]. Equivalently, one node can be a hidden terminal to C1, only if it is in the shadowed area in Figure 5.2. The corresponding probability is less than 0.0025. A more practical example would be a bidirectional highway, which has four lanes in each direction. The width of such a highway is typically less than 40 m. In this case, the probability that a node would become the hidden terminal of C1 is again less than 0.0025. In a congested scenario where there are 10 vehicles per 100 m in each lane, the expected number of hidden terminals of C1 is less than 1.6. Compared to the interference and collisions of other

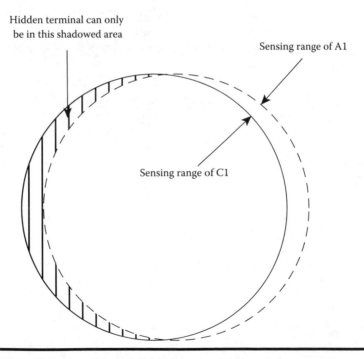

Figure 5.2   An example of in-cabin Wi-Fi hidden terminals.

nodes, the interference of the hidden terminals is very limited. Similar conclusions can be made to exposed terminals in the in-cabin Wi-Fi system. Therefore, we do not discuss hidden and exposed terminals in this chapter.

Feature 2: *(Power Loss) The vehicle cabin introduces a 5 dB to 10 dB power loss to the in-cabin Wi-Fi transmission signals* [6,7].

Impact of Feature 2: Due to this feature, the interference range of in-cabin Wi-Fi shrinks significantly. Therefore, when low transmission power is used, interference signals are restricted in a very local area. For the same deployment density, an in-cabin Wi-Fi usually causes less communication interference (hence fewer collisions) compared with a conventional Wi-Fi. However, along with the bursty traffic on different roads, the deployment density of in-cabin Wi-Fi hotspots can vary widely, ranging from a very small number to a very large number. Hence, it is not intuitive to understand the performance of in-cabin Wi-Fi communications.

Feature 3: *(Mobility) All in-cabin Wi-Fi hotspots are moving on wheels.*

Impact of Feature 3: Due to this feature, in-cabin Wi-Fi communications are highly dynamic. The mobility feature not only indicates that the speed of each vehicle is varying, but also suggests that the traffic density around a vehicle is ever changing. Each in-cabin Wi-Fi needs to adapt itself to the surrounding environments. It is worth noting that the density of in-cabin Wi-Fi hotspots can be very high on a congested road. During busy hours, there can be hundreds of vehicles at a single crossing. With the popularity of the in-cabin Wi-Fi technology, every vehicle may become a hotspot. Even considering Feature 1 and Feature 2, the loss of communication quality can be severe due to fading and interference. Although the speed of vehicles is low on a congested road, the variation of density can be highly dynamic. Thus, we need to extensively analyze in-cabin Wi-Fi communication performances under different traffic conditions.

5.2.4 Metrics

The audio and video streaming with in-cabin Wi-Fi will consume most of the network bandwidth. Goodput is the most important metric to guarantee the quality of service for passengers. We employ two goodput definitions in our work, so as to show both the node-wise and region-wise performances.

The average goodput R_{avg} is used to denote the node-wise goodput, which is defined as follows:

$$R_{avg} = \frac{1}{N} \sum_{i=1}^{N} \frac{n_{recv}^{i} \cdot S_{packet}}{T}, \tag{5.1}$$

where T is the period we are interested in, n_{recv}^{i} is the number of successfully received packets for node i during T, S_{packet} is the size of each packet, and N is the total number of nodes.

Regional goodput R_{reg} is defined from the view of a reference region Ω. It sums up all the goodputs of nodes inside Ω:

$$R_{reg} = \sum_{i \in \Omega} \frac{n_{recv}^{i} \cdot S_{packet}}{T}. \tag{5.2}$$

Denote the number of vehicles in the reference region Ω as n_Ω. Then the relationship between the average goodput and regional goodput is:

$$R_{reg} = n_\Omega \cdot R_{avg}. \tag{5.3}$$

While the average goodput indicates the quality of in-cabin Wi-Fi service for each passenger, the regional goodput reveals how the in-cabin Wi-Fi network performance scales with the traffic density.

5.3 Modeling of In-Cabin Wi-Fi Communications

In this section, we present our modeling of the in-cabin Wi-Fi system. We establish a cross-layer framework to capture the in-cabin Wi-Fi communications. In the medium access (MAC) layer, the proposed framework employs a two-dimensional (2D) Markov chain model, which is similar but different from previous models. In the physical (PHY) layer, the proposed framework considers unique features of in-cabin Wi-Fi communications (comparing with conventional Wi-Fi communications). Based on this framework and its analytical models, we calculate the average goodput and packet loss ratio (PLR).

5.3.1 MAC Layer Modeling

We discuss the MAC layer model from the view of a single in-cabin Wi-Fi device. We establish a 2D Markov chain, which is similar to but different from previous 2D Markov chain models. The improvements made in our MAC layer model are summarized as follows: (1) Unlike previous MAC layer models of IEEE 802.11, our MAC layer model works interactively with our PHY layer model, to capture the unique features of in-cabin Wi-Fi communications. (2) Unlike previous models of saturated communications (*e.g.*, the model proposed by Bianchi in Ref. [2]), our MAC layer model generalizes both saturated and nonsaturated communications. (3) Unlike previous models of nonsaturated communications (*e.g.*, models proposed by Malone et al. [8], Nguyen et al. [9], and Daneshgaran et al. [10]), our MAC layer model captures the limited retransmissions defined in the IEEE 802.11n standard [11]. (4) Unlike the models proposed by Yao et al. [12], our MAC layer model refines the definition of a virtual time slot, hence improves the model accuracy.

The 2D Markov chain for the backoff process is presented in Figure 5.3. The 2D Markov chain models proposed by Daneshgaran et al. [10] and Yao et al. [12] have similar structures to our model. In this chapter, we refine the definition of a virtual time slot, and hence obtain a more accurate model. Further, our 2D Markov chain is coupled with the PHY layer model to integrate the unique features of an in-cabin Wi-Fi system. Concretely, in our 2D Markov chain, time is divided into virtual time slots. We call each virtual time slot a time slot, if not further specified. We define the length of each time slot as the expected duration E_s that the backoff process spends in each state of the 2D Markov chain. E_s is defined in Equation 5.19. We denote each state in the chain as $\{s(t), b(t)\}$. Here, $s(t)$ stands for the backoff stage, which is determined by the number of transmission attempts of the current packet, and $b(t)$ denotes the backoff time counter, which goes down if the current time slot is sensed idle. The backoff stage $s(t)$ represents the first dimension of the 2D Markov chain, while the backoff time counter $b(t)$ is considered as the second dimension. To model the unsaturated throughput, we further append an idle state $\{idle\}$ in the Markov chain to characterize the packet arrival process for real-time Wi-Fi scenarios.

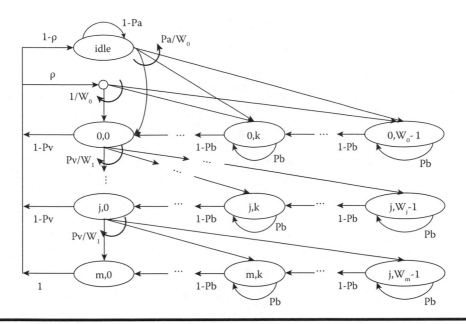

Figure 5.3   2D Markov chain for an in-cabin backoff process.

Let ρ be the probability that the queue of the in-cabin Wi-Fi device is not empty. Use W_j to denote the maximum length of contention window for backoff stage j. Let P_a be the probability that the transmitter leaves the idle state. Let m be the maximum number of retransmission attempts. Let P_b be the probability that the channel is busy at the current time slot. Let P_v denote the probability that a transmission has failed. The cause of a transmission failure is due to both collisions in the MAC layer and signal deteriorations in the PHY layer. Thus, the calculation of P_v is a joint work of both the MAC layer model and the PHY layer model. As a result, these two models are coupled.

Suppose an in-cabin Wi-Fi $A1$ starts functioning with no packet in its queue, that is, it begins from the {*idle*} state. In the next time slot, a packet arrives $A1$ with probability P_a. Thus, $A1$ begins backoff with probability P_a. In this case, it reaches the backoff Stage 0 and uniformly randomly chooses a backoff counter k from $0, 1, \ldots, W_0-1$. From state {$0,k$}, $A1$ goes to state {$0,k-1$} if the channel is not busy (with probability $1-P_b$), or stays in state {$0,k$} if the channel is busy (with probability P_b). Once $A1$ reaches state {$0,0$}, it transmits the packet at the head of its queue. If the transmission fails (with probability P_v), $A1$ enters the next backoff stage, and repeats the backoff procedure again. Upon a successful transmission (with probability $1-P_v$), $A1$ goes back to the {*idle*} state if there is no more packet in its queue (with probability $1-\rho$), or enters the backoff procedure at backoff Stage 0 if its queue is not empty (with probability ρ). If the transmission of a packet has failed m times, $A1$ drops this packet, and goes back to the {*idle*} state or one of the states in backoff Stage 0, depending on the condition of the queue.

Let $\pi_{j,k}$ be the stationary probability of state {$s(t) = j, b(t) = k$} and π_{idle} be the stationary probability of state idle. Then, we can have the following equations [13]:

$$\pi_{idle} = \frac{1-\rho}{P_a} \cdot \pi_{0,0},$$
(5.4)

$$\pi_{j,0} = (P_v)^j \cdot \pi_{0,0}, \textit{ for } j > 0, \tag{5.5}$$

$$\pi_{0,k} = \frac{W_0 - k}{(1 - P_b)W_0} \cdot \pi_{0,0}, \textit{ for } k > 0, \tag{5.6}$$

$$\pi_{j,k} = \frac{(W_0 - k) \cdot (P_v)^{j+1}}{(1 - P_b)W_j} \cdot \pi_{0,0}, \textit{ for } j, k > 0, \tag{5.7}$$

$$1 = \pi_{idle} + \sum_{j=0}^{m} \sum_{k=0}^{W_j - 1} \pi_{j,k}. \tag{5.8}$$

Thus, we can combine Equations 5.4 through 5.8 to obtain the following equation:

$$\pi_{0,0} = \left\{ \frac{W_0 P_v [1 - (2P_v^{m+1})]}{(1 - P_b)(1 - 2P_v)} + \frac{[1 - (P_v)^{m+1}]}{(1 - P_v)} + \right.$$
$$\left. \frac{W_0}{(1 - P_b)} + \frac{(1 - \rho)}{P_a} \right\}^{-1}. \tag{5.9}$$

5.3.2 PHY Layer Modeling

Our PHY layer modeling consists of two analytical models: the propagation model and the packet receiving model. To capture the unique features of in-cabin Wi-Fi communications, we extend existing empirical models accordingly.

5.3.2.1 The Propagation Model

We improve the dual-scope path loss model in Ref. [14] to capture the in-cabin Wi-Fi path loss. Concretely, we adopt a cabin path loss C to represent the unique path loss by vehicle cabins. The in-cabin Wi-Fi path loss $L(d)$ is expressed as follows:

$$L(d) = \begin{cases} -10\gamma_1 \log_{10}\left(\dfrac{d}{d_0}\right) - C, & d_0 \leq d \leq d_c, \\[2ex] -10\gamma_1 \log_{10}\left(\dfrac{d_c}{d_0}\right) - 10\gamma_2 \log_{10}\left(\dfrac{d}{d_c}\right) - C, & d \geq d_c, \end{cases} \tag{5.10}$$

where d denotes the transmission distance, d_0 is the reference distance, d_c is the equivalent transmission distance, γ_1 and γ_2 are path loss factors, and C is the cabin loss constant. For propagations within the reference distance d_0, we assume that they follow a free space propagation.

Propagations outside the reference distance d_0 follow the dual-scope model described by Equation 5.10. According to this model, propagations outside the equivalent transmission distance d_c experience more server path loss than those with d_0.

5.3.2.2 The Packet Receiving Model

We adopt empirical waterfall curves (summarized from our extensive simulations) to model the in-cabin Wi-Fi packet receiving process. In this model, the PLR is a nonincreasing function of signal-to-interference-plus-noise ratio (SINR). The empirical SINR-PLR curves in our packet receiving model are depicted in Figure 5.4. Due to the space limitation, we do not include the mathematical expressions of the SINR-PLR curves in Figure 5.4.

5.3.3 Calculation of Average Goodput and Regional Goodput

5.3.3.1 Estimating P_v

The value of P_v (*i.e.*, the probability that a transmission has failed) is determined by data rate and SINR value. Thus, we do not need to solve the MAC layer 2D Markov chain to get P_v. Instead, we take the SINR and data rate as the inputs, and estimate P_v with the SINR-PLR curves (the empirical SINR-PLR curves are illustrated in Figure 5.4). The SINR value can be estimated accurately with many existing techniques (*e.g.*, Ref. [15,16,17]).

5.3.3.2 Estimating Throughput

Define λ as the packet arrival rate and μ as the average service rate. Then the probability ρ that the queue is not empty can be expressed as follows:

$$\rho = \min\left(1, \frac{\lambda}{\mu}\right). \tag{5.11}$$

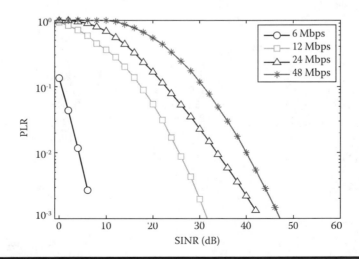

Figure 5.4   Empirical SINR-PLR curves.

Define τ as the probability that the transmitter transmits at current time slot, then the relationship between τ and P_b is:

$$\tau = \sum_{j=0}^{m} \pi_{j,0} = \left[1 - \left(P_b \right)^{m+1} \right] \cdot \pi_{0,0},$$ (5.12)

$$P_b = 1 - (1 - \tau)^{n-1},$$ (5.13)

where n is the total number of vehicles.

We can further express the probability P_{tr} that at least one node is transmitting at the current time slot as follows:

$$P_{tr} = 1 - (1 - \tau)^n.$$ (5.14)

The probability P_s that a transmission is successful can be expressed as follows:

$$P_s = n\tau(1 - \tau)^{n-1}.$$ (5.15)

The relative throughput S is defined as the fraction of time that the channel is used for transmission [2]:

$$S = \frac{E[\text{Pay load information transmitted in a time slot}]}{E_s}$$

$$= \frac{P_s T_p}{(1 - P_{tr})\sigma + P_s T_s + (P_{tr} - P_s)T_c},$$ (5.16)

where Es is expected length of a virtual time slot in our MAC layer model, σ is the basic slot length defined in the 802.11n standard [11], T_s is the round-trip time for a successful transmission, T_c is the time for a failed transmission, and T_p is the transmission time for the data section in a data packet. T_s and T_c are defined as follows [2]:

$$T_s = H + T_p + SIFS + \sigma + T_{ACK} + DIFS,$$ (5.17)

$$T_c = H + T_p + DIFS,$$ (5.18)

where H is the time for packet headers transmission, T_{ACK} is the time for ACK packet transmission, $SIFS$ is the time interval of the short interframe space, and $DIFS$ is the time interval of the distributed coordination function (DCF) interframe space. Then, the expected length of a time slot E_s is calculated as follows:

$$E_s = (1 - P_{tr})\sigma + P_s T_s + (P_{tr} - P_s)T_c.$$ (5.19)

We assume that the packet arrival follows a Poisson process. Then, we can express the probability of exiting idle state p_a as follows:

$$P_a = 1 - e^{-\lambda \cdot E_s}.$$ (5.20)

The average service rate of each node is calculated as follows:

$$\mu = \left\{ (1 - P_v) T_s \sum_{j=0}^{m} \pi_{j,0} + E_s \sum_{j=0}^{m} \sum_{k=1}^{W_j-1} \pi_{j,k} + P_v T_c \pi_{m,0} \right\}^{-1}.$$ (5.21)

By solving the system of nonlinear equations from Equation 5.9 through 5.21, we get the value of relative throughput S.

Then, we can estimate the average throughput as follows:

$$S_{average} = \frac{S \cdot r}{n},$$ (5.22)

where r is the data rate.

5.3.3.3 Obtaining PLR and Goodput

We calculate the PLR as follows:

$$PLR = P_v \cdot \pi_{m,0} = (P_v)^{m+1} \cdot \pi_{0,0}.$$ (5.23)

The average goodput, R_{avg}, is calculated as follows:

$$R_{avg} = PLR \cdot S_{average}.$$ (5.24)

5.4 Communication Model Validation

Real testbed experiments will be ideal to evaluate this research. However, it is not practical, due to following reasons: first, to make the setup emulate real-life scenarios and test scalability, we need to build a testbed with 30–1200 vehicles. This is expensive and difficult to operate; second, even if we have the testbed, it is hard to duplicate previous experiment setups. Hence, we resort to comprehensive simulation studies. In this section, we first describe the setup for the evaluation. We then validate our analytical models by comparing the theoretical results with ns-2 simulations, with a vehicle mobility module developed in-house at General Motors (GM). Due to the space limitation, we consider a typical simulation scenario as described in Tables 5.1 through 5.4.

5.4.1 Evaluation Setup

In this subsection, we describe the highway scenario, device and cabin settings, PHY layer model, and propagation model.

Urban highway scenario: We conduct our analysis in a typical bidirectional urban highway scenario, of which the length is 3000 m. The highway contains four lanes, each of which has a width of 4 m. Taking the isolation zone into consideration, the total width of the highway is 20 m. Vehicles on each lane are deployed according to a Poisson process. The average length of vehicles is set to 4 m. Each vehicle is equipped with one in-cabin Wi-Fi device. As the main purpose is to investigate in-cabin Wi-Fi communication performances under different traffic conditions, we consider a simple case that each in-cabin Wi-Fi device serves one passenger. The reference range

Table 5.1 Settings for the Highway Scenario

Description	Value
Highway length	3000 m
Highway width	20 m
Number of lanes	4
Lane width	4 m
Isolation zone width	4 m
Positions of vehicles	Poisson
Number of vehicles (N)	$N \in [30,1200]$
Vehicle length	4 m
Number of Wi-Fi devices on each vehicle	1
Number of users served by each AP	1
Distances between AP and target users	[0.1 m, 0.5 m]

Note: AP, access point.

Table 5.2 Settings for In-Cabin Wi-Fi Device

Description	Value
Center frequency	2.4 GHz
Bandwidth	20 MHz
Transmission scheme	CBR (unicast)
Network protocol	UDP
Transmission power	0, 10, 20, and 28 dBm
Buffer size	5 MB
Packet size	500 bytes
MAC protocol	CSMA/CA
Workload distribution	Poisson
Data arrival rate	340 Kbps, 2 Mbps, and 6 Mbps

Ω for the regional goodput calculation is set to 1000 m, which is a typical coverage range of an LTE base station in an urban scenario [18]. Table 5.1 summarizes all the settings for the highway model.

In-cabin Wi-Fi device settings: According to 802.11n protocol [11], we utilize the 2.4 GHz frequency and allocate 20 MHz bandwidth for transmissions. Workload arrivals of each user follow the Poisson process, which is commonly used to characterize the Internet bursty traffic [19]. The duration of each workload is an exponential random variable. The interval time between two adjacent workloads is also an exponential random variable. During each transmission interval, the workloads come in a constant bit rate (CBR) manner. The transmission scheme is unicast;

Table 5.3   Settings for the Propagation Model

Description	Value
Reference distance, d_0	1 m
Equivalent transmission distance, d_c	220 m
In-cabin constant, C	10 dB
Path loss parameter, γ_1	1.9
Path loss parameter, γ_2	4.0

Table 5.4   Data Rates in This Chapter

Data Rate	Modulation	Coding Rate
48 Mbps	64 QAM	2/3
24 Mbps	16 QAM	1/2
12 Mbps	QPSK	1/2
6 Mbps	BPSK	1/2

the network protocol is User Datagram Protocol (UDP). Table 5.2 summarizes the settings for Wi-Fi devices.

Propagation model: The empirical values of parameters in the propagation model are summarized in Table 5.3. We modify the model parameters in Ref. [14], so as to fit the model to our scenario.

Data rates: The perceivable data rates are determined by modulation and coding rates used in transmissions. We consider four defined transmissions options in IEEE 802.11n, illustrated in Table 5.4.

5.4.2 Model Validation

In order to illustrate the effectiveness of our proposed analytical model, we compare the theoretical results with simulation results, in terms of both average and regional goodputs. We further compare our proposed model with the model proposed by Yao et al. [12], so as to illustrate the improvement brought by our model. Yao's model is one of the most accurate and related models that capture the unsaturated throughput of 802.11-based communications. In this subsection, we set transmission power as 20 dBm and data arrival rate as 340 Kbps. The value of m is set to 6, which means a packet is dropped if the transmission attempt has failed seven times. As shown in Figure 5.5, we compare the modeled goodputs of our model (denoted as "Model") with the simulated goodputs (denoted as "Sim"), as well as the modeled goodputs of Yao's model (denoted as "Yao").

1. Average goodput
2. Regional goodput

In Figure 5.5a, we can observe that the average goodputs of our analytical model for all four data rates can fit the simulated average goodputs very well. Similarly, in Figure 5.5b, the regional goodputs with analytical models for all data rates can fit the simulated regional goodputs very well. In contrast, Yao's model sometimes yields a large error in modeling the

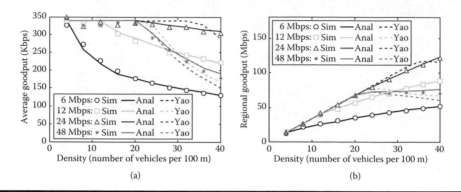

Figure 5.5   Model evaluation via (a) average and (b) regional goodputs. The simulated goodputs are denoted as "Sim" and marked by dots. The modeled goodputs of our model are denoted as "Anal" and marked by solid lines. The modeled goodputs of Yao's model are denoted as "Yao" and marked by dash lines.

goodputs in the high-density region (*i.e.*, when the vehicle density is greater than 20 vehicles per 100 m). This confirms that our model improves the modeling accuracy. The modifications made by our model over the existing models have already been summarized in Section 5.3.

5.5   Extensive Simulation Study, Observation, and Analysis

In this section, we present our simulation study of the in-cabin Wi-Fi communications. We summarize several important observations to show how in-cabin Wi-Fi performance is impacted by traffic density, transmission power, and data rate. In-depth analysis of these observations is provided. In order to obtain a certain goodput guarantee under different traffic conditions, we provide important recommendations of default settings for in-cabin Wi-Fi devices to automotive engineers. We further illustrate the impacts of transmission power and data rate on queuing delay, in order to provide useful guidelines for delay-sensitive applications.

5.5.1   Impacts of Traffic Density on Goodputs

Due to the mobility of vehicles, in-cabin Wi-Fi communication environments change a lot with traffic conditions. To illustrate the impacts of traffic density, we consider an example with transmission power of all vehicle as 20 dBm and the data rate of all vehicles as 48 Mbps. The data arrival rate of each vehicle is 340 Kbps ($\lambda = 340Kbps$). Figure 5.6 illustrates how the average goodput is affected by the traffic density.

Based on Figure 5.6, we summarize our observation as follows:

Observation 1: *The average goodput decreases with the traffic density. Meanwhile, the average goodput is degrading faster in the high-density region than in the low-density region.*

We can explain Observation 1 as follows:

1. As the density increases, the number of interference node *n* increases. When more interference nodes appear, the interference power increases, thus leading to a decrease in the SINR value. According to Equation 5.13, the probability P_b that the channel is sensed as busy

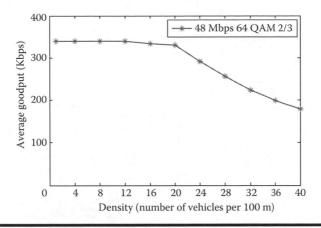

Figure 5.6   Average goodput when transmission power is 20 dBm and data rate is 48 Mbps.

increases with n. Both PHY-layer and MAC-layer performances degrade with the increasing traffic density. As a result, the average goodput of all the users drops with the traffic density.

2. When the traffic density is low, the impact of interference is low. The wireless channel capacity is larger than the data arrival rate. All the arrived packets can be transmitted immediately. Thus the average goodput shown in Figure 5.6 is approximating the data arrival rate. This explains why the average goodput is degrading very slowly in the low-density region.

3. As the traffic density increases, n and P_b increase. Due to the increasing interference, the SINR drops sharply. As a result, the capacity of the wireless channel begins to decrease, and can no longer serve all arrived packets. The average goodput begins to decrease severely with the decreasing channel capacity. This explains why the average goodput is degrading much faster in the high-density region than in the low-density region. ■

To verify the above reasoning, we further demonstrate how channel SINR is affected by traffic density. We use transmission power of 20 dBm as an example. Figure 5.7 presents how the SINR value changes according to traffic density.

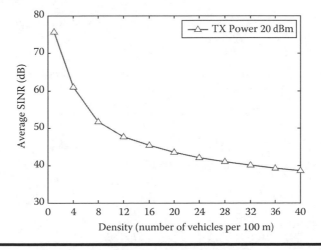

Figure 5.7   SINR value for the transmission power of 20 dBm.

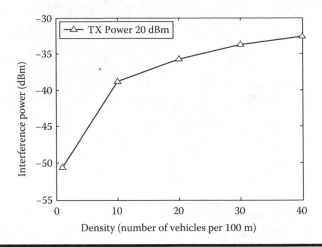

Figure 5.8   Interference power when transmission power is 20 dBm.

It is shown in Figure 5.7 that SINR decreases with traffic density. Meanwhile, SINR is decreasing faster in a lower traffic density region than in a higher traffic density region. To explain the above impact of traffic density on SINR value, we first consider how interference power changes according to traffic density. Figure 5.8 shows how interference power evolves with respect to traffic density.

It is shown in Figure 5.8 that interference power increases with traffic density. Meanwhile, interference power increases faster in a lower-density region than in a higher-density region. As traffic density increases, interference nodes increase, thus interference power increases. Meanwhile, the probability of collisions starts to converge, which means the number of simultaneously transmitting nodes converges. Thus, the interference power is increasing slowly in a higher-density region.

When the signal power and noise power is fixed, the relationship between interference power and SINR value is reciprocal:

$$\text{SINR} = \frac{S_P}{N_p + I_p},$$ (5.25)

where S_P denotes the signal power, N_P denotes the noise power, and I_p denotes the interference power. Combining the tendency of interference power and Equation 5.25, we can see why SINR changes in the way that Figure 5.7 shows.

We also analyze the impact of traffic density on regional goodput. Figure 5.9 illustrates how the regional goodput is affected by the traffic density.

The following observation can be summarized from Figure 5.9.

Observation 2: *The regional goodput increases with the traffic density. Meanwhile, the regional goodput is increasing faster in the low-density region than in the high-density region.*

We can explain Observation 2 as follows:

1. As the traffic density increases, the number of vehicles in the reference range n_Ω increases. From Equation 5.3, we can see that the regional goodput increases with n_Ω. This explains why the regional goodput increases with the traffic density.

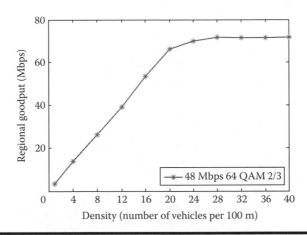

Figure 5.9   Regional goodput when transmission power is 20 dBm and data rate is 48 Mbps.

2. From Equation 5.3, the regional goodput increases with both the number of vehicles in the reference range and average goodputs of these vehicles. When the average goodput decreases quickly in the high traffic density region, the regional goodput becomes increasing slower even with a larger number of vehicles in the reference range. ■

5.5.2 Impacts of Transmission Power on Goodputs

The transmission power of an in-cabin Wi-Fi device not only determines the signal power for passengers in the vehicle, but also affects the interference level of in-cabin Wi-Fi communications. Usually, a higher transmission power provides a better signal power but introduces more interference; a lower transmission power provides a worse signal power but introduces less interference. To see the impact of the transmission power, we use a data arrival rate of 340 Kbps as an example. Figure 5.10 shows how the transmission power affects the average goodput.

1. Transmission data rate: 48 Mbps
2. Transmission data rate: 24 Mbps
3. Transmission data rate: 12 Mbps
4. Transmission data rate: 6 Mbps

From Figure 5.10, we can summarize the following observation.

Observation 3: *In a lower traffic density region, the transmission power has a very limited impact on the average goodput. The small transmission powers (0 dBm and 10 dBm) usually achieve better average goodputs than that of the large transmission powers (20 dBm and 28 dBm) in a higher traffic density region. In general, the transmission power of 10 dBm provides the best and the most robust performance in average goodput.*

We further compare the regional goodput of different transmission powers in Figure 5.11, and we can summarize the following observation.

1. Transmission data rate: 48 Mbps
2. Transmission data rate: 24 Mbps

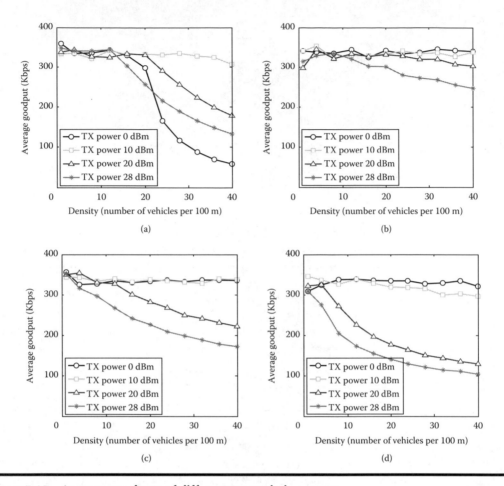

Figure 5.10   Average goodputs of different transmission power.

3. Transmission data rate: 12 Mbps
4. Transmission data rate: 6 Mbps

Observation 4: *In a lower traffic density region, the transmission power has a very limited impact on the regional goodput. The small transmission powers (0 dBm and 10 dBm) provide better regional goodputs than that of the large transmission powers (20 dBm and 28 dBm) in a higher traffic density region. In general, the transmission power of 10 dBm provides the best and the most robust performance in regional goodput.*

We can explain Observations 3 and 4 with a trade-off between PHY-layer performance and MAC-layer performance in terms of transmission power.

1. Usually, the PHY-layer performance increases with transmission power. Meanwhile, the PHY-layer performance is increasing slower in the high power region (20–28 dBm) than in the low-power region (0–10 dBm). As the transmission power increases from 0 dBm to 10 dBm, an increased transmission power provides an increased signal power, leading to a large increase in SINR and a significant improvement of the PHY-layer performance. However, as we further

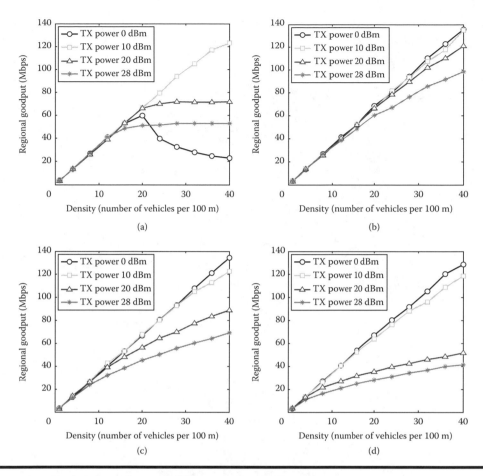

Figure 5.11   Regional goodputs of different transmission power.

increase the transmission power to over 20 dBm, the channel distortion, as well as the interference power, become the bottleneck, and this change cannot be reflected by the SINR value. In this case, we can only achieve marginal improvement in the PHY-layer performance with a large increase in transmission power.

2. Usually, the MAC-layer performance decreases with transmission power. As the transmission power increases, interference power increases. Thus, the interference range of each in-cabin Wi-Fi increases, leading to an increase in the number of interference vehicles n. According to Equation 5.13, as n increases, collisions are more likely to happen. Thus, the MAC-layer performance degrades with the increasing transmission power.

3. With the largest transmission power (28 dBm), in-cabin Wi-Fi hotspots impose a large interference power on each other, making the MAC-layer performance degenerate substantially. With the lowest transmission power (0 dBm), in-cabin Wi-Fi hotspots may fail to deal with channel fading and path loss efficiently, resulting in a huge loss in the PHY-layer performance, especially when the data rate of 48 Mbps is used. Therefore, both the largest and the smallest transmission power cannot achieve the best performance.

4. The transmission power of 10 dBm provides the best and the most robust performance in terms of average goodput and regional goodput. It achieves a good trade-off between the

PHY-layer performance and the MAC-layer performance: it provides a well enough signal power such that P_v approximates to 0; the transmission interference is limited to a local area, and the number of interference nodes n is small. ■

5.5.3 Impacts of Data Rate on Goodputs

The adaptive modulation and coding (AMC) technique [20] has been widely studied and applied in wireless communications. It is still necessary to analyze the impacts of data rate, as we have a completely different application with many distinct features. To focus on the data rate, we fix the data arrival rate as 340 Kbps. Figure 5.12 illustrates the average goodputs with different data rates.

1. Transmission power: 28 dBm
2. Transmission power: 20 dBm
3. Transmission power: 10 dBm
4. Transmission power: 0 dBm

From Figure 5.12, we summarize the following observation. ■

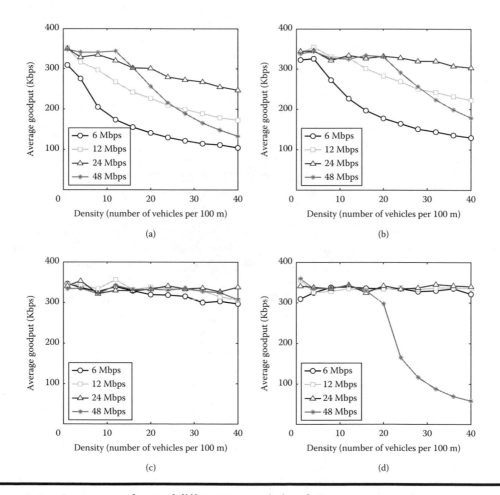

Figure 5.12   Average goodputs of different transmission data rates.

Observation 5: *In general, the data rate of 24 Mbps provides the largest and the most robust average goodput for in-cabin Wi-Fi communications. Both the highest and the lowest data rates provide poor average goodputs.*

We explain Observation 5 with a trade-off between the PHY-layer performance and MAC-layer performance in terms of the data rate.

1. The PHY-layer performance decreases with the data rate. A higher data rate corresponds to a higher order modulation, thus is more vulnerable to noise and interference. To achieve the same PLR performance, a higher data rate requires a higher SINR value (*i.e.*, a better channel condition). In other words, when the PHY-layer condition is fixed, PLR increases with the data rate. This suggests that each packet experiences a larger number of retransmissions when a higher data rate is used.
2. The MAC-layer performance increases with the data rate. A higher data rate can transmit more packets when the channel is clear. Meanwhile, the transmission duration of each packet reduces with data rate. For example, for a 500-byte packet, the transmission duration with 6 Mbps and 48 Mbps is 6.7 msec and 0.8 msec, respectively. With a higher data rate, each packet goes through the wireless channel faster, and experiences less collision in transmission.
3. According to the above explanations, there is a trade-off between the PHY layer performance and MAC layer performance in terms of the data rate. For the data rate range of 6 Mbps to 24 Mbps, the decrease of the average goodput due to the channel fading and interference is complemented by the increase of the transmission rate. Hence, the average goodput increases with the data rate in this range. For the data rate range of 24 Mbps to 48 Mbps, the impact of channel fading and interference overshadows the increase of the data rate. As a result, the average goodput decreases with the data rate in this range.
   a. Transmission power: 28 dBm
   b. Transmission power: 20 dBm
   c. Transmission power: 10 dBm
   d. Transmission power: 0 dBm   ■

Figure 5.13 illustrates regional goodputs of different data rates. We can summarize the following observation from Figure 5.13.

Observation 6: *In general, the data rate of 24 Mbps provides the largest and the most robust regional goodput for in-cabin Wi-Fi communications.*

As the explanation for Observation 6 is similar to that of Observation 5, we skip the explanation due to the space limitation. We also notice that the tendency of the goodput of 48 Mbps is different from those of 6 Mbps and 12 Mbps. The goodput of 48 Mbps decreases rapidly in the high-density region. The reason is explained as follows. For the data rates of 6 Mbps and 12 Mbps, the goodput decreases mainly due to the increasing collisions. The impact of collisions on goodput grows gradually with the traffic density. For the data rate of 48 Mbps, the goodput reduces mainly due to the decreasing channel SINR. According to the empirical packet receiving model (*i.e.*, the SINR-PLR curves in 5.4), the packet loss increases dramatically with the decreasing channel SINR, resulting in a fast-reducing goodput. Consequently, the goodput of 48 Mbps decreases much faster than that of other data rates.

Figures 5.12 and 5.13 also confirm Observation 3 and Observation 4. For example, by comparing Figure 5.12a with Figure 5.12b, c, and d, we can see that the transmission power of 10 dBm

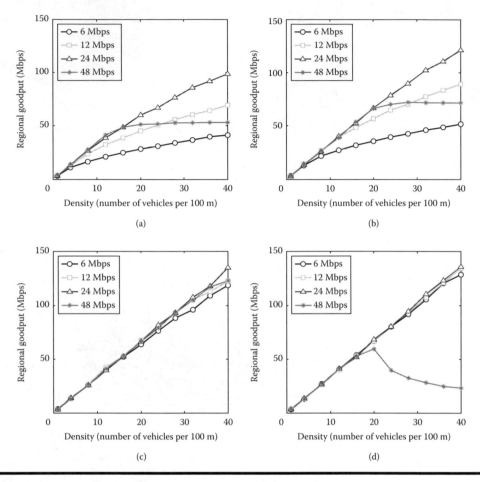

Figure 5.13 Regional goodputs of different transmission data rates.

can support all the data rates well, while other transmission powers result in performance degeneration of one or several data rates. Similarly, we can use Figures 5.10 and 5.11 to confirm that the transmission power of 10 dBm is superior to other power options.

5.5.4 More Results on the Impact of Data Rate with Increased Data Arrival Rates

We present more simulation results on the impact of data rate. To better understand the service speed supported by in-cabin Wi-Fi, we increase the data arrival rate to 10 Mbps (for high-quality video streaming service) and 48 Mbps (for saturated throughput), respectively. We fix the transmission power as 10 dBm, which gives the best goodput performance as previously discussed. Figure 5.14 presents the impact of data rate on the average goodput when the transmission power is 10 dBm and the data arrival rates are 10 Mbps and 48 Mbps, respectively.

1. Packet arrival rate: 10 Mbps.
2. Packet arrival rate: 48 Mbps.

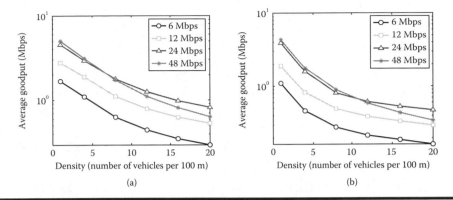

Figure 5.14 **Impact of data rate on average goodputs when transmission power is 10 dB and the data arrival rates are (a) 10 Mbps and (b) 48 Mbps, respectively.**

Figure 5.14 shows that, in general, the data rate of 24 Mbps is still the best choice for in-cabin Wi-Fi goodput performance. Although the data rate of 48 Mbps slightly outperforms the data rate of 24 Mbps by 10.0% when there is only one vehicle in each 100 m, the advantage of the data rate of 48 Mbps is limited. First of all, the data rate of 48 Mbps only performs well when traffic density is less than eight vehicles per 100 m, and the goodput gap between data rates of 48 Mbps and 24 Mbps is marginal (less than 10%) in this case. When traffic density increases, the data rate of 24 Mbps outperforms the data rate of 48 Mbps by up to 33.8%. Second, traffic density is highly dynamic and will not always stay very low. That is to say, the data rate of 48 Mbps cannot catch up with the change of traffic conditions as well as the data rate of 24 Mbps. Therefore, Figure 5.14 again confirms Observation 5.

5.5.5 Recommended Settings for In-Cabin Wi-Fi Devices[*]

With our analytical models and simulation study, we are able to answer the question raised by in-cabin Wi-Fi communications at the beginning of this chapter.

Remark 1: *(Answer to the question) Through Observations 1–6, we can conclude that performance of in-cabin Wi-Fi communications degrades quickly with the interferences from nearby Wi-Fi hotspots. However, such a degradation can be largely alleviated if we choose the transmission power and data rate appropriately.*

We recommend the default settings of transmission power and data rate for in-cabin Wi-Fi devices, so as to guarantee a good quality of service under different traffic conditions.

- ■ **Recommended transmission power:** 10 dBm. The transmission power of 10 dBm achieves a good trade-off between the signal power and the interference, and provides the largest goodput. Meanwhile, 10 dBm introduces less radiation to the human body than 20 dBm, which is the default transmission power for most of the traditional Wi-Fi devices.

---

[*] Disclaimer: The results and opinions are those of the authors and do not represent the views and opinions of GM.

■ **Recommended data rate:** 24 Mbps. The data rate of 24 Mbps achieves a good trade-off between MAC- and PHY-layer performances. Although 48 Mbps slightly outperforms 24 Mbps when traffic density is extremely low, 24 Mbps is still the most reliable choice that provides the best goodput in most of the cases.   ■

In general, our recommended settings provide the best goodput performance when the default settings of Wi-Fi hotspots are fixed for all traffic conditions. AMC schemes (*e.g.*, the one proposed by Goldsmith [20]) can be employed for in-cabin Wi-Fi communications to achieve better goodput performance. However, the design of AMC and power control is out of the scope of this chapter.

5.5.6 Impact of Transmission Power on Delay

Delay is another important metric of in-cabin Wi-Fi performance. To see the impact of the transmission power on delay, we use a data arrival rate of 340 Kbps and the data rate of 24 Mbps as an example. Figure 5.15 shows how queuing delay is affected by transmission power.

From Figure 5.15, we summarize the following observation.

Observation 7: *The small and medium transmission power (0 dBm and 10 dBm) achieve lower queuing delays than those of the large transmission powers (20 and 28 dBm). The transmission power of 10 dBm is again a good candidate for in-cabin Wi-Fi communications, in term of queuing delay.*

The reasons of Observation 7 are similar to those of observations 3 and 4. The transmission power of 10 dBm achieves the best trade-off between MAC- and PHY-layer performance. On one hand, packets with 10 dBm transmission power encounter less collision than those with 20 dBm and 28 dBm. On the other hand, packets with 10 dBm transmission power experience

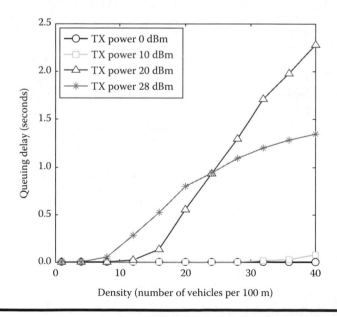

Figure 5.15   Queuing delays of different transmission powers when the data rate is 24 Mbps.

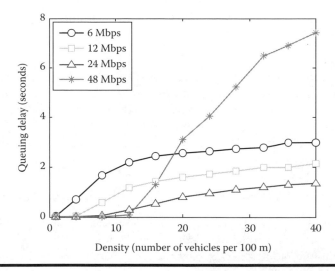

Figure 5.16 Queuing delay of different data rates when the transmission power is 10 dBm.

fewer retransmissions than those with 0 dBm. Thus, transmission power of 10 dBm achieves the smallest queuing delay. ■

5.5.7 *Impact of Data Rate on Delay*

To show the impact of data rate on queuing delay, we use a data arrival rate of 340 Kbps, and a transmission power of 10 dBm as an example. Figure 5.16 depicts the queuing delays of different data rates.

The following observation is summarized from Figure 5.16.

Observation 8: *The data rate of 24 Mbps provides the smallest queuing delay. Both the highest and the lowest data rates provide poor delay performances.*

The reasons of Observation 8 are similar to those of observations 5 and 6. The data rate of 24 Mbps achieves the best trade-off between MAC- and PHY-layer performances. At the MAC layer, the data rate of 24 Mbps can transmit packets much faster than data rates of 6 and 12 Mbps. Thus, packets with 24 Mbps stay for less time in service queues than those with 6 and 12 Mbps. At the PHY layer, SINR is large enough to provide a good communication quality for 24 Mbps. But for 48 Mbps, the channel quality degrades substantially with the increase of traffic density; that is why the queuing delay of 48 Mbps increases with the traffic density. Compared to the other data rates, the data rate 24 Mbps transmits packets rapidly, and experiences very few retransmissions. Thus, 24 Mbps has the best delay performance. ■

5.6 Related Work

Advanced technologies have been integrated in vehicles, so as to provide a driving experience that is much safer [3], cleaner [21], and easier [22,23] than ever. In this chapter, we focus on the emerging in-cabin Wi-Fi communications, especially on the performance of the network formed by surrounding in-cabin Wi-Fi devices. We review the work that is most pertinent to ours.

5.6.1 The Deployment of Wi-Fi Hotspot on Wheels

Since 2014, GM have partnered with AT&T to offer embedded 4G broadband on most of its vehicles [22]. This built-in broadband connection not only powers passengers' cellphones and laptops through high-speed Wi-Fi interface, but also provides the potential to improve driving safety. GM is not the only manufacturer providing this technology. Ford [24], BMW [25], and Audi [26] are actively deploying similar in-cabin Wi-Fi devices in their vehicles. Meanwhile, standalone mobile Wi-Fi hotspots are designed by telecommunication companies like Huawei [27]. To offer an attractive in-cabin Wi-Fi service to the customers, it is of critical importance to analyze the performance and develop a comprehensive model of in-cabin Wi-Fi communications.

5.6.2 Modeling and Measurement of Wi-Fi Communications

Analytical models of 802.11-based communications have been extensively studied. Qiu et al. [1] develop a general interference model for wireless communication, based on a Markov chain and a log-normal assumption of interference. However, this method and its extensions are not suitable for large-scale, in-cabin Wi-Fi networks, due to the fact that the sizes of their Markov chains increase quadratically with the number of vehicles. Therefore, we employ a scalable framework, which models the 802.11 DCF and enhanced distributed channel access (EDCA) mechanism. In the seminal work, Bianchi [2] develops a two-dimension (2D) Markov chain to analyze 802.11 saturated unicast throughput with unlimited retransmissions. Subsequent papers extend the study to understand the unsaturated throughput with limited retransmissions (*e.g.*, Malone et al. [8], Daneshgaran et al. [10], and Nguyen et al. [9]). Meanwhile, 802.11 broadcast communications in vehicular networks have been discussed in the literature (*e.g.*, Ma et al. [28], Misic et al. [29], Campolo et al. [30], Yao et al. [12]). While the above approaches are based on per-slot statistics, Tinnirello et al. [31] utilize channel access cycles and design an enhanced nonslot method, which further increases the modeling accuracy. However, we are still lacking an cross-layer study, which couples the unique features of an in-cabin Wi-Fi PHY layer with its MAC layer protocol.

5.6.3 Mobile and Vehicular Ad Hoc Networks

Each in-cabin Wi-Fi operates in a decentralized manner. It obtains access to the 2.4 GHz/5 GHz wireless channel through contentions with surrounding in-cabin Wi-Fi devices. Thus, an in-cabin Wi-Fi network can be considered as a mobile ad hoc network. Previous work (*e.g.*, Zhang et al. [32], Wang et al. [33], Han et al. [34]) extensively studies the capacity and delay of mobile ad hoc networks. Meanwhile, each in-cabin Wi-Fi works closely with cellular communication networks (3G/4G LTE). Together they form a hybrid ad hoc network, of which the throughput and scheduling have been widely discussed (*e.g.*, Li et al. [35], Chen et al. [36], and the references therein). Lu et al. [37] give a comprehensive survey of capacity and delay in a mobile wireless ad hoc network.

Routing in VANETs has been extensively studied. According to Dua et al. in the survey [38], routing schemes in VANETs can be categorized into topology-based routing (*e.g.*, Toutouh et al. [39] and Liu et al. [40]), clustering-based routing (*e.g.*, Pan et al. [41] and Wang et al. [42]), geography-based routing (*e.g.*, Kim et al. [43] and Lee et al. [44]), data fusion-based≈routing (*e.g.*, Wagh et al. [45] and Zhang et al. [46]) and hybrid routing (*e.g.*, Al-Rabayah et al. [47] and Minh et al. [48]). To further improve the communications performance in VANETs, distributed congestion control techniques are proposed for both vehicle-to-vehicle (V2V) communications (*e.g.*, Jiang et al. [49] and Huang et al. [50]) and vehicle-to-infrastructure (V2I) communications (*e.g.*, Chuang et al. [51]).

However, to capture the features of in-cabin Wi-Fi communications, it is necessary to develop a holistic approach covering the details in both PHY and MAC layers. Such an approach should take into consideration the previously undiscussed features of in-cabin Wi-Fi communications.

5.6.4 Measurement and Modeling of Wi-Fi Communications

Measurement study is of critical importance for us to characterize 802.11-based communications. A comprehensive overview of existing measurement tools, as well as their effectiveness in estimating wireless parameters, is presented by Dujovne et al. [52]. Nonintrusive methods, such as Jigsaw (proposed by Cheng et al. [53]), Wit (proposed by Mahajan et al. [54]) and PIE (proposed by Shrivastava et al. [55]), are designed to monitor enterprise wireless local area networks (WLANs) with little system downtime. However, these methods cannot be applied directly to an in-cabin Wi-Fi study. At this early deployment stage, a real testbed with a number of 30 to 1200 vehicles is expensive and extremely difficult to operate. For instance, only to investigate the impact of the transmission power, it is necessary to conduct multiple tests under the same traffic condition. However, it would be hard to maintain the same traffic condition for multiple tests.

5.7 Conclusion

Automotive manufacturers have begun to equip their vehicles with 4G LTE broadband abilities, and are trying to reinvent the mobile experience for all passengers. The built-in Wi-Fi hotspot, that is, the in-cabin Wi-Fi, is deployed to extend this broadband connectivity to Wi-Fi-enabled devices throughout the vehicle. The in-cabin Wi-Fi has some unique features making it different from conventional Wi-Fi. In order to evaluate the performance of in-cabin Wi-Fi communications, we establish a new evaluation platform based on analytical models and extensive ns-2 simulations. With this evaluation platform, we are able to show how in-cabin Wi-Fi communication performance is affected by traffic conditions and device settings. We also make valuable recommendations of Wi-Fi device settings. With the recommended transmission power (*i.e.*, 10 dBm) and data rate (*i.e.*, 24 Mbps), it is possible to maintain good quality of in-cabin Wi-Fi service, even for roads with high traffic densities.

There are several enhancements we would like to explore for the in-cabin Wi-Fi system. In order to improve the performance of the in-cabin Wi-Fi communications, we are currently working on the distributed scheduling of the in-cabin Wi-Fi hotspots. Meanwhile, we are interested in designing a more scalable in-cabin Wi-Fi system to support real-time applications (*e.g.*, video streaming). Another interesting research topic would be to explore the correlation between the cellular link and the Wi-Fi link.

References

1. L. Qiu, Y. Zhang, F. Wang, M.K. Han, and R. Mahajan. A general model of wireless interference. In *Proc. ACM Mobicom*, Montréal, Québec, Canada, pp. 171–182, September 09–14, 2007.
2. G. Bianchi. Performance analysis of the IEEE 802.11 distributed coordination function. *IEEE Journal on Selected Areas in Communications*, 18(3):535–547, 2000.
3. J.B. Kenney. Dedicated Short-Range Communications (DSRC) standards in the United States. *Proc. IEEE*, 99(7):1162–1182, 2011.

4. N. Balasubramanian, A. Balasubramanian, and A. Venkataramani. Energy consumption in mobile phones: A measurement study and implications for network applications. In *Proc. ACM SIGCOMM IMC,* Chicago, Illinois, November 04–06, 2009.

5. Fu-Yi Hung, S. Pai, and I. Marsic. Performance modeling and analysis of the IEEE 802.11 distribution coordination function in presence of hidden stations. In *Proc. IEEE MILCOM,* pp. 1–7, Washington, DC, October 23–25, 2006.

6. G.A. Breit, H. Hachem, J. Forrester, P. Guckian, K.P. Kirchoff, and B.J. Donham. RF propagation characteristics of in-cabin CDMA mobile phone networks. In *Proc. DASC,* volume 2, pp. 12–30, Washington, DC, October 30–November 03, 2005.

7. K.W. Hurst and S.W. Ellingson. Path loss from a transmitter inside an aircraft cabin to an exterior fuselage-mounted antenna. *IEEE Transaction on Electromagnetic Compatibility,* 50(3):504–512, 2008.

8. D. Malone, K. Duffy, and D. Leith. Modeling the 802.11 distributed coordination function in non-saturated heterogeneous conditions. *IEEE/ACM Transactions on Networking,* 15(1):159–172, 2007.

9. S.H. Nguyen, H.L. Vu, and L.L.H. Andrew. Performance analysis of IEEE 802.11 WLANs with saturated and unsaturated sources. *IEEE Transactions on Vehicular Technology,* 61(1):333–345, 2012.

10. F. Daneshgaran, M. Laddomada, F. Mesiti, and M. Mondin. Unsaturated throughput analysis of IEEE 802.11 in presence of non-ideal transmission channel and capture effects. *IEEE Transactions on Wireless Communications,* 7(4):1276–1286, 2008.

11. *IEEE 802.11n standard.* Information technology–Telecommunications and information exchange between systems Local and metropolitan area networks–Specific requirements Part 11: Wireless LAN Medium Access Control (MAC) and Physical Layer (PHY) Specifications," in ISO/IEC/IEEE 8802-11:2012(E) (Revison of ISO/IEC/IEEE 8802-11-2005 and Amendments), pp. 1–2798, November 21, 2012.

12. Y. Yao, L. Rao, X. Liu, and X. Zhou. Delay analysis and study of IEEE 802.11p based DSRC safety communication in a highway environment. In *Proc. IEEE INFOCOM,* pp. 1591–1599, Turin, Italy, April 14–19, 2013.

13. Y. Yao, L. Rao, and X. Liu. Performance and reliability analysis of IEEE 802.11p safety communication in a highway environment. *IEEE Transaction on Vehicular Technology,* 62(9):4198–4212, 2013.

14. L. Cheng, B.E. Henty, D.D. Stancil, F. Bai, and P. Mudalige. Mobile vehicle-to-vehicle narrow-band channel measurement and characterization of the 5.9 GHz Dedicated Short Range Communication (DSRC) frequency band. *IEEE Journal on Selected Areas in Communications,* 25(8):1501–1516, 2007.

15. X. Zhou and X. Wang. Channel estimation for OFDM systems using adaptive radial basis function networks. *IEEE Transaction on Vehicular Technology,* 52(1):48–59, 2003.

16. F.A. Dietrich and W. Utschick. Pilot-assisted channel estimation based on second-order statistics. *IEEE Transactions on Signal Processing,* 53(3):1178–1193, 2005.

17. C. Pandana, Y. Sun, and K.J.R. Liu. Channel-aware priority transmission scheme using joint channel estimation and data loading for OFDM systems. *IEEE Transactions on Signal Processing,* 53(8):3297–3310, 2005.

18. *3GPP LTE.*

19. M. Zukerman, T.D. Neame, and R.G. Addie. Internet traffic modeling and future technology implications. In *Proc. IEEE INFOCOM,* volume 1, pp. 587–596, San Francisco, CA, March 30–April 03, 2003

20. A. Goldsmith. Adaptive modulation and coding for fading channels. In *Proc. IEEE Inform Theory Commun Workshop,* pp. 24–26, Kruger National Park, South Africa, June 25, 1999.

21. *Tesla Motors.*

22. *General Motors built-in Wi-Fi hotspots.*

23. N. Cheng, N. Lu, N. Zhang, Xuemin (Sherman) Shen, and J. W. Mark. Vehicular wifi offloading: Challenges and solutions. *Vehicular Communications,* 1(1):13–21, 2014.

24. *Ford SYNC features.*

25. *BMW car hotspot.*

26. *Audi enables in-car Internet connectivity.*

27. *Huawei mobile Wi-Fi standalone devices.*

28. X. Ma, J. Zhang, X. Yin, and K.S. Trivedi. Design and analysis of a robust broadcast scheme for VANET safety-related services. *IEEE Transaction on Vehicular Technology*, 61(1):46–61, 2012.

29. J. Misic, G. Badawy, and V.B. Misic. Performance characterization for IEEE 802.11p network with single channel devices. *IEEE Transactions on Vehicular Technology*, 60(4):1775–1787, 2011.

30. C. Campolo, A. Vinel, A. Molinaro, and Y. Koucheryavy. Modeling broadcasting in IEEE 802.11p/WAVE vehicular networks. *IEEE Communications Letters*, 15(2):199–201, 2011.

31. I. Tinnirello and G. Bianchi. Rethinking the IEEE 802.11e EDCA performance modeling methodology. *IEEE/ACM Transaction on Networking*, 18(2):540–553, 2010.

32. C. Zhang, X. Zhu, Y. Song, and Y. Fang. C4: A new paradigm for providing incentives in multi-hop wireless networks. In *Proc. IEEE INFOCOM*, pp. 918–926, Shanghai, China, April 10–15, 2011.

33. X. Wang, W. Huang, S. Wang, J. Zhang, and C. Hu. Delay and capacity tradeoff analysis for motion-cast. *IEEE/ACM Transaction on Networking*, 19(5):1354–1367, 2011.

34. C. Han, M. Dianati, R. Tafazolli, X. Liu, and X. Shen. A novel distributed asynchronous multichannel MAC scheme for large-scale vehicular ad hoc networks. *IEEE Transaction on Vehicular Technology*, 61(7):3125–3138, 2012.

35. P. Li and Y. Fang. Impacts of topology and traffic pattern on capacity of hybrid wireless networks. *IEEE Transactions on Mobile Computing*, 8(12):1585–1595, 2009.

36. X. Chen, W. Huang, X. Wang, and X. Lin. Multicast capacity in mobile wireless ad hoc network with infrastructure support. In *Proc. IEEE INFOCOM*, pp. 271–279, Orlando, FL, March 25–30, 2012.

37. N. Lu and X. Shen. Scaling laws for throughput capacity and delay in wireless networks—A survey. *IEEE Communications Surveys Tutorials*, 99:1–16, 2013.

38. A. Dua, N. Kumar, and S. Bawa. A systematic review on routing protocols for vehicular ad hoc networks. *Vehicular Communications*, 1(1):33–52, 2014.

39. J. Toutouh, J. Garcia-Nieto, and E. Alba. Intelligent OLSR routing protocol optimization for VANETS. *IEEE Transaction on Vehicular Technology*, 61(4):1884–1894, 2012.

40. Y. Liu, J. Niu, J. Ma, L. Shu, T. Hara, and W. Wang. The insights of message delivery delay in VANETs with a bidirectional traffic model. *Journal of Network and Computer Applications*, 36(5):1287–1294, 2013.

41. M. Pan, P. Li, and Y. Fang. Cooperative communication aware link scheduling for cognitive vehicular networks. *IEEE Journal on Selected Areas in Communications*, 30(4):760–768, 2012.

42. S.-S. Wang and Y.-S. Lin. Passcar: A passive clustering aided routing protocol for vehicular ad hoc networks. *Computer Communications*, 36(2):170–179, 2013.

43. Y.-J. Kim, R. Govindan, B. Karp, and S. Shenker. Geographic routing made practical. In *Proc. NSDI*, pp. 217–230, Boston, MA, May 2–4, 2005.

44. K.C. Lee, M. Le, J. Harri, and M. Gerla. Louvre: Landmark overlays for urban vehicular routing environments. In *Proc. IEEE VTC Fall*, pp. 1–5, Calgary, AB, Canada, September 21–24, 2008.

45. A. Wagh, X. Li, R. Sudhaakar, S. Addepalli, and C. Qiao. Data fusion with flexible message composition in driver-in-the-loop vehicular {CPS}. *Ad Hoc Networks*, 11(7):2083–2095, 2013.

46. L. Zhang, D. Gao, W. Zhao, and H.-C. Chao. A multilevel information fusion approach for road congestion detection in {VANETs}. *Mathematical and Computer Modelling*, 58(56):1206–1221, 2013.

47. M. Al-Rabayah and R. Malaney. A new scalable hybrid routing protocol for VANETS. *IEEE Trans on Vehicular Technology*, 61:2625–2635, 2012.

48. L.V. Minh, Y. MingChuan, and G. Qing. End-to-end delay assessment and hybrid routing protocol for vehicular ad hoc network. *{IERI} Procedia*, 2:727–733, 2012.

49. L. Jiang and J. Walrand. Approaching throughput-optimality in distributed CSMA scheduling algorithms with collisions. *IEEE/ACM Transaction on Networking*, 19(3):816–829, 2011.

50. C.-L. Huang, Y.P. Fallah, R. Sengupta, and H. Krishnan. Adaptive intervehicle communication control for cooperative safety systems. *IEEE Network*, 24(1):6–13, 2010.

51. C.-C. Chuang and S.-J. Kao. A probabilistic discard congestion control for safety information in vehicle-to-infrastructure vehicular network. In *Proc. Int Conf Comput Ind Eng*, pp. 1–5, Awaji, Japan, July 2010.

52. D. Dujovne, T. Turletti, and F. Filali. A taxonomy of IEEE 802.11 wireless parameters and open source measurement tools. *IEEE Communications Surveys & Tutorials*, 12(2):249–262, 2010.

53. Y-C. Cheng, J. Bellardo, P. Benkö, A.C. Snoeren, G.M. Voelker, and S. Savage. Jigsaw: Solving the puzzle of enterprise 802.11 analysis. In *Proc. SIGCOMM*, pp. 39–50, Pisa, Italy, September 11–15, 2006.
54. R. Mahajan, M. Rodrig, D. Wetherall, and J. Zahorjan. Analyzing the MAC-level behavior of wireless networks in the wild. In *Proc. ACM SIGCOMM*, pp. 75–86, Pisa, Italy, September 11–15 2006.
55. V. Shrivastava, S. Rayanchu, S. Banerjee, and K. Papagiannaki. PIE in the sky: Online passive interference estimation for enterprise WLANs. In *Proc. USENIX NSDI*, pp. 25–38, Berkeley, CA, April 01, 2011.

Chapter 6

IEEE 802.15.13-Compliant Visible Light Communications in Vehicular Networks

Bugra Turan and Sinem Coleri Ergen

Koç University
Istanbul, Turkey

Omer Narmanlioglu and Murat Uysal

Ozyegin University
Istanbul, Turkey

Contents

6.1 Introduction

Intelligent transportation systems (ITS) aim to provide relevant traffic and road information in a timely and reliable manner to improve traffic flow while constituting a milestone for autonomous driving. RF-based communication schemes such as DSRC, based on IEEE 802.11p, and long-term evolution (LTE) are currently considered as main candidates for vehicular connectivity [1–3].

However, RF-based solutions suffer from spectrum scarcity and RF interference in addition to their susceptibility to intentional jamming and spoofing. Furthermore, high data rate requirements to ensure safe critical mobility information delivery is known to result with congestion and packet collisions in a wireless communication scheme with limited bandwidth.

As a complementary technology to RF-based solutions, visible light communication (VLC) is foreseen to be a low-cost, energy efficient, low-complexity, and secure vehicular connectivity scheme, providing high data rates. VLC uses modulated optical radiation of LED [4] or laser diodes to convey digital information without any noticeable effect on the human eye. Currently, LEDs are widely deployed in the headlights and taillights of many production vehicles. LED usage in automobiles is considered favorable as they provide brighter road illumination while their small sizes enable easier shape manipulation of vehicle lights. LEDs are also known to consume less power than their halogen and high-intensity discharge bulb counterparts. The low-energy consumption of LEDs with their dimming support capability, makes them energy efficient communication transmitters. Furthermore, line-of-sight (LoS) communications are targeted with the utilization of LEDs for VLC. Hence, VLC is immune to malicious jamming or spoofing, providing secure vehicular communications. As a consequence of these reasons, VLC has received increasing attention both from academia and industry (*e.g.,* Refs. [5,6] and the references therein). A detailed comparison of the VLC and IEEE 802.11p is provided in Table 6.1 [7].

IEEE 802.11p uses the Wi-Fi standard tailored to the vehicular environment, targeting low-latency local information exchange in the immediate vicinity of the vehicles (*i.e.,* up to 400 m). The physical (PHY) layer of IEEE 802.11p includes eight different data rates from 3 Mbps to 54 Mbps. However, the standard defined 6 Mbps as the default data rate, due to its highest packet delivery ratio (PDR) performance [8]. Higher transmission power levels are required for higher

Table 6.1   A Comparison of the VLC and IEEE 802.11p Performance Characteristics

	VLC	*IEEE 802.11p*
Communication mode	Point-to-point	Point-to-multipoint
Latency	Very low	Less than 50 ms
Data rate	Up to 400 Mbps	Up to 54 Mbps
Range	Up to 100 m	Up to 1 km
Frequency band	400 THz–790 THz	5.8 GHz–5.9 GHz
License	Unlicensed	Licensed
Cost	Low	High
Mobility	Medium	High
EMI	No	Yes
Power consumption	Relatively low	Medium
Coverage	Narrow	Wide
Weather conditions	Sensitive	Robust
Ambient light	Sensitive	Not affected

Source: Uysal, M., et al., *Veh. Technol. Mag. IEEE,* 10(4), 45–53, 2015.

data rates to decrease the packet duration and channel load. As higher transmission power levels increase the interference range, IEEE 802.11p data rates are limited, even for close distance vehicle-to-vehicle (V2V). Hence, providing high data rate vehicular communications [9], VLC is regarded to be a strong candidate for high-speed V2V.

Intensity modulation and direct detection (IM/DD) is typically preferred in the PHY of VLC due to the noncoherent characteristics of light-emitting diode (LED). Therefore, initial works are mainly based on simple modulation techniques such as on-off keying (OOK) and pulse position modulation (PPM) [10]. OOK and PPM have been selected as modulation schemes of the IEEE 802.15.7 standard [11] which is the first standard of VLC, published in 2011. The IM/DD scheme enables low-complexity VLC systems, as IM/DD only utilizes the amplitude of the optical signal. The IEEE has established the Task Group 802.15.13* Short-Range Optical Wireless Communications, which is currently in the process of developing a standard for VLC. This standard aims to deliver peak data rates of 10 Gbits per second. It is expected that advanced PHY techniques such as optical orthogonal frequency-division multiplexing (OFDM), multiple input multiple output (MIMO) communications, link adaptation, and relay-assisted transmission will be employed to reach this ambitious goal [12].

Several optical OFDM solutions have been proposed in the literature [13–17]. These include asymmetrically-clipped optical OFDM (ACO-OFDM) [13], direct current biased optical orthogonal frequency-division multiplexing (DCO-OFDM) [14], unipolar OFDM (U-OFDM) [15], flip OFDM [16], and enhanced unipolar OFDM (eU-OFDM) [17]. The main objective of all these different solutions is essentially the same. They aim to generate a real and unipolar signal that is able to drive the LED for communication. To obtain a real-valued signal at the output of the inverse discrete Fourier transformation equation (IDFT) block, the frame structures in these solutions are rearranged with respect to Hermitian symmetry. Among these modified OFDM schemes, ACO-OFDM [13] loses three-quarters of the spectral efficiency with respect to conventional OFDM transmission. Similarly, U-OFDM [15], also known as flip OFDM [16], loses more than three-quarters of the spectral efficiency however, some noise enhancement occurs. On the contrary, DCO-OFDM [14] uses less than half of the subcarriers for data transmission and it has the maximum achievable spectral efficiency among the proposed solutions. Alternatively, eU-OFDM [17], which doubles the spectral efficiency of U-OFDM, has been recently proposed and its spectral efficiency approaches to DCO-OFDM.

MIMO transmission techniques have been also adopted to VLC to achieve higher data rates and overcome the FOV limitations of a single LED. Several VLC studies [18–27] have already investigated MIMO techniques. In Ref. [18], three MIMO techniques, namely repetition coding (RC), spatial multiplexing (SM), and spatial modulation (SMD), are investigated. Similar analysis is conducted with the integration of OFDM in Ref. [19]. Ref. [20] provides a multiuser MIMO-OFDM analysis and Ref. [21] provides a novel constellation design for MIMO VLC. In Ref. [22], the performance of SM is investigated using suboptimal receiver techniques and the effect of channel correlation is discussed. In Ref. [23], SM is considered and joint optimization of precoder and equalizer is studied to combat the influence of channel estimation imperfections. In Ref. [24], the transmit power budget is adaptively shared among transmitters and optimal modulation size is deployed under the consideration of the same technique. A transmitter/receiver selection algorithm for MIMO VLC system is proposed in Ref. [25].

As automotive LEDs are deployed in dual configurations, a MIMO structure can be naturally employed in vehicular VLC. In our previous works [26,27], MIMO-OFDM concepts for both inter- and intravehicular communication are investigated. The goal of this chapter is to evaluate the performance of MIMO-OFDM–based vehicular VLC PHY, using red chip-based LED, based on the practical measurements from an off-the-shelf automotive taillight, that may be key to overcome the limited modulation bandwidth of vehicle LEDs and achieve higher data rates.

---

* In 2015, IEEE 802.15.13 has been initiated as a revision of IEEE 802.15.7. It took its current name in 2017.

The remaining part of this chapter is organized as follows. In Section 6.2, an overview of IEEE 802.15.13 is provided. In Section 6.3, an analytic expression of a MIMO-OFDM–based PHY solution is given, including bit error rate (BER) and data rate calculations. In Section 6.4, we explain a channel model that is obtained through experimental measurements and used for the evaluation of a MIMO-OFDM–based PHY in the context of VLC. In Section 6.5, we evaluate the performance of a MIMO-OFDM–based PHY, and we finally conclude this chapter in Section 6.6.

Notation: $\|\bullet\|^2$, $[\bullet]^H$, and $[\bullet]^T$ denote Euclidean distance, complex conjugate, Hermitian, and transpose respectively. $\otimes$ is the convolution operator. $Q(.)$ is the tail probability of standard normal distribution.

6.2 IEEE 802.15.13 Overview

The IEEE 802.15.13 Task Group has been formed to write a revision to IEEE 802.15.7. VLC. One of the features that has been accommodated aims high-speed and bidirectional mobile wireless communications using LED as the transmitter and high-speed photodetector (PD) as the receiver. This standard defines a PHY and MAC layer for short-range VLC. The standard is capable of delivering data rates sufficient to support audio and video multimedia services and also considers the mobility of the optical link, compatibility with variable light infrastructures, impairments due to noise, and interference from sources (*e.g.,* ambient light). Specifically, it targets to deliver a peak data rate of 10 Gbps. To achieve this target data rate, deployment of such advanced PHY techniques such as adaptive modulation, optical OFDM, relaying, and MIMO communications has been expected.

Optical OFDM with current specifications supports both a high-efficiency PHY mode designed to enable optimal utilization of the resources as well as a low-complexity PHY mode designed to enable high energy efficiency in mobile applications. The adaptive modulation concept has been also adopted to current specification and allows adaptive adjustment of the modulation order regarding channel conditions in order to increase transmission rates and efficiency under the consideration of specific quality of service (QoS) requirements.

In addition to adaptive modulation and OFDM concepts, beside main ambient lights, there can be secondary light sources such as desk or floor lights used for task lighting purposes in some indoor environments. This motivates the deployment of such secondary light sources as relaying terminals to boost the link reliability and/or extend coverage, therefore the PHY specification supports relaying mechanisms for the cases when dedicated relay terminals are available.

MIMO, on one hand, can leverage the additional communication capacity of multiple light sources as well as the additional communication capacity introduced by the utilization of different optical wavelengths and light polarization for communication. The use of several MIMO schemes is foreseen in order to support diversity and multiplexing. The main objective is to enable a dynamic trade-off between diversity and multiplexing, so that the optimum number of streams is always selected to maximize the throughput while operating the link reliably. MIMO transmission is further improved with the adaptation of dynamic selection of the best transmitter and receiver pairs. It is assumed that the MIMO link will be operated adaptively in a bidirectional closed-loop manner where MIMO metrics reports, regarding the forward link are provided over the reverse link.

Use cases considered for VLC with these advanced PHY solutions can be expressed as, but are not limited to

- indoor office/home applications
- data center/industrial establishments

- wireless small cell backhaul/local area network bridging
- vehicular communications including both V2V and vehicle-to-infrastructure (V2I)

Regarding the latter use case, we explain MIMO-OFDM–based PHY solutions with varying modulation orders for V2V applications in the rest of this chapter.

6.3 MIMO-OFDM–Based Physical Layer

We consider a MIMO-based transmission scheme with N_T LED and N_R PD. LEDs are assumed to be concurrently driven by the same hardware. Regarding automotive lighting with a certain required bias level, we select DCO-OFDM to transmit the information bits.

It is known that the IDFT response of complex or real-valued vector satisfying Hermitian symmetry is real. Let the DCO-OFDM frame size without CP set to N, M–ary PSK or QAM symbols (*s*) are rearranged with respect to Hermitian symmetry such that

$$\mathbf{X} = [0 s_1 s_2 s_3 ... s_{N/2-1} 0 s^*_{N/2-1} ... s^*_2 s^*_1]^T, \tag{6.1}$$

before IDFT process, hence, the output (**x**), which can be written as

$$x[n] = \frac{1}{\sqrt{N}} \sum_{k=0}^{N-1} X[k] e^{j\frac{2\pi nk}{N}}, \, n \in \{0,1,...,N-1\}, \tag{6.2}$$

where $X[k]$ is the kth element of $\mathbf{X}$ and $x[n]$ is the nth element of **x**, becomes real valued. After CP with the length of N_{CP} is appended to **x**, it is passed through a pulse shaping filter and converted to an analog signal which is denoted by $x(t)$. A DC bias (V_{bias}) is then added to shift $x(t)$ into the linear operation range of the LED, limited by turn-on voltage (V_{tov}) and maximum-allowed voltage (V_{max}) The signal driving the input of LED can be written as

$$x_S(t) = x(t) + V_{bias}. \tag{6.3}$$

Signals out of the allowed range will be clipped such that

$$\bar{x}_S(t) = \begin{cases} V_{max} & , \; if \; x_S(t) > V_{max} \\ V_{tov} & , \; if \; x_S(t) < V_{tov} \\ x_S(t) & , \quad otherwise \end{cases} . \tag{6.4}$$

In order to avoid clipping, the signal power level is kept within the linear operation range of LED, that is, $\bar{x}_S(t) = x_S(t)$

The received signal on the nth PD with N_T LED can be written as

$$y_n(t) = \sum_{m=1}^{N_T} \sqrt{\frac{P}{N_T}} R x_{S_m}(t) \otimes h_{nm}(t) + v_n(t), \; n \in \{1,..N_R\}, \tag{6.5}$$

where P is the total electrical OFDM signal power, R is the optical-to-electrical conversion coefficient (A/W) of PD, $x_{S_m}(t)$ is the transmitted signal from the mth LED, $h_{nm}(t)$ is the time-domain response of optical communication channel between the nth PD and the mth LED, and $v_n(t)$ is additive white

Gaussian noise (AWGN) term with zero mean and σ_N^2 variance at n^{th} PD. Variance is calculated by $N_0 W$ where N_0 is power spectral density and W is system bandwidth. The frequency-domain received signal can be written as

$$Y_n[k] = \sum_{m=1}^{N_T} \sqrt{\frac{P}{N_T}} R X_m[k] H_{nm}[k] + V_n[k], \quad n \in \{1,..N_R\}, \tag{6.6}$$

where $X_m[k]$ is the transmitted signal by the mth LED at k^{th} subcarrier where $k \in \{1,..N/2-1\}$, $H_{nm}[k]$ is the frequency response of optical communication channel between n^{th} PD and m^{th} LED, and $V_n[k]$ is AWGN with zero mean and σ_N^2 variance. We consider diversity and multiplexing (MIMO) modes which are explained in following subsections.

6.3.1 Repetition Code

In RC, the same information is transmitted by each LED simultaneously. The Maximum Likelihood (ML) decision rule at the kth subcarrier is formulated as

$$\hat{X}[k] = \underset{X[k]\in\Omega}{\text{argmin}} \left(\sum_{n=1}^{N_R} \left\| \frac{Y_n[k]}{\sqrt{\frac{P}{N_T}}R} - X[k]\sum_{m=1}^{N_T} H_{nm}[k] \right\|^2 \right), k \in \{1,..N/2-1\}, \tag{6.7}$$

where Ω is the set of constellation points of deployed modulation scheme. The signal-to-noise ratio (SNR) per subcarrier is given by

$$SNR[k] = \frac{\frac{PR^2}{N_T} \sum_{n=1}^{N_R} \left| \left(\sum_{m=1}^{N_T} H_{nm}[k] \right) \right|^2}{\sigma_N^2}. \tag{6.8}$$

Then, the subcarrier-based BER can then be calculated by [28]

$$\text{BER}_{RC}(k) \approx \begin{cases} Q\left(\sqrt{2SNR[k]}\right) & , \quad 2-\text{PSK} \\ \frac{2\left(\sqrt{M}-1\right)}{\sqrt{M}\log_2\sqrt{M}} Q\left(\sqrt{\frac{3SNR[k]}{(M-1)}}\right) & , \quad \text{square}-M-\text{QAM} \\ \frac{2}{\log_2(U\times J)}\left[\frac{U-1}{U}Q\left(\sqrt{\frac{6SNR[k]}{U^2+J^2-2}}\right) + \frac{J-1}{J}Q\left(\sqrt{\frac{6SNR[k]}{U^2+J^2-2}}\right)\right] & , \quad \text{rectangular}-M=U\times J-\text{QAM} \end{cases} \tag{6.9}$$

The average BER among subcarriers is expressed as

$$\bar{\text{BER}}_{RC} = \frac{2}{N-2} \sum_{k=1}^{N/2-1} \text{BER}_{RC}[k] \tag{6.10}$$

The data rate of RC-based DCO-OFDM is given by

$$\text{Data rate} - \text{RC} = \frac{1}{T_S} \frac{N/2-1}{N+N_{CP}} \log_2 M \, \text{bit}/s,$$ (6.11)

where T_S is the sampling interval, which is equal to $1/2W$ with the consideration of Nyquist rate.

6.3.2 Spatial Multiplexing

In SM, each LED disseminates different information simultaneously*. The ML decision rule at the kth subcarrier is formulated as

$$\hat{\mathbf{X}}[k] = \underset{\mathbf{X}[k] \in \Phi}{\text{argmin}} \left(\left\| \mathbf{Y}[k] - \sqrt{\frac{P}{N_T}} R\mathbf{H}[k]\mathbf{X}[k] \right\|^2 \right), \quad k \in \{1,..N/2-1\},$$ (6.12)

where $\mathbf{Y}[k]$ is the received signal vector with the dimension of N_R, $\mathbf{H}[k]$ is $N_R x N_T$ channel matrix on the kth subcarrier and Φ includes all possible combinations of transmitted signal vectors. The upper bound of subcarrier-based BER values given by ML decision rule can be calculated by Fath and Haas [18].

$$\text{BER}_{\text{SM-ML}}[k] \leq \frac{1}{M^{N_T} \log_2(M^{N_T})} \sum_{m_1=1}^{M^{N_T}} \left(\sum_{m_2=1}^{M^{N_T}} d_H\left(\mathbf{b_{m_1}},\mathbf{b_{m_2}}\right) Q\left(\sqrt{\frac{PR^2}{2\sigma_{N_T}^2} \left\| \mathbf{H}[k]\left(\mathbf{s_{m_1}} - \mathbf{s_{m_1}}\right)\right\|^2} \right) \right),$$ (6.13)

where $d_H\left(\mathbf{b_{m_1}},\mathbf{b_{m_2}}\right)$ is the Hamming distance of two-bit assignments $\left(\mathbf{b_{m_1}} \text{ and } \mathbf{b_{m_2}}\right)$ of the signal vectors $\mathbf{s_{m_1}}$ and $\mathbf{s_{m_2}}$ [18]. However, an exponential time search is needed for the ML decision rule to approach the bound, specifically M^{N_T}. Because of this complexity, we present minimum mean square error (MMSE) receiver structure. With MMSE receiver structure, the equalization process can be defined as $\mathbf{W}^{\text{MMSE}}[k]\mathbf{Y}[k]$ where

$$\mathbf{W}^{\text{MMSE}}[k] = \left(\mathbf{H}[k]^H \mathbf{H}[k] + \frac{\sigma_N^2 N_T}{PR^2} \mathbf{I} \right)^{-1} \mathbf{H}[k]^H$$ (6.14)

and $\mathbf{I}$ is identity matrix. Define $\mathbf{T}^{\text{MMSE}}[k] = \mathbf{W}^{\text{MMSE}}[k]\mathbf{H}[k]$. Under the assumption that residual signal and interference from other spatial streams at the output of MMSE receiver is approximated as Gaussian [29], signal-to-interference-plus-noise ratio (SINR) per subchannel (independent paths between LED and PD after equalization) denoted by $m \in \{1,...\min(N_R, N_T)\}$ can be written as

$$\text{SINR}_m[k] = \frac{\frac{PR^2}{N_T} \left| T_{mm}^{\text{MMSE}}[k] \right|^2}{\frac{PR^2}{N_T} \sum_{n=1, n \neq m}^{N_R} \left| T_{mn}^{\text{MMSE}}[k] \right|^2 + \sigma_N^2 \sum_{n=1}^{N_R} \left| W_{mn}^{\text{MMSE}}[k] \right|^2}.$$ (6.15)

---

* We assume that N_R is equal to or greater than N_T.

Then, BER with the use of MMSE receiver for m^{th} subchannel can be approximated as Equation 6.9 using SINR instead of SNR and denoted by $\text{BER}_{\text{SM-MMSE}_m}[k]$ at the kth subcarrier. Similar to RC, the average BER among subcarriers for SM with the use of ML and MMSE receivers is expressed as

$$\overline{\text{BER}}_{\text{SM-ML}} = \frac{2}{N-2} \sum_{k=1}^{N/2-1} \text{BER}_{\text{SM-ML}}[k], \qquad (6.16)$$

$$\overline{\text{BER}}_{\text{SM-MMSE}} = \frac{2}{\min(N_T, N_R)(N-2)} \sum_{m=1}^{\min(N_R, N_T)} \sum_{k=1}^{N/2-1} \text{BER}_{\text{SM-MMSE}_m}[k]. \qquad (6.17)$$

The data rate of SM-based DCO-OFDM is

$$\text{Data Rate} - \text{SM} = \min(N_T, N_R) \frac{1}{T_S} \frac{N/2-1}{N+N_{\text{CP}}} \log_2 M \text{ bit/s}. \qquad (6.18)$$

6.4 Channel Model for Vehicular VLC

The experimental setup includes a transmitting and a receiving vehicle located on a three-lane road at nighttime. The leading vehicle, equipped with three LED brake lights, is assumed to be transmitting information solely with its brake lights. Optical received power measurements are conducted up to 20 m for three different lanes. The leading vehicle is positioned in the middle lane, while the following vehicles are assumed to be located in all three lanes (*i.e.,* L1, L2, and L3) as depicted in Figure 6.1. A single vehicle with two receivers is considered to be at 30 different direct LoS positions for each measurement (*i.e.,* three lanes at every 2 m up to 20 m). Thereby, communication possibility with respect to received optical power is inspected for various receiver angles and distances. Moreover, both leading and following vehicles are assumed to be proceeding on a straight three-lane road. Additional challenges such as blocking and power imbalance on curvy roads are beyond the scope of this paper.

Each brake light is made up of an array of LEDs. Left and right brake lights are encapsulated under the taillight housing with optical reflector and concentrators. Hence, the Lambertian radiation pattern is not applicable to the brake lights under interest. Taillights separation distance is 70 cm, while the third brake light is 1 m away from each taillight's inner corner. Left and right brake lights transmit optical power (P_{LED}) of 10 dBm. Each following vehicle is assumed to be equipped with two receivers at the height of 80 cm, adjacent to the left and right headlights with a separation distance of 1.33 cm. Vehicles in nearby lanes are considered to be 1.5 m separated from each other. A test field is selected to be free of additional roadside objects, for the purpose of preventing reflection effects on the received optical power. Received optical power measurements are conducted separately for each brake light. Radiation from a single light source is measured while the other two lights are blocked.

Received power (P_{PD}) (see Figure 6.2) is measured using Thorlabs PM 200 optical power meter with S142C PD and FGL610 colored glass filter in order to avoid artificial light interference below 610 nm wavelength. As depicted in Figure 6.2, taillights have a symmetrical illumination pattern, consistent with the regulation [30]. The optical channel can be extracted by the following equation:

$$h_{\text{opt}}[\text{dB}] = P_{\text{PD}} - P_{\text{LED}}. \qquad (6.19)$$

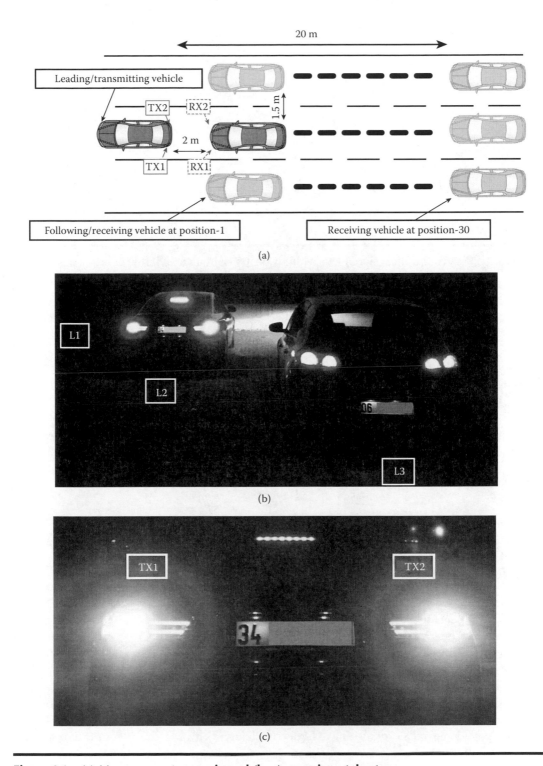

Figure 6.1 (a) Measurement scenario and (b–c) experimental setups.

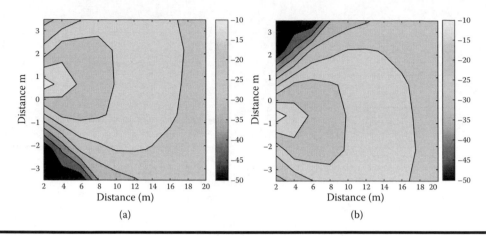

Figure 6.2   Received optical signal levels [dBm] for (a) right and (b) left LED brake lights.

In addition to the channel gain characteristics (Equation 6.19) obtained by experimental results, the effects of LED sources should be further taken into account in the channel modeling. The frequency response of the LED is assumed to be [31]

$$H_{\text{LED}}(f) = \frac{1}{1+j\dfrac{f}{f_{\text{cut-off}}}},$$
(6.20)

where $f_{\text{cut-off}}$ is the LED 3–dB cutoff frequency. Hence, the time-domain electrical channel response can be written as

$$h(t) = h_{\text{opt}} \int_{-\infty}^{\infty} H_{\text{LED}}(f)e^{j2\pi ft}\,df.$$
(6.21)

6.5 Performance Evaluations

In this section, we demonstrate the BER and data rate performance of MIMO-OFDM–based PHY using the channel model described in Section 6.4. Simulation parameters are given in Table 6.2.

In order to make a fair comparison between RC and SM MIMO modes, we consider the different deployment of modulation sizes that yield equal data rates (see Table 6.3). Since we consider a 2 × 2 MIMO system, the SM mode doubles the data rate with respect to the RC mode. Therefore, when 2–PSK is deployed for SM MIMO mode, 4–QAM is selected for RC mode. Similarly, 4–QAM and 16–QAM deployed SM are compared with 16–QAM and 64–QAM deployed RC, respectively. It should be noted that the MMSE receiver type is used for 16–QAM deployed SM mode due to high complexity with a ML decision rule (256 searches for each symbol).

In Figure 6.3, we present the BER and data rate performance of a MIMO-OFDM–based transmission on Lane 2 at different inter-vehicular distances. For 2 m inter-vehicular distance (see Figures 6.3a and 6.3b), it can be observed that RC with 4–QAM gives slightly better performance

Table 6.2   Simulation Parameters

Parameter	Value
LED 3–dB modulation optical bandwidth ($f_{cut-off}$)	2 MHz
System bandwidth (W)	20 MHz
Power spectral density of noise (N_0)	10^{-22} W/Hz
Responsivity of PD (R)	1.0 A/W
Number of subcarriers (N)	64
Length of CP (N_{CP})	8
Number of LEDs (N_T)	2
Number of PDs (N_R)	2

Table 6.3   Modulation Orders and MIMO Types for Different Spectral Efficiency Levels

Modulation	MIMO Type	Receiver Type	Data Rate
4–QAM	RC	ML	34 Mbps
2–PSK	SM	ML	34 Mbps
16–QAM	RC	ML	68 Mbps
4–QAM	SM	ML	68 Mbps
64–QAM	RC	ML	103 Mbps
16–QAM	SM	MMSE	103 Mbps

than SM with 2–PSK at the target BER of 10^{-3}. Comparing RC with 16–QAM and SM with 4–QAM, SM outperforms RC with a gain of approximately 3–dB. Moreover, when modulation orders are increased to 64–QAM for RC and 16–QAM for SM, multiplexing provides a 5–dB gain in diversity. In direct proportion to these BER improvements, for lower modulation orders, RC achieves higher data rate values at lower transmitted bit energy levels. On the contrary, benefits of SM over RC become clear for higher modulation orders. When the distance of 4 m is considered (see Figure 6.3c and 6.3d), it can be observed that RC with both 4–QAM outperforms SM with 2–PSK with a gain of 2–dB. The main reason of this additional improvement is because SM suffers from higher channel correlation and channels between LED and PD has higher correlation (see Figure 6.2). On the contrary, SM with 4–QAM and 16–QAM gives still better results than RC with 16–QAM and 64–QAM, respectively, as a consequence of RC suffering from higher modulation orders. When we set the distance between LED and PD to 6 m (see Figure 6.3e and 6.3f), due to more robust channel correlation, RC with 4–QAM and 16–QAM outperforms SM with 2–PSK and 4–QAM, respectively, with the gain of approximately 7–dB and 3–dB. However, SM with 16–QAM achieves better BER performance with a gain of 3–dB responding to RC with 64–QAM.

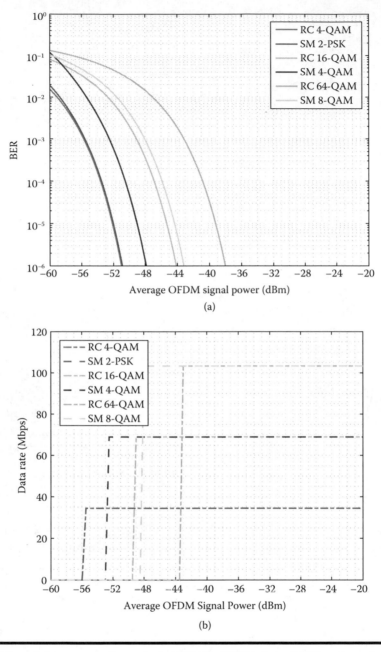

Figure 6.3   BER and data rate performance of MIMO-OFDM–based PHY on Lane 2 at the distance of (a)–(b) 2 m.   *(Continued)*

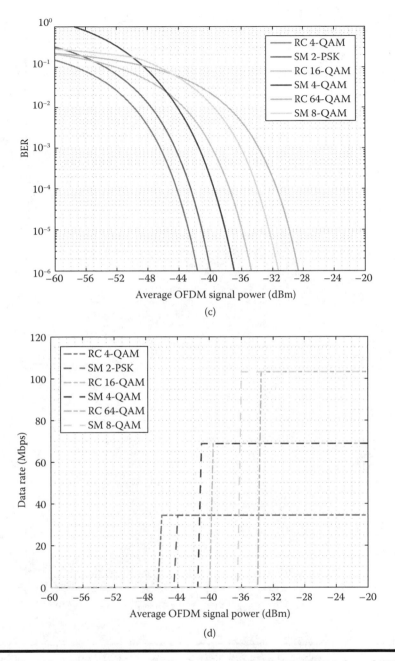

(c)

(d)

Figure 6.3 (Continued) **BER and data rate performance of MIMO-OFDM–based PHY on Lane 2 at the distance of (c)–(d) 4 m.** *(Continued)*

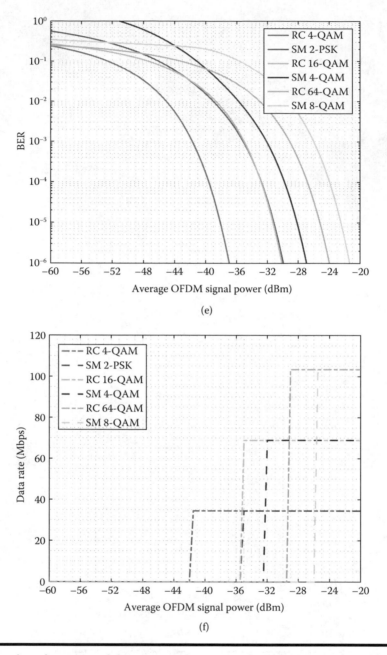

Figure 6.3 (Continued)   BER and data rate performance of MIMO-OFDM–based PHY on Lane 2 at the distance of (e)–(f) 6 m.

6.6 Conclusion

VLC is foreseen to be a complementary technology for vehicular communications. VLC is free from legacy considerations such as spectrum scarcity, RF interference, and security. Furthermore, wide deployment of LED lights in automobiles, makes VLC a low-cost vehicular communication scheme. VLC, with its potential to provide high data rates with minimum latency, is believed to be a reliable complementary vehicular communication solution. In this chapter, we proposed a MIMO-OFDM–based vehicular VLC scheme to fulfill higher data rate requirements, while taking advantage of dual automotive light configuration. Furthermore, we evaluated the performance of the proposed MIMO-OFDM vehicular VLC scheme based on practical vehicle LED taillight optical power measurements. Simulation results, based on measured channel model indicates that, RC MIMO suffers from higher modulation orders, whereas SM MIMO suffers from channel correlation. Thus, it can be concluded that, to utilize MIMO-OFDM, rapidly changing channel model due to high mobility should be carefully considered, and the MIMO scheme should accordingly selected.

Acknowledgment

The work of M. Uysal was supported by the Turkish Scientific and Research Council (TUBITAK) under Grant 215E311.

References

1. L. Cheng, B. E. Henty, D. D. Stancil, F. Bai, and P. Mudalige, Mobile vehicle-to-vehicle narrowband channel measurement and characterization of the 5.9 GHz dedicated short range communication (DSRC) frequency band, *Selected Areas in Communications, IEEE Journal on*, vol. 25, no. 8, pp. 1501–1516, 2007.
2. J. B. Kenney, Dedicated short-range communications (DSRC) standards in the United States, *Proceedings of the IEEE*, vol. 99, no. 7, pp. 1162–1182, 2011.
3. G. Araniti, C. Campolo, M. Condoluci, A. Iera, and A. Molinaro, LTE for vehicular networking: A survey, *IEEE Communications Magazine*, vol. 51, pp. 148–157, 2013.
4. S. Arnon, J. Barry, G. Karagiannidis, R. Schober, and M. Uysal, *Advanced optical wireless communication systems*, Cambridge university press: New York, 2012.
5. D. Karunatilaka, F. Zafar, V. Kalavally, and R. Parthiban, LED based indoor visible light communications: State of the art, *IEEE Communications Surveys Tutorials*, vol. 17, pp. 1649–1678, third quarter 2015.
6. H. Elgala, R. Mesleh, and H. Haas, Indoor optical wireless communication: Potential and state-of-the-art, *IEEE Communications Magazine*, vol. 49, pp. 56–62, September 2011.
7. M. Uysal, Z. Ghassemlooy, A. Bekkali, A. Kadri, and H. Menouar, Visible light communication for vehicular networking: Performance study of a V2V system using a measured headlamp beam pattern model, *Vehicular Technology Magazine, IEEE*, vol. 10, no. 4, pp. 45–53, 2015.
8. D. Jiang, Q. Chen, and L. Delgrossi, Optimal data rate selection for vehicle safety communications, in *Proceedings of the fifth ACM international workshop on VehiculAr Inter-NETworking*, pp. 30–38, San Francisco, California, September 15, 2008.
9. Y. Goto, I. Takai, T. Yamazato, H. Okada, T. Fujii, S. Kawahito, S. Arai, T. Yendo, and K. Kamakura, A new automotive VLC system using optical communication image sensor, *IEEE Photonics Journal*, vol. 8, no. 3, pp. 1–17, 2016.
10. T. Komine and M. Nakagawa, Fundamental analysis for visible-light communication system using LED lights, *Consumer Electronics, IEEE Transactions on*, vol. 50, no. 1, pp. 100–107, 2004.
11. IEEE Standard for Local and Metropolitan Area Networks Part 15.7: Short-Range Wireless Optical Communication Using Visible Light, in IEEE Std 802.15.7-2011, pp. 1–309, September 6, 2011.

12. TG7r1, *Technical considerations document, doc: IEEE 802.15-15/0492r3*, July. 2015. [Online; accessed 19–November–2017]. Available: https://mentor.ieee.org/802.15/dcn/15/15-15-0492-03-007a-technical-considerations-document.docx.

13. J. Armstrong and A. J. Lowery, Power efficient optical OFDM, *Electronics Letters*, vol. 42, pp. 370–372, March 2006.

14. J. Armstrong, OFDM for Optical Communications, *Journal of Lightwave Technology*, vol. 27, pp. 189–204, February 2009.

15. D. Tsonev, S. Sinanovic, and H. Haas, Novel Unipolar Orthogonal Frequency Division Multiplexing (U-OFDM) for optical wireless, in *2012 IEEE 75th Vehicular Technology Conference (VTC Spring)*, Yokohama, Japan, pp. 6–9, May 2012.

16. N. Fernando, Y. Hong, and E. Viterbo, Flip-OFDM for unipolar communication systems, *IEEE Transactions on Communications*, vol. 60, pp. 3726–3733, December 2012.

17. D. Tsonev, S. Videv, and H. Haas, Unlocking spectral efficiency in intensity modulation and direct detection systems, *IEEE Journal on Selected Areas in Communications*, vol. 33, pp. 1758–1770, September 2015.

18. T. Fath and H. Haas, Performance comparison of MIMO techniques for optical wireless communications in indoor environments, *Communications, IEEE Transactions on*, vol. 61, no. 2, pp. 733–742, 2013.

19. M. O. Damen, O. Narmanlioglu, and M. Uysal, Comparative performance evaluation of MIMO visible light communication systems, in *2016 24th Signal Processing and Communication Application Conference (SIU)*, Zonguldak, pp. 525–528, May 2016.

20. Q. Wang, Z. Wang, and L. Dai, Multiuser MIMO-OFDM for visible light communications, *Photonics Journal, IEEE*, vol. 7, no. 6, pp. 1–11, 2015.

21. Y.-J. Zhu, W.-F. Liang, J.-K. Zhang, and Y.-Y. Zhang, Space-collaborative constellation designs for MIMO indoor visible light communications, *Photonics Technology Letters, IEEE*, vol. 27, no. 15, pp. 1667–1670, 2015.

22. C. He, T. Q. Wang, and J. Armstrong, Performance of optical receivers using photodetectors with different fields of view in a MIMO ACO-OFDM system, *Journal of Lightwave Technology*, vol. 33, pp. 4957–4967, December 2015.

23. K. Ying, H. Qian, R. J. Baxley, and S. Yao, Joint optimization of precoder and equalizer in MIMO VLC systems, *IEEE Journal on Selected Areas in Communications*, vol. 33, pp. 1949–1958, September 2015.

24. K. H. Park, Y. C. Ko, and M. S. Alouini, On the power and offset allocation for rate adaptation of spatial multiplexing in optical wireless MIMO channels, *IEEE Transactions on Communications*, vol. 61, pp. 1535–1543, April 2013.

25. P. F. Mmbaga, J. Thompson, and H. Haas, Performance analysis of indoor diffuse VLC MIMO channels using angular diversity detectors, *Journal of Lightwave Technology*, vol. 34, pp. 1254–1266, February 2016.

26. B. Turan, O. Narmanlioglu, S. Ergen and M. Uysal, Broadcasting Brake Lights with MIMO-OFDM Based Vehicular VLC, IEEE Vehicular Networking Conference (VNC), Columbus, Ohio, pp. 8–10 December 2016.

27. B. Turan, O. Narmanlioglu, S. Ergen and M. Uysal, On the Performance of MIMO OFDM-Based Intra-Vehicular VLC Networks, IEEE 84th Vehicular Technology Conference (VTC), Montréal, Canada, pp. 18–21 September 2016.

28. K. Cho and D. Yoon, On the general BER expression of one-and two-dimensional amplitude modulations, *Communications, IEEE Transactions on*, vol. 50, no. 7, pp. 1074–1080, 2002.

29. E. Eraslan, B. Daneshrad, and C. Y. Lou, Performance indicator for MIMO MMSE receivers in the presence of channel estimation error, *IEEE Wireless Communications Letters*, vol. 2, pp. 211–214, April 2013.

30. U. Nation, *ECE Regulation No. 48, Uniform provisions concerning the approval of vehicles with regard to the installation of lighting and light-signaling devices*, pp. 52–54, 2015. [Online; accessed 19-November-2017]. Available from https://www2.unece.org/wiki/download/attachments/25265573/SLR-04-01-Rev.2e.docx?api=v2, vol. Rev.1/Add.47/Rev.6.

31. L. Grobe and K. D. Langer, Block-based PAM with frequency domain equalization in visible light communications, 2013 IEEE Globecom Workshops (GC Wkshps), Atlanta, GA, 2013, pp. 1070–1075.

V2X APPLICATIONS

Chapter 7

Toward Platoon-Aware Retiming of Traffic Lights in a Smart City

Dan C. Marinescu
University of Central Florida
Orlando, Florida

Ryan Florin and Stephan Olariu
Old Dominion University
Norfolk, Virginia

Contents

7.1 Summary

This chapter promotes the vision of PaTras—a cyber-physical system for platoon-aware traffic light rescheduling in the smart cities (SCs) of tomorrow. PaTras will enhance urban mobility by exploring methods and management strategies that increase system efficiency and improve individual mobility through information sharing, avoid congestion of key transportation corridors through cooperative navigation systems, and handle nonrecurring congestion.

PaTras will explore how SCs can leverage new opportunities available at the nexus of intelligent infrastructure, data analytics, connected vehicles, and engaged drivers that are the hallmark of SCs. The problems posed by the realization of PaTras are very challenging given the dynamic nature of the urban traffic environment in which coherent decisions about the timing of traffic lights must be made. To address these challenges, PaTras leverages knowledge from multiple areas of computer science and applies them to transportation systems. Coalition formation and self-organization principles will be applied to the creation of platoons and platoon algebra, processor coscheduling algorithms and real-time scheduling algorithms will guide the efforts for traffic lights rescheduling to ensure a virtual green wave for platoons, and simulations of the system will involve insights in cloud computing.

PaTras is built around, and depends on, the seamless integration of computational algorithms and the physical components. The algorithms that affect the aggregation and disassembly of urban platoons and other forms of route guidance will be performed in cyberspace. The vehicles in the physical space will benefit from route guidance and, in turn, will close the loop by providing feedback to the cyber system. PaTras will transcend the way people interact with engineered systems and will result in a better utilization of the road infrastructure in SCs by decreasing travel times and unnecessary idling at traffic lights and, consequently, by reducing fuel consumption and carbon emissions.

7.2 Introduction and Motivation

Urban transportation is key to sustaining economic vitality and to promoting a high quality of urban life. With the ever-increasing number of vehicles and vehicle-miles traveled, the surface transportation system in our cities is confronted with a number of significant issues ranging from increased levels of congestion to traffic delays and carbon emissions [1–3]. Indeed, in 2015, the US-DOT reported annual congestion costs in excess of $100 billion, congestion-related accident costs of well over $300 billion, as well as a continuously degrading air quality that fuels serious concerns about climate change [4]. Not surprisingly, improving urban transportation infrastructure management and utilization ranks high among the US-DOT's strategic priorities for 2015–2019 and beyond [5]. Intelligent transportation systems (ITS) and connected vehicles have been recognized by the US-DOT as key enabling technologies in SCs [4].

Traffic signals assign the right of way at intersections, but they can be a source of significant delays and unnecessary pollution if the cycle and phase lengths are inappropriate for current traffic conditions. One solution is the use of *adaptive traffic signal control* (ATSC) systems to dynamically optimize flow along a corridor. While some ATSC systems have been in existence for over 20 years, they have not been deployed beyond the corridor level. The main reason for this is the dynamic nature of urban traffic in which coherent decisions about the timing of traffic lights are very hard to make [6,7]. Under current practices, the metropolitan traffic management centers (TMC) only have fragmentary information about traffic conditions and trends, and this lack of awareness often leads to poor traffic signal timing regimens that, in turn, lead to congestion, increased periods of idling at traffic lights, pollution, and lost worker productivity [3,4].

One important trend that we are beginning to see is the emergence of *autonomous vehicles*. The penetration rate of different levels of autonomous vehicles is expected to increase exponentially, and the citizens of SCs will have access to high-quality and timely traffic information; they will very likely follow the traffic guidance offered to them. Projects aiming to reduce congestion and to offer the driving public high-quality information that will allow them to reduce their trip time and its variability are now feasible.

In ITS, a platoon is defined as a set of vehicles traveling at the same speed and sharing common group dynamics [8,9]. The wealth of real-time traffic data collected in conjunction with perceived traffic trends will make *urban platooning* and platoon-aware traffic signal retiming possible. In turn, this will offer new opportunities for enabling connected urban spaces, for drastically reducing travel times and their variability as well as for limiting unnecessary idling and the resulting carbon emissions.

The goal of this chapter is to promote the vision of PaTras—a cyber-physical system for platoon-aware retiming of TRAffic lights in a smart city (SC). PaTras is specifically designed to manage platoons, groups, or equivalent classes of vehicles with similar destinations, rather than individual vehicles. This view of the traffic in an SC will reduce travel time and fuel consumption, and eliminate congestion or reduce its effect. It can also reduce red-light-running hazards by combining early platoon splitting with extended green lights for the platoons and eliminate dilemma zones at intersections to reduce angle collisions.

In addition to being platoon aware, PaTras is also aware of the advantages of the future SC. As autonomous vehicles become more popular, the driving public will be restricted from SCs. This is because the autonomous vehicle can make decisions faster, thus can safely maneuver with less headway distance, and needs less room in the lane. Additionally, they may not require visible road signs or visible road markers. Instead, this information can be communicated by the infrastructure. Such a model will allow for a dynamic assignment of lanes and directions that is not possible with a human driver.

A fair number of significant results related to connected vehicles and ATSCs have been reported in recent years, including the ones from the PATH project [10–15], the PAMSCOD system [16] and several other projects discussed in Section 7.3.4.

In a sharp departure from previous approaches, PaTras adopts a quantum view of the traffic and promotes the formation of urban platoons and the retiming of relevant traffic lights that will allow these platoons to traverse a series of intersections without stopping. The combination of platooning and platoon-aware retiming of traffic lights will significantly improve mobility and reduce trip times and unnecessary idling at traffic lights. The first novelty that distinguishes PaTras from previous efforts is that in PaTras the formation of platoons is a deliberate activity and that traffic light retiming is platoon-aware. In PaTras, we take the view that platoons are *coalitions* of vehicles negotiated by smart vehicular agents that, acting in the best interest of the vehicles they represent, negotiate short- or long-term associations that will benefit the group of vehicles in terms of reducing the trip time of all the vehicles in the platoon, as well as the time these vehicles spend idling at traffic lights. Coalition formation algorithms are at the heart of platoon formation and subsequent platoon management.

The second novelty of PaTras is that, once formed, the various platoons are analogous to jobs whose duration is the time it takes a given platoon to clear the next intersection. This reduction allows the problem of traffic signal retiming to be perceived as an instance of multiple processor coscheduling, where the output of one processor becomes, in time, the input of the next one. We plan to extend and adapt preliminary results that Marinescu and his coworkers have obtained [17–20] to produce a platoon-aware retiming of consecutive traffic lights that will produce a green wave, allowing platoons to traverse a number of traffic lights without stopping or reducing their speed.

The third novelty is that in PaTras there is a permanent interaction between platoons and the dedicated traffic control infrastructure to allow for exceptional event handling, *e.g.*, lane blockage due to an accident or preemptive traffic lights setting by an emergency vehicle. Thus, PaTras will reduce red-light-running hazards (RLR) by combining early platoon splitting with extended green lights for the platoons and eliminate dilemma zones at intersections to reduce angle collisions. This could lead to fuel savings, as vehicles are advised early to slow down, as well as increased traffic safety.

Finally, unique to SCs is the ability to dynamically change the road infrastructure. PaTras can work outside of such an environment, but the benefits of platoons are enhanced by having fully autonomous drivers and the ability leverage changes in road infrastructure.

One of the fundamental factors that will make the platoon-aware retiming of traffic lights feasible in SCs is the penetration rate of connected vehicles and of different levels of autonomous vehicles, which is expected to increase exponentially [21,22]. As already noted, PaTras is built around, and depends on, the seamless integration of computational algorithms and the physical components.

PaTras is a visionary system that enables development and validation of innovative elements of an advanced traffic management system for the SCs of the future. PaTras explores how SCs can leverage new opportunities available at the intersection of the intelligent infrastructure, data analytics, connected vehicles, and engaged drivers that are the hallmark of SCs. Our vision is that the wealth of real-time traffic data collected in conjunction with perceived traffic trends will make urban platooning and platoon-aware traffic signal retiming possible. In turn, this will offer new opportunities for enabling connected urban spaces, for drastically reducing travel times and their variability as well as for limiting unnecessary idling and the resulting carbon emissions.

PaTras functions at the nexus of urban platooning and platoon-aware retiming of traffic signals. This represents a transformative, next-generation computing paradigm that is not possible with currently available technology and practices. Under current practices, coordinated urban navigation relies on route guidance provided on overhead panels and on data provided by probe vehicles [23,24] or inferred from cellphone traces [25]. PaTras goes well beyond the state of the art by actively coordinating the formation of urban platoons of vehicles that share similar destinations. Once platoons are formed, the roadway infrastructure is configured to reduce congestion. Then, the phases of the traffic lights are configured so as to allow the platoons to move as a unit through intersections. PaTras investigates issues related to dynamic resource discovery, availability prediction and management in a challenging wireless environment, and provides insight into the coordination between tomorrow's vehicles and their onboard computational, storage, and networking resources. Enabling this coordination faces many challenges, including the requirement of wireless communications and the mobility of individual vehicles which affects the dynamics of groups of vehicles. We anticipate that PaTras will have a profound and lasting societal impact.

The remainder of this chapter is organized along the following lines: Section 7.3 offers a review of SCs cities and ITS techniques that will be used in PaTras; Section 7.4 introduces the PaTras system architecture; next, Section 7.5 discusses in detail the challenges of PaTras and the various approaches for its efficient implementation; finally, Section 7.6 offers concluding remarks.

7.3 Technical Background

PaTras straddles a number of research areas ranging from traffic estimation and prediction, to traffic signal optimization, to platoon management and security. The main goal of this section is to offer a succinct synopsis of relevant recent developments in these areas.

7.3.1 Smart Cities

Apparently, the term *smart city* (SC) was introduced in the early 1990s to illustrate how urban development was turning toward technology, innovation, and globalization [4]. While SCs have been defined in myriad ways [26–28], it is striking that all of these definitions have two characteristics in common: they assume an *anticipatory* governance and management style in conjunction with broad *engagement* and *participation* of the citizens [27]. In fact, empowering the responsible citizens of SCs by increased access to high-quality information is one of the defining dimensions of an SC [29,30]. The role played by the citizens is poised to increase since, according to recent statistics, 70% of the US population already resides in big cities. It is, therefore, of a fundamental importance to *provide user-centric services that connect and sustain urban spaces* and this is precisely what PaTras aims to do for transportation in SCs.

In our view, SCs differ from present-day cities in three major respects. First, SCs will be instrumented with, and will actively rely on, *intelligent infrastructure*—these are smart devices that can sense the environment, send and receive data, and are networked together and with other networked elements in the SC. The intelligent infrastructure is apt to provide real-time traffic data on which timely management decisions can be based. Second, SCs will make extensive use of strategies and techniques to incentivize and to engage its connected citizens. These strategies and techniques will influence short- and long-term driver behavior.

Additionally, it is expected that vehicle transportation will be a service offered by the SC. Autonomous vehicles will be a shared resource; vehicles will bring passengers from source to destination, then return to the road in a holding pattern waiting for its next transport. They will be called upon much like an Uber or Lyft of today. Passengers will be picked up from an origin and dropped off at the requested destination. At peak times, the entire fleet of vehicles may be "roaming" the streets to minimize the average wait time of a ride. As the load decreases, vehicles will exit the streets into existing parking garages where they can wait.

Since autonomous vehicles are not owned by citizens, there are no dedicated parking spots for certain vehicles. In the parking garage, there is no need for wide parking spots to allow for the doors to open, as the passengers will already have exited prior to parking. Vehicles can be parked close together, maximizing the use of garage space, allowing for reuse of some of the current parking architecture.

Furthermore, the roads of tomorrow's SCs will be off-limits to any other types of vehicles. By restricting the cars to the shared resource, street parking is eliminated, allowing additional room for pedestrians or additional lanes. Also, since cars will be autonomous, lanes may be narrower, or perhaps nonexistent. Lanes and direction of travel of the lanes may be altered as the traffic flow changes.

PaTras is a small subsystem of the much larger SC. A certain amount of network, computational, and storage support on the scale of an SC is expected; however, it is not required.

7.3.2 Traffic Estimation and Prediction

In the 1990s, the US-DOT has initiated the development of traffic estimation and prediction (TrEP) [2] to enhance the mobility of our urban transportation system. Two TrEP systems, namely DynaMIT [31,32] and Dynasmart [33], were developed, and several demonstration projects were conducted at various locations [34–37]. Two key elements governing the success of TrEP are the following" (i) How accurately can the system estimate and predict short-term traffic conditions in real-time? and (ii) How well can the system model driver behavior to properly use this information to achieve better mobility and sustainability? Over the years, many researchers have enhanced the accuracy and performance of TrEP by considering weather conditions [35,37], by enhancing origin and destination (OD) demand, and by presenting/demonstrating calibration of supply parameters [38]. However, little research was conducted on exploiting the wealth of onboard sensor data available in present-day vehicles. Suitably data-mined and aggregated, these data can contribute tremendously to enhancing the accuracy of TrEP systems. At this time, mostly microwave vehicle detecting systems collect vehicle speed, type, counts, lane occupancy, and other parameters for all lanes. By using recent advances in communications, location, and computing technologies, one could take advantage of the information and communication technology applications in collecting traffic conditions, developing and disseminating route guidance to drivers, and receiving feedback from drivers about their route selection [22,39–41].

7.3.3 Connected Vehicles

The past decade has seen a rapid convergence of systems that exploit the ability of vehicles to communicate with each other and with the surrounding infrastructure, promising to revolutionize the way we drive by creating a safe, secure, and robust computing environment that will eventually pervade our highways and city streets.

Aligned with the Connected Vehicles Program initiated by US-DOT [42], an idea that exploits the prevalence of smartphones is to supplement legacy traffic monitoring with traffic incident reports submitted by the driving public. A recent implementation of this idea has led to 511 Traffic that offers an at-a-glance view of road conditions in a given geographic area [43]. Unfortunately, 511 Traffic is a centralized system that accumulates and aggregates traffic-related feeds at TMCs and, due to inherent delays, often displays stale traffic information [44]. The Mobile Millennium project at UC Berkeley exploits information collected by probe vehicles to infer information about the traffic [23,45]. Relying solely on traffic data collected by probe vehicles seems to work best in urban environments that experience high concentration of vehicles and less well on highways where there may be no "critical mass" of probe vehicles [24].

Waze [25] is a GPS-based geographical navigation application program for smartphones with GPS support and display screens which provides traffic status information from driver-submitted location-dependent traffic status information along with travel times and route details. Waze is a centralized system where the entire communication takes place over the mobile telephone network. Waze can be easily hacked and a number of security attacks can be easily mounted against Waze, including Sybil attacks, as well as an assortment of denial of service (DoS) attacks. For example, in March 2014, a hacking attempt was successfully made by students from the Technion-Israel Institute of Technology to fake a traffic jam [25].

7.3.4 Traffic Signal Optimization

While over 90% of traffic signals in the US are actuated signals (*i.e.*, duration of green times depends on the traffic condition based on fixed-point sensors), the optimization of the traffic signal systems currently occurs off-line at either the isolated intersection or corridor level. As recently as 2012, the Federal Highway Administration estimated that more than 75% of the country's 330,000 traffic lights were operating with outdated or uncoordinated signal timing plans [46].

Adaptive signal control that adjusts traffic signal timing plan on the fly has mixed results. One of the key challenges is predicting vehicle arrivals at the intersection. The state-of-the-art approach is to use a rolling horizon based on vehicle information from upstream fixed-point sensors. There are two key elements in optimizing traffic signal timing plans: (i) How accurate a tool is used in the evaluation of traffic signal timing plan during optimization? and (ii) How good the optimization engine is? For the former, the state of practice relies on microscopic simulation modeling tools including VISSIM, Synchro, CORSIM, and TRANSYT-7F [47–49]. Several studies have shown limitations of these approaches, as microscopic modeling does not properly capture the stochastic variability such as turn pocket and/or link overflow as well as the interactions of mixed vehicles including heavy vehicles and trucks. For the latter, as the traffic signal optimization formulated to minimize vehicular delay cannot be solved by analytical approaches, many heuristic optimization approaches (*e.g.*, hill-climbing, and genetic algorithm) that do not guarantee a globally optimal solution were used. However, due to their complexity, these methods have not been deployed beyond the corridor level [50].

Connected vehicles and traffic in SCs have been the focus of active research for several years [51,52]. Data fusion is critical for adaptive control systems [53]. A 2013 review of ATSCs [11] lists nine systems in operation in several countries. The Sydney Coordinated Adaptive Traffic System, with some 200 installations, has a centralized architecture and adjusts background fixed-time plans based on measurements of the degree of traffic saturation. The Split Cycle Offset Optimization Technique with some 50 installations is a decentralized system; it uses data from detectors located upstream of each intersection to estimate the size and shape of traffic platoons for each signal cycle, and then adjusts the timings such as the platoon to clear the intersection during the green time thus minimizing vehicle delays and stops. The IntelliDriveSM system is presented in [54].

The PATH project [10–15] is closely related to PaTras. PATH proposed several strategies for mobility including *queue spill back avoidance, dynamic lane allocation* [12,15], and *dynamic all-red extension* [14]. Their study analyzes the reduction of RLR hazards through minimization of vehicle arrivals during the yellow interval and platoon segmentation [13]. This study is focused on a microscopic view of the traffic and views the platoons as ad hoc high-density formations. The PAMSCOD system [16] uses a mixed-integer linear program (MILP) to compute future optimal signal plans based on the real-time arterial platoon request data and traffic controller status. MILP is computationally intensive for realistic cases, and platoons are ad hoc groupings of vehicles created to reduce the computing time.

7.3.5 Vehicle Coalitions

Recent studies have shown that various forms of route guidance strategies can help increase urban mobility, eliminate congestion or drastically curtail its effects, and reduce fuel consumption and carbon emissions due to idling at traffic lights [34,36,38]. One such strategy that holds the promise of reducing travel times and fuel consumption on highways is *platooning* [55–61]. We look at a platoon as a coalition of vehicles. Informally, a *coalition* is a group of agents who cooperate for achieving a well-defined goal and expect to be rewarded for their accomplishments. A coalition could last for an extended period of time, or for a relatively short time; coalition stability is very important for the former, while the efficiency of coalition formation is critical for the latter. The lifetime of platoons is rather short, so we are not concerned with coalition stability.

A *coalitional game* is a pair $G = (A,v)$, with A a set of players/agents and v a real-valued function associated with each coalition $C \in A$. The worth of a coalition, $v(C) \in R$, quantifies the benefits obtained by the players collaborating with one another. Coalition formation requires solutions to two problems: (i) the decision process for joining a coalition and (ii) the policies and mechanisms for reward sharing among the members of a coalition.

The agents could play a noncooperative or a cooperative game when the goal is to optimize a common objective function. We now present the definitions of frequently used terms in coalition games. Consider a set A of agents A_i with $n = |A|$; a *coalition* C is a nonempty set of A. The entire set A is called the *grand coalition*. A *coalition structure* is a *partition* of the set of all agents into disjoint, nonempty coalitions $S = \{C_1, C_2, \ldots C_m\}$.

7.4 The PaTras System Architecture

The goal of this section is to define the architectural components of the PaTras system. The specific ways in which the various architectural components of PaTras interact with each other to enable the proposed platoon-aware retiming of traffic light will be discussed in Section 7.5.

7.4.1 The Vehicle Model

In its 2006 ruling [62], the NHTSA mandated that starting September 2010 an *Event Data Recorder* (EDR) be installed in vehicles with an unloaded weight of less than 5,000 lbs. The EDR is responsible for recording mobility attributes including acceleration, deceleration, lane changes, and the like. Each such transaction is associated with an instantaneous GPS reading. All of the vehicle's subassemblies feed their readings into the EDR [63,64].

Each vehicle will have a DSRC-compliant radio for vehicle-to-vehicle (V2V) and vehicle-to-infrastructure (V2I) communications [65]. Such radios, mounted on the windshield, are already required in several states and are motivated by Congress' decision to have a national electronic toll-collection mechanism in place by 2018 [66]. These radios allow vehicles to stay connected with other vehicles and to various forms of fixed infrastructure made available by the SC. In particular, vehicles traveling as a platoon will keep and update, in real-time, the previous and next vehicles in the platoon as well as the current headway distance [67] to these vehicles. In addition, each vehicle contains an onboard server-class computer, a GPS receiver, and a digital map. The GPS receiver and the digital map provides the EDR with real-time trip information such as the vehicle's geographic location, current lane, speed, delay at intersections, stopping time, and number of stops.

7.4.2 Intelligent Vehicular Agents

Each vehicle is associated with two intelligent agents responsible for interacting with the SC-provided infrastructure for the explicit purpose of providing platoon-aware guidance to the vehicle. One of these intelligent agents, colocated with the EDR of each vehicle, is referred to as the *local agent* (LA); the second intelligent agent, referred to as the *remote agent* (RA), is colocated with the municipal TMC or resides in the SC-managed municipal cloud. While the vehicle is on the road, the two vehicular agents are exchanging traffic-related information. This exchange of information allows the vehicle to have a global view of relevant traffic attributes from its current location to the trip destination. The granularity of this information is such that the vehicle has a more accurate perception of the traffic parameters within a horizon of size τ. The accuracy of the information decreases progressively as the look-ahead horizon is increased.

As will be discussed in subsection 7.5.1, the RAs of several vehicles may enter coalitions intended to promote the formation of a platoon involving vehicles with a similar destination and sharing similar spatial and temporal characteristics.

It is important to realize that the TMC can harvest, in the obvious way, real-time traffic information from a large vehicle population that participates in the traffic. Suitably aggregated with historical data, this sort of real-time information should allow the servers of the SC-managed cloud to acquire an accurate picture of the following:

- The current traffic parameters (average speed, density, and flow) on various links within the SC
- The short-term variation in traffic parameters
- The gradient of change of various traffic parameters over time

The TMC can use this aggregated information to predict traffic parameters within a given time horizon and to effect changes in the timing of various traffic signals to meet the dynamically changing demand on the road infrastructure. In addition to changing the timing of traffic signals, the TMC may set up a regimen of dynamic lane assignments dedicated to platoon traffic.

7.4.3 TMC Cloud and TMC Cloudlets

The main computing infrastructure of the SC is called a smart city community cloud and is based on several interconnected warehouse-scale computers (WSCs) [68,69]. One WSC is dedicated to traffic management and used exclusively by the municipal TMC. We shall refer to this as the *TMC cloud*. For tractability purposes, we will adopt a hierarchical view of the entire urban area, whereby the latter will be perceived as partitioned into *cells* as discussed in subsection 7.5.2. Several rack arrays of the WSC supporting the TMC cloud will be dedicated to computational tasks executed in various such cells. We refer to the resulting subcloud as a *TMC cloudlet*.

Each LA of a vehicle in the traffic will upload to the TMC cloud the time the trip started as well as the link-by-link OD information weighted by the probability that the link will be taken. By aggregating this information with information about the average speed on each link, the TMC acquires a very good idea of the load on each link over a time horizon of length τ. This link-by-link traffic information is shared with each intersection adjacent to the given link. Observe that this information is far richer and more accurate than the conventional traffic data used, under current practices, by state-of-the-art TrEP systems.

7.4.4 The Platoon Model

The platoon was defined in Ref. [8] as "a collection of vehicles that travel together, actively coordinated in formation." In our interpretation, a platoon is a set of vehicles traveling at the same speed, sharing a *common destination region* of the urban area, as well as common group dynamics. Each platoon has a *platoon leader*, which is the first vehicle in the platoon, and a number of platoon members. The RA of the platoon leader is the *platoon RA* and will be in charge of negotiating, on behalf of the platoon, with upcoming traffic light controllers (TLCs). As mentioned before, the vehicles forming a platoon are in constant V2V communications with each other and with the platoon leader.

Important system platoon design parameters include the following:

■ Geographical destination area of vehicles in the platoon
■ Current size of a platoon
■ Average headway distance within the platoon, also known as *intraplatoon* headway distance
■ Common acceleration, deceleration, and instantaneous speed maintained by all the vehicles in the platoon

These platoon parameters are negotiated by the RAs of the vehicles forming the platoon and are communicated to the TLCs within a given time horizon τ. Conversely, the TLCs may suggest changes to some of the platoon parameters, such as speed, in such a way that the platoon will clear an intersection without stopping. The vehicles forming a platoon are in constant V2V communications with each other and are being updated by the platoon leader. As new vehicles join the platoon, or vehicles leave the platoon, the changed platoon parameters, such as platoon size, will be communicated to all the concerned vehicles. In the presence of several platoons it makes sense to talk about the *interplatoon distance* which is the distance separating two consecutive codirectional platoons. Platoon formation and disassembly will be discussed in detail in subsection 7.5.1.

7.4.5 The Traffic Light Model

We assume that the municipality of the SC has instrumented the urban area under its jurisdiction by deploying high-bandwidth access to individual traffic lights and supporting infrastructure. Each TLC contains a DSRC-compliant radio transceiver for communicating with neighboring vehicles, an LTE or 4G cellular broadband connection to communicate with the TMC and a dedicated server-class computing device.

Using DSRC, each TLC transmits periodically to the platoon RAs and the nonplatooned vehicles approaching the intersection

- Identification information to alert approaching vehicles
- The timing and duration of its current and next green phases
- On a red phase, the length of the queue of vehicles accumulated at the traffic light

By communicating with the RAs of the approaching platoons and nonplatooned vehicles as well as by monitoring legacy devices (including cameras, ILDs, and infrared sensors), each TLC acquires accurate knowledge about the traffic in its vicinity. This knowledge is refined by aggregation with similar traffic data provided by adjacent TLCs and the TMC. When individual TLCs become aware of an imminent trend or a traffic incident, they disseminate this information to the approaching vehicles, neighboring TLCs, and the TMC.

As discussed in subsection 7.4.6, each TLC is aware of the traffic attributes that it will see in a window of size τ, including the locations, size, and speed of various platoons embedded in such traffic. The TLC will coordinate with the various platoon RAs suggesting speed changes that will allow the platoon(s) to pass through the intersection without stopping or else suggest a reduced speed that will allow the platoon to clear the intersection on the next green phase without idling at the red light.

7.4.6 Historical Traffic Database

We assume that the TMC has collected, over a statistically significant time period, traffic-related parameters (including flow, density, and average speed) on a link-by-link basis for the entire metropolitan area. This information that is updated daily is maintained in a database that we call the historical traffic database (HTD) and stored in the TMC cloud. The granularity of the information in the HTD is five minutes and is qualified by day of the week and adjusted seasonally.

The organization of the HTD is inspired by Google's *BigTable*; this database includes data collected over many months, possibly years, and its size could reach petabytes. The data are mined off-line to compute history-based recommendations for traffic light settings and platoon guidance for different traffic conditions, time of day, day of week, and traffic intensity. Reinforced learning is used to select a subset of near-optimal solutions and rank them based on appropriate performance metrics.

Access to the HTD is granted to all the RAs and to all the TLCs within the smart city. Each TLC supplements the information in the HTD, at the same granularity, by the average number and size of the platoons for each of the directions N–S and E–W that it expects to see. As a result, by consulting the HTD and its local database, and by suitable interpolation, for each time interval $[t, t + \tau]$, each TLC has a fairly good idea of the average traffic flow, vehicle density, average speed, and number and size of the platoons that it expects to see, in each of the N–S and E–W directions, in the time interval $[t, t + \tau]$. This allows it to set the durations of the green and red phases accordingly.

We note that the current state of the traffic may vary considerably from the average values in the HTD and the platoon parameters stored by individual TLCs. Adjustments to the green and red phases that maintain the durations of these phases between prescribed values can be performed by the TLC without informing the TMC. However, in the case of a traffic incident that is blocking one of the links (or at the end of emergency vehicle preemption that resulted in congested traffic flow), the concerned TLC must inform the TMC so that a suitable retiming of traffic lights in the vicinity of the congested area can be undertaken in conjunction with provisioning route guidance to vehicles and platoons that would otherwise add to the existing congestion. The responsibility of suggesting alternate routes resides with the TMC.

7.5 PaTras—Challenges

In this section, we address the challenges involved in the design of PaTras and show how the challenges of PaTras stated above can be realized. There will be many challenges involved in making PaTras a reality. Security and privacy are major concerns when multiple entities cooperate, and PaTras is no exception [70]. However, in PaTras we are focusing our efforts on platoon formation and management, and on platoon-aware traffic light retiming in PaTras. We note that some of the challenges of providing scalable and robust security and privacy in the context of SCs has been studied in Refs. [67,70,71].

7.5.1 Challenge 1: Enhance Urban Mobility by Exploring Methods and Management Strategies that Increase System Efficiency and Improve Individual Mobility through Information Sharing

This first challenge is concerned with how PaTras will set up and manage platoons in order to increase the throughput of the urban transportation system and to improve individual mobility. PaTras will combine detailed knowledge of real-time traffic flow data with stochastic predictions within a given time horizon to help the formation of urban platoons containing vehicles with a similar destination and group dynamics.

7.5.1.1 Communication Requirements for Platoon Formation and Management

Due to the US government's Connected Vehicle program, DSRC/WAVE is becoming the *de facto* standard for vehicular communications. Thus, the secure communications between the LA and RA of each vehicle and between TLCs and both LAs and RAs can be implemented by secure V2V and V2I communications as discussed in Refs. [70,71].

Intraplatoon communication may be implemented either by direct V2V communications between the various LAs of the cars in the platoon or by RA-to-RA communication followed by RA-to-LA communication. It is not yet known which alternative is better.

Platoon to traffic light communications may be implemented as communications between the TLC and the platoon RA. The smart city will likely provide broadband Ethernet connection between the TMC and the various TLCs. Problems may arise if this communication will be conducted over wireless channels, due to restrictions in bandwidth. Interplatoon communications will be conducted solely through platoon RA to platoon RA wired communications.

7.5.1.2 Platoon Formation and Management

The stated goal of PaTras is to improve travel times through a judicious platoon-aware retiming of traffic lights and the creation of green waves of traffic. In order to achieve this goal, platoon formation must be both flexible and efficient. In support of this, we recursively partition the metropolitan area into a number of subareas termed *cells*. The cells play a number of roles: first, they are essential to platoon formation; second, they help with the creation of green waves; and third, as is explained in subsection 7.5.2, they help in determining routes.

Once a cell is created, the TMC spawns a TMC cloudlet in charge of the newly created cell. The TMC cloudlet in charge of a cell is the locus of computation and control for the cell. We note that the servers located at the various traffic lights in the cell can contribute, on a crowdsourcing basis, to enhancing the compute power of the TMC cloudlet.

In Figure 7.1, cells are shown as squares; however, there may be alternate variations on the size and shape of the cells. If the load on a certain cell is greater than can be handled, the cell can be split. Additionally, if the load on a cell is light, then it may be merged with another cell. Differing cell shapes may also be used; there also may be a benefit to overlapping cells to handle edge cases.

In conjunction with the TLCs in the cell the TMC cloudlet will compute, using the coprocessor paradigm discussed in subsection 7.5.2, a suitable retiming of the green phases of these traffic lights that maximizes the throughput (measured in the number of vehicles) that traverses the cell per time unit.

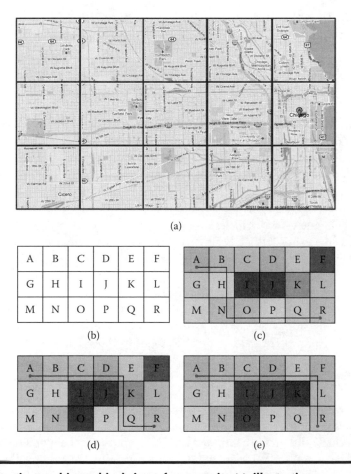

Figure 7.1 **Illustrating our hierarchical view of a smart city (a); illustrating an area of metropolitan Chicago partitioned into cells (b); a generic metropolitan area partitioned into 15 cells (c); a sample route from cell A to cell R (d); an updated sample route from cell A to cell R (e).**

To understand how the platoons are formed, refer to Figure 7.1b that shows the metropolitan area partitioned into a number of cells. We can define a binary relation on the set of vehicles that are in the same cell at time *t* by stating that two vehicles are *D-related* if they share the same *destination cell*. It is clear that the binary relation defined is an equivalence relation. As an example, consider the set of vehicles that, at time *t*, are in cell A and whose destination is somewhere in cell R. These vehicles are D-related and are very likely to benefit from traveling "together," as a platoon, through the various other cells to their (common) destination cell. Accordingly, the RAs of these vehicles will negotiate the ad hoc formation of a platoon. If the traffic is sparse, this platoon may contain a small number of vehicles. In such a case, it could be merged with the platoon of the vehicles in cell A which destination is in one of the cells L or Q. This aggregation of platoons can be achieved by using various algorithms for coalition formation devised by Marinescu [19].

When the platoons formed as discussed above reach their destination cell, the vehicles will leave the platoon and the platoon will disintegrate. Similarly, following the example above, at the appropriate moment the vehicles whose destinations are in one of the cells L or Q will leave the platoon and will continue their journey by either joining another platoon or traveling solo.

7.5.1.3 Algebra of Platoon Formation and Management

Platoons can be thought of as vehicle coalitions discussed in Section 7.3.5. The platoon algebra consists of a set of vehicles and operations such as formation, joining, splitting, and dispersion. These operations should be associative and subsets of them can be applied multiple times. The set itself changes dynamically as the vehicles in a platoon share a common path for different distances and periods of time.

The platoon formation process is distributed in space and time; it closely resembles the process a river uses to collect water from its tributaries. The process applies to the traffic at peak hours in existing cities. It is likely that the process will be significantly less complex for the SCs of the future and traffic dominated by self-driving vehicles. The principles guiding the platoon algebra are as follows:

1. Apply self-organization principles to the coalition formation process—rely on decisions made jointly by the LAs of vehicles.
2. Identify as early as possible coalitions capable to reach a *critical mass* and make early decisions to allow coalitions to proceed with minimum fuel consumption and reach intersections along the way at the right time.
3. Use a hierarchical decision-making process involving local cells and groups of local cells for guiding platoons to the joining point with the traffic on a relevant arterial.
4. Identify the cells on the path of the platoons on the main artery and inform the respective TMC cloudlets of the size and timing of platoons to coordinate scheduling.
5. Attempt to ensure a green wave for the platoons on the main artery.
6. Give priority to platoons with vehicles traveling the longest distance together.

The process starts when the LA contacts its counterpart, the RA located in the origin cell, and communicates the destination of the vehicle and the path to it. To ensure anonymity, the cell assigns random vehicle IDs prefixed with the cell-ID, identifies sets of potential vehicle coalitions V_i based on the location of the vehicles, and distributes this information to all vehicles in the set. Vehicles $v_{ij} \in V_i$ use V2V communication to coordinate and decide to join a platoon.

Platoon joining can be seamless when merging lanes are available. Platoon splitting can be foreseen or triggered by exceptional events. In the first case, a platoon can be split when more lanes are available or when the green light at an intersection cannot accommodate the size of the platoon. In such cases, the LAs are informed in advance and advised to change their speed accordingly. It is considerably more difficult to deal with exceptional events such as accidents. Traffic dispersion will be coordinated by the cell at the end of the common path of a vehicle.

7.5.2 Challenge 2: Avoid Congestion of Key Transportation Corridors through Cooperative Navigation Systems

It is key to investigate congestion avoidance techniques that rely heavily on the cooperation between platoons and various TLCs. As a result of this cooperation, the TLCs will intelligently retime the traffic lights in a platoon-aware manner. This will increase traffic throughput and also will reduce unnecessary idling at traffic lights. In turn, reducing unnecessary idling will limit the environmental impact of urban transportation. Indeed, abrupt acceleration and unnecessary idling at traffic lights have been identified as the main contributors to urban traffic-induced pollution [52,57,72,73].

7.5.2.1 Road Utilization

Given that roads in the SC are only utilized by autonomous vehicles, the road infrastructure can be dynamic. Currently, lanes, directions, and signage are all mostly static, with very few exceptions to this rule. Road infrastructure decisions can be made at the TMC cloudlet level and communicated to the RAs, allowing for lane direction to change or entire street direction to change to better utilize the roadway infrastructure.

As described previously, each cell has its own TMC cloudlet. It knows the OD of each vehicle on the road. To best use the road infrastructure, the cloudlets first must determine the traffic demand through the cell. Once the demand is known, certain roads will be chosen as primary routes and others as secondary routes, and so forth. In a Manhattan square-type city, there may be several primary routes due to the block nature of the roads. In cities where roads are less neat, there may be fewer routes due to the road layout. To aid in finding routes to cover the traffic, lane direction may be changed or even entire roads may be changed into one way to accommodate the traffic demand.

As a very simple example, if there is an increase of eastbound traffic and a decrease in westbound traffic, more of the lanes on the roads will be used for eastbound traffic.

7.5.2.2 Interaction between Platoons and TLCs

In PaTras, the tasks inherent to platoon-aware traffic light retiming are performed in a distributed fashion by a number of computational agents as described next. This latter approach also exploits the principle of locality. Namely, it is only seldom that drivers traverse the entire metropolitan area; in most cases, the OD of a driver's trip are located in a few adjacent cells and therefore global guidance is not really necessary. All that is needed is local route guidance through a reduced number of cells.

Imagine an arterial in a municipal area that traverses a number of consecutive cells. By interacting with the platoon RAs in their own cell and in the immediately previous cells, the TLCs can acquire full information about the size, location, and speed of all the platoons that they will see in a time horizon of length τ and can schedule the timing of their green phases to implement a green wave for the platoons. One of the unintended consequences of platoon-aware retiming of traffic lights is that platoons may be segmented. Consider a platoon that approaches a traffic light that cannot accommodate the entire platoon on its current green phase. This might happen because, in order to guarantee a modicum of fairness, green phases must be limited by a system-defined maximum value. Now imagine that, as the platoon approaches the traffic light, the traffic light controller realizes that it cannot accommodate the whole platoon on the next green phase (even when stretched to its maximum length). In such a case, the traffic light controller will inform the RA of (1) the number of vehicles in the platoon that can make it through the intersection on the current green phase and (2) the beginning of the next green phase. With this information at hand, the RA of the platoon will inform the LA of the leading vehicle in the platoon which, in turn, will propagate the information via V2V communications to the members of the platoon. As a result, the vehicles that can make it through the traffic light will maintain their speed, the others will reduce their speed in such a way that they will reach the traffic light on its next green phase. This will avoid idling at the traffic light and will reduce the carbon emissions involved.

7.5.2.3 Platoon-Aware Traffic Light Retiming as an Instance of Processor Coscheduling

There is a striking similarity between the traffic light retiming problem presented earlier and processor coscheduling. Consider an application requiring concurrent execution of all processes in a process group. When only a subset of processes are scheduled for execution, these processes may attempt to communicate with the subset of processes that are not executing, which will cause them to block. Coscheduling can be formulated as the problem of concurrent scheduling of all processes in a process group to minimize the completion time for the entire process group. Coscheduling was researched in late 1980s for running compute-intensive tasks on a network of workstations. The problem is of renewed interest now for scheduling concurrent threads of an application displaying course-grain parallelism on multicore processors or on multiple instances of a computer cloud.

The analogy with processor scheduling allows the application of coscheduling algorithms to traffic light retiming. In this analogy, the traffic light in intersection (i, j) of a Manhattan network controlling the traffic in two directions, north–south (N–S) and east–west (E–W), can be represented by a processor with two cores running asynchronously, one core assigned to the N–S traffic and the other to the E–W traffic. When one core is running, the other is blocked. Each core can be in one of the three states: running, corresponding to the green light; context-switching, corresponding to yellow light; and blocked, corresponding to the red light. The service time of a core will represent the time for a platoon to pass through the intersection controlled by the traffic light.

A set of processors interconnected by a Manhattan network will represent the traffic lights in a city with $n \times m$ nodes, $n_{i-1,j}$, $n_{i,j}$, $n_{i+1,j}$, $1 \le i \le n$, $1 \le j \le m$ adjacent in the N–S direction and nodes $n_{i,j-1}$, $n_{i,j}$, $n_{i,j+1}$ adjacent on the E–W direction. We assume that at a given time of day the traffic intensity for one of the directions either N or S and either E or W is dominant, and thus, the traffic in that direction should be optimized. Coscheduling of the N–S cores in direction S could represent a green wave when the N–S processor core in node $n_{i,j}$ starts running $\delta_{i-1,i}^{N-S,N}(t)$ seconds after node $n_{i-1,j}$; $\delta_{i-1,i}^{N-S,N}(t)$ is the travel time between the two intersections at time t. The service time of each core along the path is determined by the number of vehicles in a platoon traveling in that direction. We propose to adapt the algorithm developed by Atallah et al. [17] to this formulation of coscheduling.

If traffic intensity at each intersection is known, then $S_{N-S,N}(t)$, the solution for the N–S direction from N to S at time t, will automatically impose a solution $S_{E-W/N-S,N}(t)$ for the E–W traffic. Similarly, $S_{E-W,E}(t)$, the solution for the E–W direction from E to W, will impose a solution for the N–S direction, $S_{N-S/E-W,E}(t)$. Call $\bar{T}_{N-S,N}(t)$ the average time required for all cars to cross all intersections in the congested area in the first case and $\bar{T}_{E-W,E}(t)$ the corresponding time for the second solution. We can then choose $S_{N-S,N}(t)$ over $S_{E-W,E}(t)$ if $\bar{T}_{N-S,N} < \bar{T}_{E-W,E}$.

Traffic lights cannot be held indefinitely and the vehicle speed may depend on several factors including atmospheric conditions. Therefore, $\delta_{i-1,i}^{N-S,N}(t)$ must be bounded, $\delta Min_{i-1,i}^{N-S,N}(t) \le \delta_{i-1,i}^{N-S,N}(t) \le \delta Max_{i-1,i}^{N-S,N}(t)$ so the core service time may be calculated using the data about the current traffic. The upper limit may force platoons to split.

A more effective solution may need to balance $S_{N-S,N}(t)$ and $S_{E-W,E}(t)$ and adjust the timings of the traffic lights and require the platoons to adjust their speed in order to catch the green wave. Instead of calculating traffic light times in real time, a predetermined set of traffic signal templates can be used. The current traffic load in the cell is analyzed against a predetermined template; the matching template is used for the traffic signals.

7.5.3 Challenge 3: Handling Nonrecurring Congestion

The crux of this challenge is how to mitigate the effects of nonrecurring congestion should it occur. Since most of our city streets and arterials operate close to capacity [46], nonrecurring congestion is triggered by change fluctuations in traffic flow, such as caused by an incident, lane closure, or an emergency vehicle preempting the green phase of a traffic light. The major tasks that pertain to this challenge are (1) detecting signs of imminent congestion, (2) establishing the required communications and cooperation between the cells in the area of interest (AoI) and the TMC, and (3) investigating computational models for transient traffic signal retiming intended to dissipate congestion as efficiently as possible and setting up a temporary route guidance system for preventing platoons and other vehicles from adding to the congestion in the AoI.

7.5.3.1 Detecting Signs of Imminent Congestion

One of the challenges in PaTras is determining when to alert the TMC of conditions that indicate imminent congestion. As mentioned earlier, we focus here on nonrecurrent congestion rather than normal rush hour congestion. During rush hour, the main problem is that there is more demand for road bandwidth than the maximum capacity the road can serve. Nonrecurrent congestion, such as during a traffic incident, is often caused by a *decrease* in the capacity of the roadway. Once vehicles are past the bottleneck area, the congestion dissipates. Our goal is to retime the signals such that vehicles may quickly pass through the bottleneck area or take an alternate route to avoid it altogether.

The 2010 edition of the Highway Capacity Manual (HCM-2010) [74] defines various levels of congestion in terms of the volume-to-capacity ratio (VtC, for short). Specifically, a roadway or street where the VtC is between 0.85 and 0.95 is considered to be "near capacity," while a value of VtC in the range 0.95 to 1 indicates that the roadway is at capacity. Of course, a VtC larger than 1 indicates over capacity where we expect congestion to manifest itself in various degrees. HCM is based on static traffic analysis, while in SCs, data should be collected and processed in real time. To the best of our knowledge, under present-day practices, the TMC does not proactively monitor the VtC and, as a result, congestion is often detected too late, precluding meaningful congestion-preventive actions. One such preventive action would be to start retiming the traffic lights when the VtC reaches a critical level. One of the goals of PaTras is to address the issue of taking preventive action, well ahead of experiencing stopped traffic.

7.5.3.2 Establishing the Required Communications and Cooperation with the TMC

With legacy data collection resources, it is difficult or even infeasible to proactively monitor the local VtC along each city block. We now propose a possible way of achieving this goal under a platoon-centric data collection regimen. Specifically, every traffic light actively seeks input from approaching platoons. Specifically, the RA of each such platoon, when approaching an intersection, will drop off with the next traffic light controller with the current platoon parameters: GPS location of head vehicle, platoon size, average time headway, and speed. By aggregating information received from the platoons and isolated vehicles, the traffic light controller has an accurate view of the traffic that it will see in the next τ time units. Once this is done, it is a simple matter to estimate the current VtC.

By using simple data aggregation, the TLC keeps track of the most recent values of the VtC as well as the gradient of the change. The reason for the latter is that sudden steep increases in the VtC (even at low initial values) may indicate rapid traffic buildup that, if left unchecked, may result in congestion.

The TMC may use its own previously deployed sensors (ILDs, cameras, police reports, etc.) to detect the VtC at a particular intersection, but as PaTras-enabled vehicles can communicate with the cluster controllers, they can provide additional real-time information about the conditions at a particular intersection. Average speed is a difficult metric to use on signalized arterial roads because vehicles may stop or reduce their speed at intersections due to signal timings, without any congestion being experienced. To overcome this limitation, Hall and Vyas [75] suggested that a "congestion alarm" be set whenever a vehicle must wait more than one cycle before being able to pass through an intersection. This condition is called a signal cycle failure and indicates that not all vehicles queued at an intersection during a red phase were able to depart during the next green phase. Zheng et al. [76] developed an algorithm for detecting signal cycle failures using video image processing. They advised keeping track of the end of the queue during a red phase. If the last vehicle does not clear the intersection during the green phase, then there is a cycle failure.

We suggest two possible approaches. First, one can produce a robust probabilistic model that uses platoon idling time to determine probabilistically signal cycle failure. Second, one can use an approach similar to the one in Ref. [76], however using the vehicles or platoons themselves to report signal cycle failures instead of relying on infrastructure to detect this. There are several ways to approach this. First, we can assume that the range of communications of a cluster controller will cover all of the vehicles between it and the previous signal. In this case, the TLC can broadcast the beginning of the signal phases (red, yellow, and green). The vehicle at the end of the queue will hear the red broadcast. If it has not passed through the intersection before hearing the next red broadcast, it will alert the cluster controller that there has been a cycle failure. The problem becomes more challenging if we assume that the communication range of a cluster controller will not reach all vehicles in its block. In this case, we may have to rely on the vehicle at the end of the queue communicating with the previous cluster controller or performing some heuristics to determine when it has been in the block for the duration of an entire cycle. In both cases, our evaluation of the congestion detection technique will consider its accuracy, including the number of false positives generated. Random fluctuations in arrival volumes may cause isolated cycle failures (*e.g.*, if emergency vehicle preemption shortens a green phase), but the queue may be cleared on subsequent cycles. We will set a threshold on repeated cycle failures before we indicate to the TMC that signal optimization may be needed.

Once imminent congestion is detected by a TLC, using a procedure to be described in subsection 7.5.3, and the TMC has been alerted, the TMC identifies the boundaries of the AoI where traffic rerouting is needed in order for congestion to be prevented or dissipated. The various TMC cloudlet servers in conjunction with the TMC cloud may use this aggregated information to predict traffic parameters within a time horizon and to effect changes in the timing of various traffic signals to meet the dynamically changing demand on the municipal road infrastructure. In turn, having collected and aggregated traffic data from the platoon RAs and TLCs in its own cells, each TMC cloudlet has a fairly good idea of the rate of change (*i.e.*, gradient) of various traffic parameters that are of interest to route guidance. Periodically, the TMC cloudlets are reporting the status of the traffic flow in their own cell to the TMC cloud. Of course, more rapid changes trigger more frequent updates to the TMC cloud.

Referring to Figure 7.1c, imagine a platoon formed in cell A whose destination is in cell R. Being aware of the fact that, at the moment, cells I and J are heavily loaded, the TMC sets up a tentative, high-level route for the platoon in the form of the sequence ABHNOPQR, indicating the sequence of cells through which the platoon is to be guided. We refer to this sequence as the high-level route of the platoon.

The sequence ABHNOPQR, along with the ID of the platoon, is then sent to the TMC cloudlets in charge of the respective cells. Each individual TMC cloudlet is responsible for guiding the platoon to an "entry point" into the next cell. For example, the TMC cloudlet in charge of cell A will decide, based on the local traffic conditions in cell A (perhaps, in consultation with the TMC cloudlet in charge of cell B) to which of the entry points into B the platoon should be guided. Once there, the TMC cloudlet in charge of cell B takes over and, in consultation with the TMC cloudlet in charge of cell H, will guide our platoon to an entry point into H.

What happens if, once in B, the platoon RA in consultation with the TMC cloudlet decides to enter cell C instead of H? As soon as the TMC cloudlet in charge of C realizes that the platoon will enter a cell that differs from the original high-level sequence ABHNOPQR, the TMC is informed of the new cell sequence, *e.g.*, CDEKQR will be devised and communicated to the respective TMC cloudlets for implementation. This continues as described until finally, our platoon has reached its destination cell. We recall that for each driver in the platoon, once the destination has been reached, the LA and RA update the driver's compliance record that will be used for data mining for patterns in the compliance history of the driver.

Now, imagine that later in the day, another platoon starts out in cell A heading to a destination in cell R. At the moment, cell O is heavily congested as shown in Figure 7.1d and so the old route ABHNOPQR will be suboptimal. Knowledge about the loads of the other cells prompt the TMC to tailor the optimal high-level sequence ABCDEKQR of cells for the trajectory of the platoon. As before, this sequence will be communicated to the cloudlets in charge of the respective cells and the process described before is repeated.

As mentioned before, the shape and position of a cell may influence how routes are chosen. It is likely that square cells are not the optimal shape. Rather, these cells may have irregular shapes depending on the roads themselves.

7.5.3.3 Building a Model of Retiming of Traffic Lights and Route Guidance to Dissipate Congestion

While the ITS community has devoted substantial effort to changing the timing of traffic lights at the corridor level, to the best of our knowledge, the problem of retiming traffic lights at the scale of a wider urban area is still very much uncharted territory. The principle reason behind this state of affairs is, no doubt, the combinatorial explosion inherent in the process of transiting from small-scale to large-scale, complex problems. This is clearly the case when the traffic that flows from several corridors compete for unsharable resources—time and road bandwidth.

Though congestion may begin in a small area, in time it may spill over and affect side streets. Once the signal timings have been changed, the vehicles in the area of highest concentration of congestion will begin to move to other regions of the roadway network. The flow of vehicles leaving the congested area may well generate secondary congestion events over a larger area. If new congestion events result from the original retiming, the TMC will restart the process of setting up crowdsourcing of TLCs in each of the cells local to each congested area. Thus, the wide-area resolution of congestion can be perceived as a multistage process, with each of the successive stages addressing a less acute instance of congestion than the previous one. Therefore, a severe instance of congestion is reduced, in stages, to several less acute instances, spread over a much larger area. The framework that we have established in PaTras lends itself well to this recursive process.

To get a handle on the problem, first the traffic is modeled at the *macroscopic* level, as an instance of the classic diffusion problem. With insight gleaned from this macroscopic view, the model is reverted to the *microscopic* problem of platoon-aware retiming the traffic signals in the AoI.

The benefit of the macroscopic view just outlined is that it helps suggest that the wide-area resolution of congestion can be perceived as a multistage process, each of the successive stages addressing a less acute instances of congestion than the previous one. An obvious corollary of the previous discussion is that, by diffusion, one severe instance of congestion is reduced, in stages, to several, less acute instances of congestion, spread over a much larger area. This latter insight has a direct bearing on our strategy for mitigating congestion and, of course, on the attendant retiming of traffic lights.

7.6 Concluding Remarks

Transportation is a key element for sustaining economic growth and a high quality of urban life. Unfortunately, the transportation systems in our cities are confronted with increased levels of congestion, traffic delays, and accidents estimated to cost the economy in excess of $400 billion in 2015. The continuous degradation of air quality due to carbon emission fuels serious concerns about climate change. While existing traffic light retiming strategies and route guidance systems have provided drivers with convenience and limited savings in travel time, their impacts were not fully realized. Smart cities and connected vehicles technologies are truly multidisciplinary topics. The vision of PaTras demonstrates the usefulness of implementing urban platoons and a companion platoon-aware traffic light retiming regimen that will lead to reducing trip times and unnecessary idling at stop lights. Through the complex interaction of intelligent agents and TLCs, platoon speeds and sizes as well as the timing of traffic lights are coordinated in a way that results in a superior throughput, in terms of the total number of vehicles that pass through intersections. This translates into reducing total travel time and unnecessary idling at stop lights. It is our vision that the resulting traffic management system will be significantly better than the state of the art, contribution to a better utilization of our roadway infrastructure, increasing fuel savings, and reducing carbon emissions.

References

1. AAA. Crashes vs. congestions: What's the cost to society?, Cambridge Systematics Inc., Bethesda, Maryland, November 2011.
2. R.P. Roess, E.S. Prassas, and W.R. McShane. *Traffic Engineering*. 4th edition, Prentice-Hall, Upper Saddle River, NJ, 2011.
3. Texas Transportation Institute. Urban mobility report 2012. http://mobility.tamu.edu/ums/report/ [Accessed May 21, 2016].
4. USDOT. *The smart/connected city and its implications for connected transportation*, ITS Joint Program Office-HOIT, Washington, DC, 2014.
5. USDOT. 2015–2019 Strategic Plan Intelligent Transportation Systems (ITS). 2015. http://www.its. dot.gov/strategicplan.pdf.
6. N.J. Garber and L.A. Hoel. *Traffic and Highway Engineering*. Cengage Learning, Toronto, Canada, 2009.
7. J.M. Hutton, C.D. Bokenkriger, and M.M. Meyer. Evaluation of an adaptive traffic signal system: Route 291 in Lee's Summit, Missouri. Department of Transportation, MO, 2010.
8. M. Selinger. *Hitting All the greens*. Roads and Bridges, pp. 52–55, January 2010.
9. L. Wang, W.-W. Guo, X.-M. Liu, and D. Wu. Two-way green wave optimization control method of artery based on partitioning method. *Advances in Mechanical Engineering*, 8(2):1–8, 2016.
10. J.-Q. Li, K. Zhou, S. Shladover, and A. Skabardonis. Estimating queue distance under the connected vehicle technology: Using probe vehicle, loop detector, and fused data. *Transportation Research Record: Journal of the Transportation Research Board*, 2356:17–22, 2013.
11. A. Skabardonis. Advanced traffic signal control algorithms. 2013. http://www.dot.ca.gov/newtech/ researchreports/reports/2013/final_report_task_2157a.pdf
12. G. Wu, K. Boriboonsomsin, L. Zhang and M. J. Barth. Simulation-based benefit evaluation of dynamic lane grouping strategies at isolated intersections. In *Proceedings of the 15th International IEEE Conference on Intelligent Transportation Systems*, Anchorage, AK, pp. 1038–1043, September 2012.
13. L. Zhang, J. Hu, K. Zhou, and W.-B. Zhang. Signal timing optimization based on platoon segmentation by using high precision microwave data. In *Proceedings of 92nd TRB Annual Meeting*, Washington DC, January 2013.
14. L. Zhang, L. Wang, K. Zhou, and W.-B. Zhang. Dynamic all-red extension at signalized intersection: A framework of probabilistic modeling and performance evaluation. *IEEE Transactions on Intelligent Transportation Systems*, 12(4):166–179, 2011.
15. L. Zhang and G. Wu. *Dynamic Lane Grouping at Individual Intersections: Problem Formulation and Performance Analysis*. Transportation Research Board, Irvine CA, July 23, 2012.
16. Q. He, K.L. Head, and J. Ding. PAMSCOD: Platoon-based arterial multi-modal signal control with online data. *Transportation Research C, Emerging Technology*, 20(1):164–184, 2012.
17. M.J. Atallah, C.L. Black, D.C. Marinescu, H.J. Siegel, and T.L. Casavant. Models and algorithms for co-scheduling compute-intensive tasks on a network of workstations. *Journal of Parallel and Distributed Computing*, 16(4):319–327, 1992.
18. L.B. Boloni and D.C. Marinescu. Robust scheduling of metaprograms. *Journal of Scheduling*, 5(5):395–412, 2002.
19. D.C. Marinescu. *Complex Systems and Clouds: A Self-Organization and Self-Management Perspective*. Morgan Kaufman, Elsevier, 2016.
20. D.C. Marinescu and J.R. Rice. Synchronization of non-homogeneous *parallel computations*. In G. Rodrigues (Eds.), *Parallel Processing for Scientific Computing*, SIAM, Los Angeles, CA, pp. 362–367, 1989.

21. M. Aeberhard, S. Rauch, M. Babram, G. Tanzmeister, J. Thomas, Y. Pilat, F. Homm, W. Huber, and N. Kaempchen. Experience, results and lessons learned from automated driving on Germany's highways. *IEEE Intelligent Transportation Magazine*, 7(1):43–57, 2015.

22. T. Litman. Autonomous vehicle implementation predictions: Implications for transport planning. *Presented at the 2015 Transportation Research Board Annual Meeting*, Washington DC, January 11–15, 2015.

23. A.M. Bayen, J. Butler, and A.D. Patire. *Mobile Millenium Final Report.* Technical Report UCB-ITS-CWP-2011-6. California Center for Innovative Transportation, September 2011.

24. A. Hofleitner, R. Herring, P. Abbeel, and A. Bayen. Learning the dynamics of arterial traffic from probe data using dynamic Bayesian network. *IEEE Transactions on Intelligent Transportation Systems*, 13(4):1679–1693, December 2012.

25. WAZE. http://en.wikipedia.org/wiki/Waze

26. D.V. Gibson, G. Kozmetsky, and R.W.E. Smilor. *The Technopolis Phenomenon: Smart Cities, Fact Systems, Global Networks.* Rowman and Littlefield, Savage, MD.

27. C. Harrison and I.A. Donnelly. The theory of smart cities. In *Proceedings of the 55th Annual Meeting of the International Society for the Systems Sciences, (ISSS'2011)*, Hull, U.K., 2011.

28. A.M. Townsend. *Smart Cities: Big Data, Civic Hackers, and the Quest for a New Utopia.* W. W. Norton, New York, 2013.

29. D. Hatch. *Singapore Strives to become 'The Smartest City': Singapore is Using Data to Redefine What it Means to be a 21st-Century Metropolis*, Governing, pp. 22–26, February 2013.

30. P. Olson. Why Google's Waze is trading user data with local governments. 2014. http://www.forbes.com/sites/parmyolson/2014/07/07/why-google-waze-helps-local-governments-track-its-users/

31. M. Ben-Akiva, M. Bierlaire, H. Koutsopoulos, and R. Mishalani. Dynamit: A simulation-based system for traffic prediction. In *Proceedings of DACCORD Short Term Forecasting Group*, Delft, The Netherlands, February 1998.

32. M. Ben-Akiva, H.N. Koutsopoulos, C. Antoniou, and R. Balakrishna. Traffic simulation with DynaMIT. In B. J., editor, *Fundamentals of Traffic Simulation*, Springer Science, Berlin, pp. 363–398, 2010.

33. V. Alexiadis, K. Jeannotte, A. Chandra. Traffic Analysis Toolbox Volume I: *Traffic Analysis Tools Primer. Technical Report No. FHWA-HRT-04-038*, National Highway Traffic Safety Administration, Washington DC, July 2004.

34. J. Lee and B. Park. *Evaluation of Vehicle Infrastructure Integration (vii) Based Route Guidance Strategies Under Incident Conditions.* Transportation Research Board 2086:107–114, 2008.

35. B. Park, T. K. Jones, and S. O. Griffin. *Traffic analysis toolbox volume xi: Weather and traffic analysis, modeling and simulation*, Federal Highway Administration, Washington, DC, 2010.

36. B. Park and J. Lee. Assessing sustainability impacts of route guidance system under cooperative vehicle infrastructure environment, in *IEEE International Symposium on Sustainable Systems and Technology*, Phoenix, AZ, pp. 1–6, 2009.

37. D. Samba and B. Park. Incorporating inclement weather impacts on traffic estimation and prediction. In *Proceedings of the 18th ITS World Congress*, Orlando, FL, October 16–20, 2011.

38. M. Ben-Akiva, S. Gao, L. Lu, and Y. Wen. Combining disaggregate route choice estimation with aggregate calibration of a dynamic traffic assignment model. *Networks and Spatial Economics*, 15(3):559–581, 2015.

39. G. Castignani, T. Derrmann, and T. Engle. Driver behavior profiling using smartphones: A lost cost platform for driver monitoring. *IEEE Intelligent Transportation Magazine*, 7(1):91–102, 2015.

40. T. Gindele, S. Brechtel, and R. Dillmann. Learning driver behavior models from traffic observations for decision making and planning. *IEEE Intelligent Transportation Magazine*, 7(1):69–79, 2015.

41. M. Kamargianni, M. Ben-Akiva, and A. Polydoropoulou. Incorporating social interaction into hybrid choice models. *Transportation*, 41:1263–1285, 2014.

42. U.S. Department of Transportation. Connected vehicle program. 2012. http://www.its.dot.gov/connected_vehicle/connected_vehicle.htm

43. 511 Traffic. Latest traffic news. http://traffic.511.org/LatestNews

44. N.H.T.S. Administration. An analysis of recent improvements to vehicle safety. 2012. http://www-nrd.nhtsa.dot.gov/Pubs/811572.pdf

45. UC Berkeley, Mobile Millennium. Project. http://traffic.berkeley.edu/ [Accessed May 2, 2016].
46. NHTSA National Highway Traffic Safety Administration. An analysis of recent improvements to vehicle safety. 2012. http://www-nrd.nhtsa.dot.gov/Pubs/811572.pdf
47. PTV. Vissim. http://www.english.ptv.de/cgi-bin/traffic/traf_vissim.pl
48. Trafficware Ltd. *Synchro 7 User Manual.* 2006. http://www.trafficware.com
49. US Department of Transporation, Federal Highway Administration. *Corridor Simulator (CORSIM/TSIS).* 2006. http://ops.fhwa.dot.gov/trafficanalysistools/corsim.htm
50. T. Dynus. Dynamic urban systems for transportation (DynusT). 2006. http://wiki.dynust.net/doku.php?id=bibliography [Accessed May 2, 2016].
51. A. Stevanovic. *Adaptive Traffic Control Systems: Domestic and Foreign State of Practice.* Technical Report Synthesis 403. National Cooperative Highway Research Program, Washington, DC, 2010.
52. A. Winkler. Smart car meets smart lights. In *Proceedings of ITS America Conference,* Washington, DC, May 21–23, 2012.
53. N.-D. El Faouzi, H. Leung, and A. Kurian. Data fusion in intelligent transportation systems: Progress and challenges—A survey. *Information Fusion,* 12:4–10, 2011.
54. B. Smith, R. Venkatanarayana, B. Park, N. Goodall, J. Datesh, and C. Skerrit. *Intellidrivesm Traffic Signal Control Algorithms.* 2011.
55. M.A. Khan and L. Boloni. Convoy driving through ad-hoc coalition formation. In *Proceedings of the 11th IEEE Real Time on Embedded Technology and Applications Symposium,* RTAS '05, San Francisco, CA: IEEE Computer Society, pp. 98–105, 2005.
56. J. Larson, K.-Y. Liang, and K. H. Johansson. A distributed framework for coordinated heavy-duty vehicle platooning. *IEEE Transactions on Intelligent Transportation Systems,* 16(1):419–429, 2014.
57. K.-Y. Liang, J. Martesson, and K. H. Johansson. Heavy-duty vehicle platoon formation for fuel efficiency. *IEEE Transactions on Intelligent Transportation Systems,* 17(4):1051–1061, 2016.
58. P. Meisen, T. Seidl, and K. Henning. A data-mining technique for the planning and organization of truck platoons. In B. Jacob, P. Nordengen, A. O'Connor, and M. Bouteldja, editors., *International Conference on Heavy Vehicles (HVParis 2008), Heavy Vehicle Transport Technology,* Wiley, New York, pp. 389–402, 2008.
59. J. Ploeg, D.P. Shukla, N. van de Wouw, and H. Nijmejer. Controller synthesis for string stabilityof vehicle platoons. *IEEE Transactions on Intelligent Transportation Systems,* 15(2):854–865, 2014.
60. T. Tank and J.-P. Linnartz. Vehicle-to-vehicle communications for AVCS platooning. *IEEE Transactions on Vehicular Technology,* 46(2):528–536, 1997.
61. R. Zheng. Study on emergency-avoidance braking for the automatic platooning of trucks. *IEEE Transactions on Intelligent Transportation Systems,* 15(4):1748–1757, 2014.
62. NHTSA National Highway Traffic Safety Administration. 2006 ruling. http://www.injurysciences.com/Documents/NHTSA\%20Issues\%20Final\%20Rules\%20for\%20Automotive\%20EDRs.pdf
63. D.J. Gabauer and H.C. Gabler. Comparison of roadside crash injury metrics using event data recorders. 2008. http://www.sciencedirect.com/science/article/pii/S000145750700139X
64. D. Sapper, H. Cusack, and L. Staes. *Evaluation of Electronic Data Recorders for Incident Investigation, Driver Performance, and Vehicle Maintenance: Final Report.* US-DOT, Office of Research, Washington DC, 2009.
65. SIRIT-Technologies. *DSRC Technology and The DSRC Industry Consortium (dic) Prototype Team.* White Paper. 2005. http://www.itsdocs.fhwa.dot.gov/research_docs/pdf/45DSRC-white-paper.pdf
66. O. Abari, D. Vasisht, D. Katabi, and A. Chandrakasan. Caraoke: An e-toll transponder network for smart cities. In *Proceedings SIGCOMM'15,* London, UK, pp. 297–310, August 17–21, 2015.
67. G. Yan and S. Olariu. A probabilistic analysis of link duration in vehicular ad hoc networks. *IEEE Transactions on Intelligent Transportation Systems,* 4(12):1227–1236, 2011.
68. L.A. Barroso, J. Clidaras, and U. Hölzle. *The Datacenter as a Computer: An Introduction to the Design of Warehouse-Scale Machines.* 2nd edition. Morgan & Claypool, San Rafael, CA, 2013.
69. D.C. Marinescu. *Cloud Computing, Theory and Applications.* Morgan Kaufman, Waltham, MA, 2013.
70. G. Yan, D. Wen, S. Olariu, and M. C. Weigle. Security challenges in vehicular cloud computing. *IEEE Transactions on Intelligent Transportation Systems,* 4(1):6–16, 2013.

71. G. Yan, S. Olariu, J. Wang, and S. Arif. Towards providing scalable and robust privacy in vehicular networks. *IEEE Transactions on Parallel and Distributed Systems*, 25(7):1895–1906, 2014.

72. Y. Jiang, S. Li, and D.E. Shamo. A platoon-based traffic signal timing algorithm for major-minor intersection types. *Transportation Research B, Methodology*, 40(7):543–562, 2012.

73. R.K. Kamalanathsharma and H.A. Rakha. Leveraging connected vehicle technology and telematics to enhance vehicle fuel efficiency in the vicinity of signalized intersections. *Journal of Intelligent Transportation Systems*, 20(1):33–44, 2016.

74. P. Ryus, M. Vandehey, L. Elefteriadou, R. Dowling, and B. Ostrom. Highway Capacity Manual, TR News 273, Transportation Research Board, Washington DC, pp. 45–48, March–April 2011.

75. R.W. Hall and N. Vyas. Buses as a traffic probe: Demonstration project. *Transportation Research Record*, 1731:96–103, 2000.

76. J. Zheng, Y. Wang, N. L. Nihan, and M. Hallenbeck. Detecting cycle failures at signalized intersections using video image processing. *Computer-Aided Civil and Infrastructure Engineering*, 21(6):425–435, 2006.

Chapter 8

Messages for Cooperative Adaptive Cruise Control Using V2V Communication in Real Traffic

Xiao-Yun Lu and Steven E. Shladover

University of California
Berkeley, California

Aravind Kailas

Volvo Trucks
Greensboro, North Carolina

Osman D. Altan

Turner Fairbank Highway Research Center
McLean, Virginia

Contents

8.1 Introduction

Automated vehicle technologies have been implemented in practice progressively for at least a decade in passenger cars and commercial vehicles. A predecessor technology was cruise control (CC) that regulates the vehicle speed to the driver's desired/selected speed regardless of traffic immediate in front, so the driver needs to adjust the set speed for a proper following distance. The first stage of automation is adaptive cruise control (ACC) that incorporates forward remote sensor detection such as radar/lidar for front target detection, tracking, and relative distance/speed estimation. ACC involves both speed and relative distance control. Although ACC advances significantly beyond CC, it still cannot maintain string stability for multiple vehicles in tandem with a short enough following distance, *i.e.,* if three or more vehicles in tandem are in ACC mode and if the leader vehicle speed fluctuates due to traffic ahead, this fluctuation will be amplified toward the upstream which will eventually cause stability problems for the whole string. The third stage, which is still underway for research and tests, is cooperative ACC (CACC) which is basically ACC plus extra information from V2V (vehicle-to-vehicle) DSRC (dedicated short range communication). Thanks to the delay reduction achieved with V2V information, CACC is able to maintain string stability even if the overall traffic speed fluctuates. This will also significantly improve safety. Although the BSM (basic safety message) Part I [1,2] was set as a standard to support cooperative collision warnings, it is not adequate for CACC control. It is therefore necessary to come up with a minimum set of messages that satisfy the needs of CACC as well as active safety (to enhance vehicle and driver safety with automatic control technologies).

PATH has developed and field tested passenger car and heavy-duty truck CACC on freeways with other traffic, and in support of that work, we have defined a set of V2V messages to support CACC functionality. Although more messages passed between vehicles will likely lead to better performance of CACC in general, there should be a minimum set of messages that are adequate for both CACC maneuverability and safety. It makes sense to minimize the size of such messages due to potentially significant overhead of V2V messaging in a practical traffic system since hundreds of vehicles may be within V2V communication range of the subject vehicle. This chapter suggests such a message set. This set includes messages for maneuvers of individual vehicles within a CACC string, as well as for the coordination of vehicle maneuvers among multiple CACC strings in the same lane and different lanes in real traffic.

The messages include the following data:

- Data for longitudinal control CACC and platooning
- Data for lateral control (this is for future development although not implemented yet)
- Data for maneuvers of individual vehicles within a platoon (or string)
- Data for coordinated maneuvers among multiple platoons (or strings), including exchange of vehicles between two platoons (or strings)
- Data for fault detection and management for safety and maintaining platoon operation under anomalous conditions

Some of the messages are already included in BSM I and BSM II, but some are newly added for control and coordination purposes. This chapter explains the data sets sorted by their functionalities.

For communication purposes, the data to be transferred are encoded to the needed data types such as integers and then built into the communication packets. At the receiving end, the packets are resolved and decoded.

8.2 Data for Control and Active Safety

Society of Automotive Engineers standard SAE J1939 [3] is the vehicle bus recommended practice used for communication and diagnostics among vehicle components. It originates in the widely adopted diesel powered car, bus and heavy-duty truck industry in the United States and is now widely used in other parts of the world. One driving force behind this is the increasing adoption of the engine electronic control unit (ECU), which provides one method of controlling exhaust gas emissions within US and European standards. Control data include those from onboard sensors and J-1939 or other control area network (CAN) bus and control commands that would directly affect the interactions among the vehicles in the platoon. The minimum set of data used for control usually will depend on the control design method. The set of data listed here are those PATH has used for platooning of passenger cars, buses, and trucks to keep practical string stability [4,5,6]. This represents 133 bytes plus 3 bits.

8.3 Data for Coordination of Maneuvers within the Platoon

This set of data is used for the coordination of the maneuvers of individual vehicles within a platoon, which is different from the platoon behavior as a collective. The data for coordination are usually defined by the control system designer. They are not from sensors. The control designer could define a particular meaning of a number which could represent a particular maneuver. To avoid confusion for the communication between vehicles of different makes, it is necessary to standardize this set of data. Broadcasting the current maneuver status is very important for control of individual vehicles and safety so that all the vehicles in the same string know what the others are doing right now. Obviously, one of the control strategies for the subject vehicle is to avoid any space–time conflict with other vehicles in the same string for safety, maneuvering efficiency, and string stability.

Parameters for maneuver coordination would include the coordination and indication of the maneuver (dynamic interaction) of an individual vehicle in the platoon. This information is also useful from a control viewpoint for the immediately following or preceding platoon to handle the dynamic interaction between platoons. The latter would include, but not be limited to, the time adjustment between platoons, and exchange of vehicles between platoons in the same lane (joining the front platoon from the back) and adjacent lanes (lane change). This represents 16 bytes of data.

8.4 Data for Fault Detection and Handling

Parameters to represent the current health condition of the control system are very important information for other vehicles in the same platoon and other platoons nearby (in the same lane or adjacent lane) to make correct decisions for safe maneuvers. This information should include the fault types and the means for handling the fault. The outcome of the handling would be a proper control mode or relevant maneuver to avoid a collision.

It is noted that the vehicle fault mode is represented with a single long integer. The reason is explained as follows. There could be many possible faults/errors that could affect automated vehicle control, to name a few: V2V communication drops, lidar/radar detection, other sensors (speed, gyro, road grade, GPS, …), network switch, CAN and interface, control computer, control software including database, DVI (driver vehicle interface), engine torque control, engine retarder control, torque converter, transmission retarder control, and foundation brake control. Since each component would affect platooning in different aspects and at different levels, all such information should be built into the vehicle fault mode parameter. To achieve this, one could use a bit map with each bit (assumed a conventional sequence of order for all possible faults) corresponding to a specific fault. Then, this bit map could be converted into a long integer (8 byte) which can represent the fault status of 63 different components. To avoid confusion in the fault mode definition, it is necessary to have a standard that defines the threshold of fault. As examples, if intervehicle communication continuously drops for longer than 1s, it is considered as a communication fault; if the distance estimation discrepancy is over 10% of the actual distance, it is considered as a distance measurement fault; if the relative speed estimation error is over 10% of the actual relative speed, it is considered to have a relative speed measurement error. This quantification should be specified for all parameters that are critical for the control. This represents 12 bytes plus 1 bit.

8.5 Data for Coordination between Platoons

This data set is for use in coordinating maneuvers between platoons, including transfers of individual vehicle between platoons, as well as platoon actions as an entity. It could be used for V2V communications between the leaders of the two platoons. However, due to power limits and limited range of V2V communication, it may also be passed between the last vehicle in the lead platoon and the leader of the following platoon. Since each vehicle would have a chance to be the leader or the last vehicle in a platoon, for convenience, it would be necessary to include this 21-byte set of data in the V2V communication packet.

8.6 Concluding Remarks

Communication data are critical for connected automated vehicles. On the one hand, it is desirable to pass as much information as possible between vehicles and between the vehicle and the roadside coordination manager. The latter will be necessary if the market penetration of connected automated vehicles is high, but may not be necessary when the market penetration is low. On the other hand, more information passing would mean more communication overhead, considering so many vehicles are broadcasting and receiving within the DSRC range. For control performance and safety, it is necessary to have a minimum communication data set. The suggested data sets above are initial suggestions based on the experience in connected automated vehicle control research at California PATH for the past 30 years. Different vehicle types, including both light- and heavy-duty vehicles, have been taken into consideration.

If we assume the following data size—short int: 1 byte; int: 4 bytes; long int: 8 bytes; float: 4 bytes; long float: 8 bytes—the total packet size to contain the suggested data will be 182 bytes plus 4 bits.

References

1. *SAE J2735: Dedicated Short Range Communications (DSRC) Message Set Data Dictionary,* SAE International: Warrendale, PA.
2. *SAE J2945/1: On-Board System Requirements for V2V Safety Communications,* SAE International: Warrendale, PA.

3. *SAE J1939*: https://en.wikipedia.org/wiki/SAE_J1939 (Accessed on 28 May 2017).
4. X.Y. Lu and K. Hedrick, 2004, Practical string stability for longitudinal control of automated vehicles, *Int. J. Vehicle Syst. Dynam. Suppl.,* Vol. 41, pp. 577–586.
5. X.Y. Lu and S.E. Shladover, Integrated ACC and CACC development for heavy-duty truck partial automation, *to be presented at 2017 American Control Conference, (ACC-17)*, Seattle, WA, May 24–26.
6. X.Y. Lu and J.K. Hedrick, 2002, A panoramic view of fault management for longitudinal control of automated vehicle platooning, *Proc. of the 2002 ASME Congress*, DOI: 10.1115/IMECE2002-32106. New Orleans, LA, November 17–22.

ANTENNAS FOR V2X

Chapter 9

MIMO in Vehicular Communication Networks: Channel Modeling, Cooperative Relaying, and Resource Allocation

Christos N. Efrem and Athanasios D. Panagopoulos
National Technical University of Athens (NTUA)
Athens, Greece

Contents

9.1 Introduction

9.1.1 Intelligent Transportation Systems

The main application of vehicular communication networks is intelligent transport systems (ITS). ITS in the concept of a smart city is an emerging technology that aims to reduce road accidents and help drivers to avoid traffic jams. In particular, vehicles could be aware of the local and nearby traffic through V2V (vehicle-to-vehicle) and V2I (vehicle-to-infrastructure) communication. Although current vehicular networking may help the achievement of traffic safety and traffic efficiency, there still exist major technical challenges that should be addressed, due to high mobility of vehicles and high relative speeds between nodes.

As can be seen in Figure 9.1, vehicular networking applications are divided into three main categories: (1) active road safety applications, (2) traffic efficiency and management applications, and (3) infotainment (information and entertainment) applications. The first category of applications is intended to decrease the probability of traffic accidents (*e.g.*, lane change assistance, cooperative forward collision warning, emergency vehicle warning, wrong way driving warning, and traffic condition warning). The second category aims at the improvement of vehicle traffic flow and traffic coordination such as speed management, and the provision of local information and maps (*e.g.*, cooperative navigation). Finally, the last category includes cooperative local services (*e.g.*, media downloading) and global Internet services like parking zone management [1].

The implementation of a vehicle ad hoc network (VANET), to reduce road traffic accidents, constitutes another part of ITS which is a special case of mobile ad hoc networks (MANET). In general, according to Figure 9.2, VANET consists of vehicles and city infrastructure—roadside units (RSUs). From Figure 9.2, we can distinguish between two types of communication: V2V and V2I/I2V (vehicle-to-infrastructure and vice versa) which are achieved through wireless standards (*e.g.*, IEEE 802.11p and wireless access for vehicular environments (WAVE). Each vehicle plays the role of transmitter, receiver, and router for broadcasting, and also is equipped with an onboard unit (OBU) for communication and a global positioning system (GPS) to be aware of its own position as well as for tracking other nearby vehicles. The most challenging aspect in VANET is routing because of the high mobility of vehicles. According to Bhoi and Khilar [2], routing protocols in VANET are partitioned into five categories: (a) topology-based routing, (b) position/geography-based routing, (c) cluster-based routing, (d) geo cast-based routing, and (e) broadcast-based routing.

In Ref. [3], the authors propose the use of long-term evolution (LTE) architecture to support ITS applications and services for safer mobility on roads. Although the 802.11p standard has significant advantages like low cost and easy development, it has some drawbacks, namely limited capacity, poor scalability, lack of quality of service (QoS) guarantees (*e.g.*, delay and packet loss), and intermittent connectivity. Because of its high data rates, wide coverage, and high penetration, LTE could be an efficient solution for ITS. In addition, multiple-input-multiple-output (MIMO)

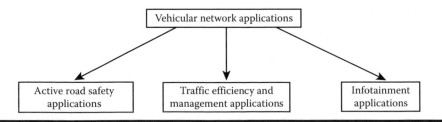

Figure 9.1 Vehicular networking applications.

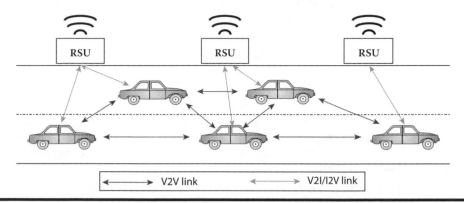

Figure 9.2 Intelligent transportation system (ITS) configuration.

antenna systems in LTE can improve the spectral efficiency even in a high mobility environment and offer multicast and broadcast transmission capabilities as well. By exploiting LTE in ITS, two main types of safety messages would be successfully realized: (1) cooperative awareness message (CAM) such as emergency vehicle warning, slow vehicle indication, intersection collision warning, speed limits notification; and (2) decentralized environmental notification message (DENM) like emergency electronic brake light, wrong way driving warning, stationary vehicle accident, and signal violation warning.

9.1.2 Mobility Models

For reliable simulation results related to ITS/VANETs, appropriate and realistic mobility models are needed because routing protocols are affected by the mobility of vehicles/nodes. Macromobility models describe the road topology, street characteristics, traffic lights and signs, nodes flow, density, and distribution, while micromobility models characterize the movement of vehicles and their interaction with nearby vehicles such as overtaking and acceleration/deceleration. Generally, the mobility depends on many factors related to street construction (intersections, single lane or multiple lanes, and one-way or two-way streets), block size, traffic control mechanism, interdependent vehicular motion (how the movement of one vehicle affects its surrounding vehicles and vice versa), and average speed (how the network topology, movement of nodes, changes over time). In a paper by [4], two types of models based on the movement of nodes are presented: (a) random node movement (movement of nodes in any direction at any velocity) and (b) real-world mobility models based on real-world data of road traffic.

Moreover, the authors in Ref. [5] provide a comprehensive survey and comparison of mobility models available for VANETs. A general framework for realistic mobility models is illustrated and the main categories of criteria for classification of mobility models are proposed as well.

9.1.3 MIMO Antenna Systems and Cooperative Diversity

MIMO antenna systems offer high data rates (gigabit wireless link) and good QoS in non-line-of-sight (NLOS) environments. These performance improvements are due to the particular MIMO characteristics, namely array gain, diversity gain, spatial multiplexing gain, and interference reduction [6].

Array gain is achieved through signal processing at the transmitter (precoding) and the receiver (combining), and leads to an increased signal to noise ratio (SNR) and a decreased bit error rate (BER) as well. Additionally, it requires channel state information (CSI) both in the transceiver and receiver.

Diversity is a novel technique to reduce the impact of fading in wireless links. The key element of diversity techniques is that the signal is transmitted over multiple radio paths in time, frequency, or space. MIMO antenna systems is a special case of a spatial (or antenna) diversity scheme which provides high diversity gain, especially when the radio paths are independent or uncorrelated to a great extent. This can be achieved via either beamforming (if the CSI is available) or space–time coding (if the CSI is not available).

Spatial multiplexing gain leads to a significant increase in channel capacity, maintaining the transmitted power at the same levels. What is more, MIMO systems can reduce the cochannel interference due to the frequency reuse in wireless networks and therefore the cell capacity increases. If CSI is available, transmitter beamforming achieves interference reduction via minimizing the interference power sent to neighbors other than the intended receiver. On the contrary, receiver beamforming (combining) minimizes signals from neighbors other than the intended transmitter.

In Ref. [7], the application of MIMO technique in broadband fixed wireless access (BFWA) at the frequency band of 10–66 GHz, where the rain attenuation plays an important role in total losses, is analyzed. The proposed MIMO/BFWA applications offer capacity enhancement in the spatial multiplexed scheme and interference reduction (SIR enhancement) in the diversity scheme with receiver antenna selection.

Furthermore, the applicability of MIMO techniques in V2V communications is examined in the paper by Attia et al. [8]. The most important reasons to incorporate MIMO into vehicular communications are the following. The first one is to meet safety requirements and acceptable user experience for infotainment applications. For example, spatial multiplexing could support high data rate applications (*e.g.*, media streaming). In the case of safety applications, MIMO can provide reliable communications for short warning messages. Secondly, because of the highly dynamic V2V channel (movement of nodes and multipath fading), the MIMO technique could be highly beneficial. Last but not least, the broadband support of MIMO creates a tremendous opportunity to introduce bandwidth-hungry applications in VANETs such as data streaming.

Due to the increased demand for capacity in next-generation 5G networks (including V2V communications), mmWave systems that operate in frequencies above 10 GHz, where the main fading mechanism is rain attenuation which can be modeled as a random variable that follows the lognormal distribution, could be a potential solution [9]. In Figure 9.3, a cooperative diversity model is illustrated. Particularly, the source node broadcasts a signal to both the destination node

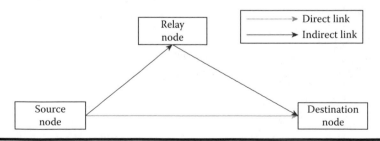

Figure 9.3 Cooperative diversity model.

and the relay node. The relay node regenerates the received signal (using the decode-and-forward scheme) and then retransmits it to the destination node. The destination node combines the direct signal and the signal received through the relay node, using either the selection combining (SC) or the maximal ratio combining (MRC) technique. In the former case, the SNR of the final combined signal corresponds to the SNR of the link with the highest SNR, while in the latter case the SNR of the final signal is equal to the sum of the received SNRs. The article [9] presents a model to evaluate the outage probability of the system that depends on the SC scheme as well as the correlation between the rain fading channels.

9.1.4 Vehicular Network Connectivity

A major issue in vehicular communications is the preservation of the required connectivity between vehicles (V2V connectivity) and between vehicles and infrastructure (V2I connectivity) as well. Connectivity is of great importance for the design of reliable and fault-tolerant communication networks. In order to achieve high availability and energy efficiency, transmission over a long distance from source to destination node is attained through multihop transmissions of smaller distances.

Two nodes (transmitter and receiver) are considered to be connected if the signal to interference-plus-noise ratio (SINR) at the receiver is greater than or equal to the SINR threshold value that depends on the receiver's sensitivity. In general, the link availability is given by the following formula:

$$P_{avail} = P[SNIR \geq SNIR_{th}]$$ (9.1)

A vehicular network is defined as connected when every random vehicular node can communicate with every other node inside the network in a multihop way, as a consequence of the connectivity. The main weakness of this definition is it cannot be used directly for connectivity evaluation in random networks and generally in cases where no spatial information for network nodes and accurate link estimations exist [10]. Therefore, an open issue is to propose metrics for vehicular networks in order to approximate the expected value of connectivity.

9.2 Channel Modeling

The development of vehicular communications requires accurate channel models so as to evaluate the system's performance. Channel modeling in a V2V environment is quite complex due to high relative velocities between the transmitter and the receiver in conjunction with a dynamic variation of nodes' position. Another significant characteristic is the high Doppler spread/shift that results in a statistically nonstationary channel.

In Ref. [11], a presentation of the physical radio channel propagation of an ultra high frequency (UHF) radio frequency identification (RFID) system for a vehicle-to-infrastructure (V2I) scattering environment is given. The presented results are based on a three-dimensional (3-D) ray launching (RL) technique and can help identify the optimal position of nodes in order to minimize power consumption and interference, and also enhance the system's performance.

According to Borhani and Patzold [12], a V2V channel model is used, making the assumption that the local scatterers move with random velocities in random directions. Afterward, the autocorrelation function (ACF), power spectral density (PSD) and the Doppler spread of the channel are calculated and validated by measurement data. Moreover, it is shown that the Gaussian mixture (GM) model can accurately describe the velocity distribution of relatively fast-moving scatterers, while the exponential distribution can reliably model the speed of relatively slow-moving scatterers.

Assuming a path loss model for VANET simulators, the authors in Ref. [13] examine the value of the model parameters in two cases: line-of-sight (LOS) and non-line-of-sight (NLOS) paths. The model parameters are derived from thorough channel measurements at 700 MHz and 5.9 GHz, in different V2V scenarios, taking into account traffic densities and speeds.

In Ref. [14], a geometry-based stochastic channel model (GSCM) for MIMO systems based on extensive channel measurements performed at 5.2 GHz in highway and rural environments is proposed. Besides the LOS component, it is observed that a significant amount of energy comes from scatterers such as cars, houses, and traffic signs. The impact of all these scatterers makes the commonly adopted wide-sense stationary uncorrelated scattering (WSSUS) assumption invalid. What is more, the Doppler spread of the channel alters quickly over time due to high mobility of nodes. Using available channel measurements, the authors suggest a channel model that is able to cope with the non-WSSUS characteristic of a V2V environment. Moreover, four different propagation environments (highway, rural, urban, and suburban) are studied in Ref. [15] so as to derive path loss models for vehicular networks based on extensive data sets. It is concluded that the path loss exponent is small in these V2V scenarios.

The author [16] deals with the modeling of a V2V channel, taking into consideration statistical nonstationarity. Furthermore, key differences between the V2V and cellular radio channels are pointed out, and channel dispersion and time variation are also examined. The dispersion results from multipath propagation and the time variation is caused by the mobility of nodes. More realistic channel models for vehicular communications often incorporate the impact of antennas (*e.g.*, in MIMO systems), filters, diplexers, and other parts of conventional radio transceivers. According to this paper, there exist two types of V2V channel models: the tapped delay line (TDL) model and the GSCM.

An extension of the two ray-tracing model, the multiray tracing model, is proposed by the authors in Ref. [17] in order to accurately study packet loss ratio and end-to-end delay in urban VANET environments. In particular, they investigate both LOS and NLOS channel modeling.

Vehicular communication systems are envisioned to operate at 5.9 GHz and over short distances (10–500 m), while recent cellular systems operate at 700–2100 MHz over long distances (up to tens of kilometers). In vehicular communications, buildings, vehicles (both static and mobile), and vegetation along with foliage have a major impact on the radio propagation. The most recent developments in channel modeling for vehicular communications can be found in [18].

As shown in Figure 9.4, vehicular channel models are classified based on three main criteria: (1) propagation mechanisms, (2) implementation approaches, and (3) properties of the model. Subsequently, each criterion is further divided into subcases, according to the same figure.

Large-scale propagation has to do mainly with the log-distance path loss model (the associated path loss exponent is estimated from empirical measurements), whereas *small-scale fading* is caused by multipath propagation and Doppler spread due to the mobility of vehicles and the movement of scatterers. Small-scale fading is usually modeled as a random variable that follows well-known distributions (*e.g.*, Weibull, Nakagami, and Gaussian distributions).

Geometry-based deterministic (GBD) models primarily concern ray-tracing methods that require a detailed description of the propagation environment. Geometry-based stochastic (GBS) models do not require any information about the environment, are simple and scalable, but they are less accurate. On the contrary, nongeometry-based (NG) models estimate the channel characteristics in a specific environment and adjust the parameters of the path loss, shadowing, and the

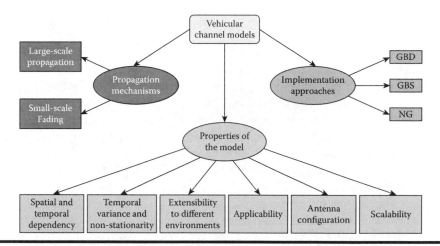

Figure 9.4   Classification of vehicular channel models.

small-scale fading appropriately. The TDL model belongs to this type of model in which each tap represents signals received from several propagation paths with different delays and Doppler shifts.

Moving on to the properties of the channel model, spatial and temporal dependency refers to variations of received signal power that are highly correlated in both time and space. In addition to that, temporal variance and nonstationarity stem from the movement of vehicles at high speeds and the presence/absence of LOS component. Extensibility to different environments describes if the model can be generalized to other propagation environments beyond those that were used to derive the model. Applicability shows the ability of models to be implemented in specific applications. Antenna configuration has to do with the incorporation of different antenna systems (*e.g.* MIMO) in the channel model. Finally, scalability is related to the channel model's performance (complexity) in large-scale simulations.

In Ref. [18], guidelines are provided in order to choose the appropriate channel model, relying on the type of application, geographic information (if it is available), computational complexity, and accuracy of the model. Particularly, stochastic models provide simple and scalable applicability, because there is no need for environmental information, but they produce less accurate calculations. Ray-tracing models achieve higher accuracy, at the expense of increased computational complexity, and require detailed information about the propagation environment that is difficult to obtain. However, some simplified GBD models offer an adequate degree of accuracy and lower complexity than the ray-tracing models.

9.3 Cooperative Techniques

Generally, cooperative MIMO antenna systems and cooperative relay techniques offer increased performance and low energy consumption as well, by taking advantage of spatial and temporal diversity gains.

In the case that roadside units (RSUs) are embedded in the traffic signs, it is quite difficult to have more than one antenna due to the constraints of space, cost, and energy consumption. Consequently, conventional MIMO systems cannot be applied directly to support I2V communication. However, MIMO techniques can be implemented in a distributed manner through a cooperative MIMO scheme in which a group of transmitters, equipped with a single antenna in each one, transmits the information to another group of receivers, thus creating a virtual MIMO system.

In Refs. [19,20], cooperative schemes such as cooperative MIMO, cooperative relay, and multihop single-input-single-output (SISO) transmission are studied for I2V communication. It is highlighted that cooperative MIMO and cooperative relay techniques outperform traditional multihop SISO techniques in terms of capacity, interference, and energy consumption. Moreover, it is pointed out that cooperative relay techniques are more efficient than the SISO technique, but are less efficient than the cooperative MISO techniques concerning energy consumption.

A cooperative relay technique for I2V communication is illustrated in Figure 9.5. The transmission from source node to destination node can be realized by a two-time slot transmission. In the first time slot, signal is transmitted by the source node to the destination node and the relay node at the same time. In the second time slot, the relay node retransmits the signal that is received in the first time slot. At the destination node, the receiver combines the received signals using a diversity technique, for example MRC or equal gain combining (EGC). Due to the fact that received signals follow different paths, the probability of having deep fades in both of them becomes very small. There exist three categories of relay techniques based on the forwarding strategy of the relay node: (1) amplify and forward, (2) decode and forward, and (3) re-encode and forward.

According to Laneman et al. [21], Skraparlis et al. [22], and Sakarellos et al. [23], in a decode-and-forward scenario, there exist two types of relaying protocols: the fixed relaying (FR) protocol and the selection relaying (SR) protocol. In the former case, the relay node always retransmits the received signal and therefore this protocol is restricted by the source-relay link. In the latter case, the relay node retransmits the received signal only when the signal is correctly decoded. In other words, if the relay node cannot decode the received signal, it does not retransmit it to the destination node and the transmission consists only of the direct link between the source and the destination node. The outage probability in each of these two protocols is given as follows:

$$P_{out}^{FR} = P\left[\gamma_2 < \gamma_{th}\right] + P\left[\gamma_2 \geq \gamma_{th}\right] \cdot P\left[\gamma_1 + \gamma_3 < \gamma_{th}\right] \tag{9.2}$$

$$P_{out}^{SR} = P\left[\gamma_2 < \gamma_{th}\right] \cdot P\left[\gamma_1 < \gamma_{th}\right] + P\left[\gamma_2 \geq \gamma_{th}\right] \cdot P\left[\gamma_1 + \gamma_3 < \gamma_{th}\right] \tag{9.3}$$

In the previous expressions, γ_1, γ_2, and γ_3 are the SNRs in dBs of the corresponding links, as shown also in Figure 9.5, and γ_{th} is the SNR threshold of both relay and destination nodes.

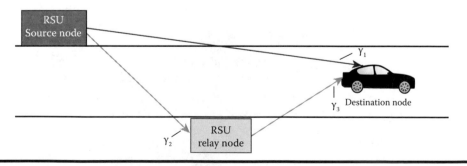

Figure 9.5 Cooperative relay technique for I2V communication.

Figure 9.6 shows the basic idea of cooperative MIMO technique, using N cooperative transmission nodes and M cooperative reception nodes. The whole procedure takes place in three stages. The first stage is called local data exchange, the second cooperative MIMO transmission, and the third cooperative reception. In the local data exchange stage, a source node (S) broadcasts the information to the other N-1 cooperative transmission nodes. Subsequently, in the cooperative MIMO transmission step, the N cooperative transmission nodes transmit, at the same time, the information to the M cooperative reception nodes (including the destination node) using the space–time block code (STBC) technique. In the final phase (cooperative reception), the M-1 cooperation reception nodes retransmit the signal to the destination node. After that, the destination node combines the received signals in order to reconstruct the initially transmitted information.

A multihop SISO transmission for I2V communication is presented in Figure 9.7. This technique is considered the simplest one, where the information is transmitted from source node to destination node (vehicle) through multiple RSUs (cooperative nodes).

The cooperative MIMO technique takes advantage of the diversity gain so as to reduce the transmission energy consumption. As shown in Figure 9.8, the source node cooperates with two of its nearby RSUs, implementing a cooperative MISO scheme to transmit a message to a vehicle (destination node). In addition to this, a cooperative MIMO transmission for I2V communication is illustrated in Figure 9.9.

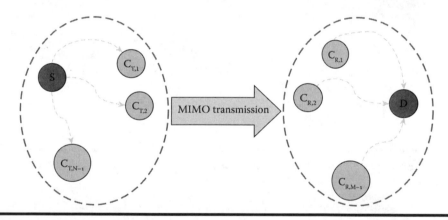

Figure 9.6 Cooperative MIMO technique—transmission from source node (S) to destination node (D) with N cooperative transmission nodes {S, $C_{T,1}$, $C_{T,2}$, ... , $C_{T,N-1}$} and M cooperative reception nodes {D, $C_{R,1}$, $C_{R,2}$, ... , $C_{R,M-1}$}.

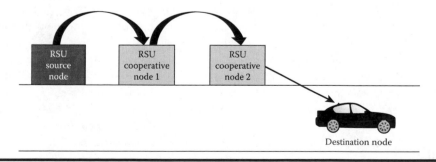

Figure 9.7 Multihop SISO transmission for I2V communication.

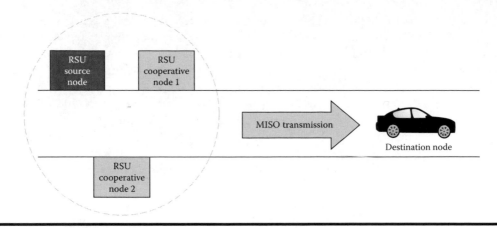

Figure 9.8 Cooperative MISO transmission for I2V communication.

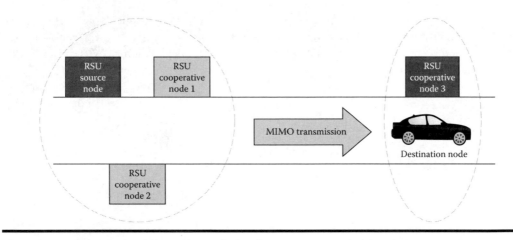

Figure 9.9 Cooperative MIMO transmission for I2V communication.

Multihop transmission is preferred when the distance between source node and destination node is long. Figure 9.10 presents a two-hop cooperative MIMO transmission in which I2V communication is mediated by a set of cooperative nodes (3 and 4).

The performance of vehicular relaying in an LTE-advanced (LTE-A) downlink scenario, employing an STBC-MIMO antenna system at both source (base station) and destination (vehicle), is studied in Ref. [24]. By implementing an appropriate linear precoding, the proposed amplify-and-forward relay technique can achieve a significant improvement on pairwise error probability (PEP) and an increased diversity gain.

The same conclusions are derived in Ref. [25], where the authors investigate the performance of a V2V MIMO cooperative communication system over doubly selective (time/frequency) fading channels. In particular, they assume that source and destination nodes (vehicles) are equipped with two antennas, while the relay node (vehicle) has only one antenna.

An asynchronous cooperative relaying for V2V communication, exploiting polarization diversity without requiring synchronization at the relay node, is examined in Ref. [26]. Because of the dynamic topology of nodes in vehicular networks (fast movement of vehicles), the synchronization, that is generally assumed for cooperative communication protocols, is quite difficult

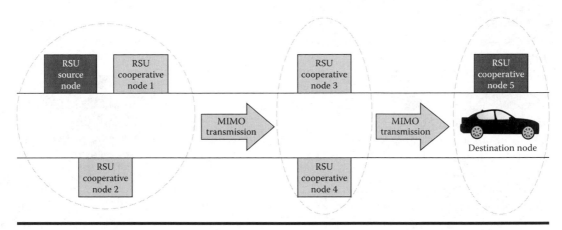

Figure 9.10   Multihop cooperative MIMO transmission for I2V communication.

to attain. The authors suggest a full duplex amplify-and-forward relay node which is equipped with dual polarized antennas. A comparison between the proposed scheme, the noncooperative SISO system, and the conventional amplify-and-forward relaying is conducted via simulations. Finally, it is concluded that the proposed technique outperforms the other two techniques in terms of capacity (data rate), BER, and total energy consumption.

9.4 Resource Allocation

The intelligent transportation system along with its applications requirements has imposed immense challenges in terms of system capacity and packet latency. As a consequence, a careful and appropriate radio resource management (RRM) strategy is needed to achieve high through-put and low delay in vehicular networks. Generally, resource allocation is expressed as an optimization problem of the following form:

$$\mathbf{max}\left\{\mathit{Throughput}\right\} \quad or \quad \mathbf{min}\left\{\mathit{Power\ consumption}\right\}$$

$$\mathbf{subject\ to} \quad \mathit{QoS\ constraints}$$

In Ref. [27], a traffic classification technique based on similarity of content is presented so as to reduce the system traffic load and ensure high QoS for drivers. Particularly, a time division multiple access (TDMA)-based scheme is proposed in which shared content of different users are transmitted via broadcast, instead of unicast.

What is more, a subchannel allocation problem in V2V communications is of great importance. A low outage probability and high outage capacity can be attained by formulating the subchannel allocation in vehicular communications as a maximum matching problem on random bipartite graphs [28]. The proposed distributed algorithm has sublinear complexity of order $O(N^{2/3})$, where N is the number of subchannels. Moreover, it is highlighted that the proposed strategy requires only one-bit broadcasting for each subchannel, thus leading to reduced signaling overhead.

Based on cognitive radio (CR) architecture, the authors in Ref. [29] propose the prioritized optimal channel allocation (POCA) scheme in which the vehicles are classified into primary providers (PPs) that transmit safety messages and secondary providers (SPs) that transmit non-safety information. POCA aims to achieve optimal load balance between channel availability and utilization subject to the throughput constraints of PPs. In the case of a distributed CR network, a POCA algorithm uses a dynamic programming technique, while in a centralized CR network the problem is solved by means of linear programming theory.

An efficient dynamic resources allocation scheme (DRAS) for V2I communication (vehicle-to-RSUs), based on IEEE 802.11p/WAVE standard, is investigated in Ref. [30]. In this strategy, buses are introduced as moving relay nodes in order to offload the RSUs in the vehicular network. Finally, it is pointed out that the improvement of QoS becomes greater as the number of vehicles increases.

The authors in Ref. [31] introduce a novel spectrum resource allocation strategy on a TV band for a cognitive radio-based high-speed vehicle network (CR-HSVN). Specifically, an optimization problem (0–1 integer programming) is formulated to maximize the TV white spaces that high-speed vehicles use in CR-HSVN. The performance of the proposed algorithm is examined through simulations and it is shown that very good spectrum utilization and fairness of the network can be attained.

Smart antenna technology offers significant benefits for V2V2I (vehicle-to-vehicle-to-infrastructure) communication, according to Pyun et al. [32]. For example, both vehicles and RSUs could use multiple antennas/directional beams so as to transmit multiple information at the same time. In such a scenario, the V2V and V2I communications coexist, thus resulting in cochannel interference and this in turn leads to performance degradation. A resource allocation strategy for the smart antenna-based vehicular network is suggested in Ref. [32], which offers capacity enhancement.

Afterward, in Ref. [33], two joint beam and power allocation algorithms that take into account the interference between directional beams are presented, in order to maximize the overall system capacity. The first algorithm is an iterative resource allocation (IRA) technique, and the second one is a heuristic resource allocation (HRA) scheme of lower complexity with insignificant performance degradation compared to the first one.

In general, joint optimization results in much higher system performance, at the expense of increased computational complexity. A joint power and subcarrier allocation scheme for vehicular networks, considering delay QoS constraints, is investigated in Ref. [34]. In this article, orthogonal frequency division multiplexing (OFDM) and perfect channel state information (CSI) are assumed. In addition, the optimization problem (minimization of total power consumption) is converted to a convex optimization problem that is solved by means of convex optimization theory.

The authors in Ref. [35] present a resource allocation policy for multicast service in cellular vehicular networks with QoS constraints. In order to cope with the high mobility of vehicles, the scheduling period is divided into multiple segments, assuming that the channel condition for one vehicle is stable in one segment. Two types of multicast services are investigated: (1) the multicast service to cover all the receivers, and (2) the multicast service with adaptive receivers. Regarding the first type, the problem is formulated as a resource scheduling strategy with fixed modulation and coding scheme (MCS) and then a k-commodity packing-based approximation algorithm is proposed. Concerning the second type of multicast service, the authors consider that the number of valid MCSs is limited and they propose a greedy heuristic algorithm to search for the appropriate MCS assignments. Finally, the efficiency of the algorithms is evaluated through simulations.

9.5 Summary and Future Research Directions

In this chapter, we firstly deal with the general concept of an ITS and mobility models in vehicular networks. Afterwards, the basic principles of MIMO antenna systems and cooperative diversity are presented. An overview of the state-of-the-art channel models for V2V environments is then provided. Subsequently, we examine various cooperative techniques in vehicular communications, including cooperative MIMO and cooperative relay schemes. Finally, a review of recent resource allocation policies is given.

Channel modeling in MIMO vehicular networks is very challenging because of the high relative velocities between the transmitter and the receiver. This dynamic topology (mobility) of nodes gives rise to high Doppler spread that, in turn, results in a statistically nonstationary channel. Channel estimation, that is, accurate acquisition of CSI, signal processing techniques, and cross-layer optimization are crucial for attaining the advantages of MIMO antenna systems.

Cross-layer optimization in MIMO networks can be classified into bottom-up and top-down architecture. In a bottom-up architecture, MIMO technique is exploited at the higher layers of the protocol stack so as to take advantage of the channel dynamics and minimize interference. On the contrary, in a top-down architecture the vehicular application adapts and optimizes the underlying layers of the protocol stack, including the MIMO physical layer, in order to satisfy the QoS requirements. One research direction related to bottom-up architecture is to develop efficient link/MAC protocols that decide the optimal MIMO mode based on physical layer information (*e.g.*, scattering and interference) and vehicle sensors (*e.g.*, speed and acceleration). Another research direction related to top-down architecture is to design suitable mechanisms that help V2V applications to adapt the MIMO signal processing mode (diversity, multiplexing, and beamforming) and the link and network layer protocols in order to meet the specific requirements in terms of QoS [8].

In addition, more in-depth investigation is required in order to examine the capability of LTE to support vehicular applications. For instance, D2D (device-to-device) communications could be an attractive solution for local data exchange among vehicles, and therefore many issues should be addressed to make D2D communication suitable for vehicular communications. Furthermore, the decision about the communication mode (cellular or D2D) should consider the interference to nearby nodes and guarantee reliable transmission of traffic messages. Another example is the design of efficient LTE packet schedulers to support vehicular applications that will be able to achieve high spectrum efficiency, throughput, and fairness [3].

Further research efforts are also required in channel modeling for vehicular communications. In particular, further studies (measurements and modeling) are needed to investigate channel characteristics for vehicles other than personal cars, for example, commercial vans, trucks, scooters, and public transportation vehicles such as buses. These types of vehicles have different dimensions and road dynamics. Additionally, signal propagation measurements in other environments such as multilevel highways, tunnels, parking garages, bridges, and roundabouts should be performed, so as to supplement the existing channel models in vehicular environments. Last but not least, the incorporation of ITS into future fifth-generation (5G) networks leads to the convergence of the research efforts in channel modeling. Channel models for D2D communications, for example, could take advantage of the existing V2V channel models and vice versa [18].

Finally, the network connectivity imposes its own challenges, since it is a crucial parameter for system performance, especially in spectrum sharing, routing, and cooperative techniques.

References

1. Karagiannis, G., Altintas, O., Ekici, E., Heijenk, G., Jarupan, B., Lin, K., & Weil, T. (2011). Vehicular networking: A survey and tutorial on requirements, architectures, challenges, standards and solutions. *IEEE Communications Surveys & Tutorials*, 13(4), 584–616.
2. Bhoi, S.K., & Khilar, P.M. (2014). Vehicular communication: A survey. *IET Networks*, 3(3), 204–217.
3. Araniti, G., Campolo, C., Condoluci, M., Iera, A., & Molinaro, A. (2013). LTE for vehicular networking: A survey. *IEEE Communications Magazine*, 51(5), 148–157.
4. Madi, S., & Al-Qamzi, H. A survey on realistic mobility models for vehicular ad hoc networks (VANETs). In *2013 10th IEEE International Conference on Networking, Sensing and Control (ICNSC)*, pp. 333–339, Evry, France: IEEE, April 10–12, 2013.
5. Harri, J., Filali, F., & Bonnet, C. (2009). Mobility models for vehicular ad hoc networks: A survey and taxonomy. *IEEE Communications Surveys & Tutorials*, 11(4), 19–41.
6. Paulraj, A.J., Gore, D.A., Nabar, R.U., & Bolcskei, H. (2004). An overview of MIMO communications-a key to gigabit wireless. *Proceedings of the IEEE*, 92(2), 198–218.
7. Liolis, K.P., Panagopoulos, A.D., Cottis, P.G., & Rao, B.D. (2009). On the applicability of MIMO principle to 10-66GHz BFWA networks: Capacity enhancement through spatial multiplexing and interference reduction through selection diversity. *IEEE Transactions on Communications*, 57(2), 530–541.
8. Attia, A., ElMoslimany, A.A., El-Keyi, A., ElBatt, T., Bai, F., & Saraydar, C. (2012). MIMO vehicular networks: Research challenges and opportunities. *Journal of Communications*, 7(7), 500–513.
9. Sakarellos, V.K., Skraparlis, D., Panagopoulos, A.D., & Kanellopoulos, J.D. (2012). Cooperative diversity performance in millimeter wave radio systems. *IEEE Transactions on Communications*, 60(12), 3641–3649.
10. Pitsiladis, G.T., Panagopoulos, A.D., & Constantinou, P. (2012). A Spanning-Tree-Based connectivity model in finite wireless networks and performance under correlated shadowing. *IEEE Communications Letters*, 16(6), 842–845.
11. Azpilicueta, L., Vargas-Rosales, C., & Falcone, F. (2016). Intelligent vehicle communication: Deterministic propagation prediction in transportation systems. *IEEE Vehicular Technology Magazine*, 11(3), 29–37.
12. Borhani, A., & Patzold, M. (2013). Correlation and spectral properties of vehicle-to-vehicle channels in the presence of moving scatterers. *IEEE Transactions on Vehicular Technology*, 62(9), 4228–4239.
13. Fernández, H., Rubio, L., Rodrigo-Peñarrocha, V. M., & Reig, J. (2014). Path loss characterization for vehicular communications at 700 MHz and 5.9 GHz under LOS and NLOS conditions. *IEEE Antennas and Wireless Propagation Letters*, 13, 931–934.
14. Karedal, J., Tufvesson, F., Czink, N., Paier, A., Dumard, C., Zemen, T., & Molisch, A.F. (2009). A geometry-based stochastic MIMO model for vehicle-to-vehicle communications. *IEEE Transactions on Wireless Communications*, 8(7), 3646–3657.
15. Karedal, J., Czink, N., Paier, A., Tufvesson, F., & Molisch, A.F. (2011). Path loss modeling for vehicle-to-vehicle communications. *IEEE Transactions on Vehicular Technology*, 60(1), 323–328.
16. Matolak, D.W. (2014). Modeling the vehicle-to-vehicle propagation channel: A review. *Radio Science*, 49(9), 721–736.
17. Tabatabaei, S.H., Fleury, M., Qadri, N.N., & Ghanbari, M. (2011). Improving propagation modeling in urban environments for vehicular ad hoc networks. *IEEE Transactions on Intelligent Transportation Systems*, 12(3), 705–716.
18. Viriyasitavat, W., Boban, M., Tsai, H.M., & Vasilakos, A. (2015). Communications: Survey and challenges of channel and propagation models. *IEEE Vehicular Technology Magazine*, 10(2), 55–66.
19. Nguyen, T.D., Berder, O., & Sentieys, O. (2011). Energy-efficient cooperative techniques for infrastructure-to-vehicle communications. *IEEE Transactions on Intelligent Transportation Systems*, 12(3), 659–668.
20. Abraham, T.S., & Narayanan, K. Cooperative communication for vehicular networks. In *2014 International Conference on Advanced Communication Control and Computing Technologies (ICACCCT)*, pp. 1163–1167, Ramanathapuram, India: IEEE, May 8–10, 2014.

21. Laneman, J.N., Tse, D.N., & Wornell, G.W. (2004). Cooperative diversity in wireless networks: Efficient protocols and outage behavior. *IEEE Transactions on Information Theory*, 50(12), 3062–3080.
22. Skraparlis, D., Sakarellos, V.K., Panagopoulos, A.D., & Kanellopoulos, J.D. (2009). Outage performance analysis of cooperative diversity with MRC and SC in correlated lognormal channels. *EURASIP Journal on Wireless Communications and Networking*, 2009, 5.
23. Sakarellos, V.K., Skraparlis, D., Panagopoulos, A.D., & Kanellopoulos, J.D. (2011). Cooperative diversity performance of selection relaying over correlated shadowing. *Physical Communication*, 4(3), 182–189.
24. Feteiha, M.F., & Hassanein, H.S. Cooperative vehicular ad-hoc transmission for LTE-A MIMO-downlink using Amplify-and-Forward relaying. In *2013 IEEE Global Communications Conference (GLOBECOM)*, pp. 4232–4237 Atlanta, GA: IEEE, December 09–13, 2013.
25. Feteiha, M. F., & Uysal, M. (2015). On the performance of MIMO cooperative transmission for broadband vehicular networks. *IEEE Transactions on Vehicular Technology*, 64(6), 2297–2305.
26. Sohaib, S., & So, D. K. (2013). Asynchronous cooperative relaying for vehicle-to-vehicle communications. *IEEE Transactions on Communications*, 61(5), 1732–1738.
27. Xin, X., Zheng, K., Liu, F., Long, H., & Jiang, Z. An efficient resource allocation scheme for vehicle-to-infrastructure communications. In *2013 8th International ICST Conference on Communications and Networking in China (CHINACOM)*, pp. 40–45, Guilin, China: IEEE, August 14–16, 2013.
28. Bai, B., Chen, W., Letaief, K.B., & Cao, Z. (2011). Low complexity outage optimal distributed channel allocation for vehicle-to-vehicle communications. *IEEE Journal on Selected Areas in Communications*, 29(1), 161–172.
29. Chu, J.H., Feng, K.T., & Lin, J.S. (2015). Prioritized optimal channel allocation schemes for multi-Channel vehicular networks. *IEEE Transactions on Mobile Computing*, 14(7), 1463–1474.
30. Jiang, T., Alfadhl, Y., & Chai, K.K. Efficient dynamic scheduling scheme between vehicles and roadside units based on IEEE 802.11 p/WAVE communication standard. In *2011 11th International Conference on ITS Telecommunications (ITST)*, pp. 120–125, St. Petersburg, Russia: IEEE, August 23–25, 2011.
31. Jiang, T., Wang, Z., Zhang, L., Qu, D., & Liang, Y.C. (2014). Efficient spectrum utilization on TV band for cognitive radio based high speed vehicle network. *IEEE Transactions on Wireless Communications*, 13(10), 5319–5329.
32. Pyun, S.Y., Cho, D.H., & Son, J.W. Downlink resource allocation scheme for smart antenna based v2v2i communication system. In *2011 IEEE Vehicular Technology Conference (VTC Fall)*, pp. 1–6, San Francisco, CA: IEEE, September 05–08, 2011.
33. Pyun, S.Y., Lee, W., & Cho, D.H. (2016). Resource allocation for Vehicle-to-Infrastructure communication using directional transmission. *IEEE Transactions on Intelligent Transportation Systems*, 17(4), 1183–1188.
34. Zhang, H., Ma, Y., Yuan, D., & Chen, H.H. (2011). Quality-of-service driven power and sub-carrier allocation policy for vehicular communication networks. *IEEE Journal on Selected Areas in Communications*, 29(1), 197–206.
35. Zhou, H., Wang, X., Liu, Z., Zhao, X., Ji, Y., & Yamada, S. QoS-aware resource allocation for multi-cast service over vehicular networks. In *2016 8th International Conference on Wireless Communications & Signal Processing (WCSP)*, pp. 1–5, Yangzhou, China: IEEE, October 13–15, 2016.

Chapter 10

Antenna Placement and Beam Direction Effects on V2V Communication Channels

Milad Mirzaee, Nischal Adhikari, and Sima Noghanian

University of North Dakota
Grand Forks, North Dakota

Contents

10.1 Introduction

The US Department of Transportation (USDOT) recently released *Intelligent Transportation Systems (ITS) Strategic Research Plan*, by the Research and Innovative Technology Administration (RITA). USDOT has recognized that the interaction between vehicles (vehicle-to-vehicle or V2V) and between vehicles and the roadway (vehicle-to-infrastructure or V2I) holds the potential to address safety issues and other difficult transportation challenges. The research on V2V and V2I is branded as IntelliDrive®. The key enabling technology for V2V communication is dedicated short-range communication (DSRC). The Federal Communication Commission (FCC) has allocated a frequency band at 5.9 GHz (5.850–5.925 GHz) to DSRC to be used for ITS applications which has various applications including vehicular safety, traffic management, toll collection, and information delivery.

There has been a tremendous amount of effort and investment from the government and private organizations to develop means for highly efficient V2V and V2I communications. V2V communications might be used to communicate information like position, speed, and direction in cooperative awareness messages, in order to derive an environmental picture, which could be used for prediction of movement [1]. This contributes to safer travel and a reduction in fatalities as the result of road accidents [2]. The challenge in developing an efficient network between the cars is to understand the nature of random channels that changes with the location and radiation patterns of antennas, surroundings, and obstacles between the transmitting and receiving vehicles.

The antenna as a key component plays an important role in the performance of V2V communications. In the past few years, various antenna configurations have been developed for vehicular communication applications [3–6]. Some other aspects of V2V communication are antenna placement, antenna pattern, channel modeling, and channel measurement [7–9]. In situations when the wireless channel is considered to be affected by multiple scattering and reflections, where the direct line-of-sight (LOS) does not exist, or it coexists with multipath components, multiple-input-multiple-output (MIMO) technology is proposed to take advantage of multipath effects. MIMO systems increase channel capacity compared to single-input-single-output (SISO) systems and offer better performance in non-line-of-sight (NLOS) scenarios.

In most of the MIMO systems, including V2V systems, omnidirectional antennas are utilized. Only a limited number of studies are focused on the use of directional antennas in V2V communication. In Ref. [10], a measurement setup is proposed and used to study the effect of steering directional beams to improve vehicle-to-vehicle communication. The setup is based on commercially available phased array antennas. An omnidirectional beam is compared with a switched beam system with eight beams of approximately 45 degrees half-power-beam-width (HPBW) at 2.4 GHz. Two scenarios were considered. The first one is the highway scenario with two to three lanes and LOS. The second one is a dense residential neighborhood where scattering is reached and a lot of time LOS does not exist. The study showed that using the directional beams improved the signal-to-noise-ratio (SNR) by 11 dB in a suburban environment, and 14 dB in the highway scenario, providing 50% to 80% improvement in the range. The beam direction was estimated using the global positioning system (GPS) coordinates of the transmitter and receiver, and directing the beam to the other car.

In Ref. [11], the effect of antenna position on the vehicle in the frequency band of 5 GHz is studied. Although the focus of this paper is on the antenna position, one conclusion was that the antenna pattern does not remain omnidirectional as it is mounted at various locations on the vehicle. Therefore, the patterns become more directional. This effect is discussed in more detail in Ref. [12]. This paper refers to the effect of finite metallic roof and nonmetalized sunroof or railings, as well as the inclination angle of the roof, that can affect the antenna patterns. Another effect is caused by the mutual coupling and effect of other antennas such as GPS and satellite radio antennas. This paper mentions the effect of the rooftop that might be tilting the omnidirectional antenna's main beam from horizontal (90 degrees from normal to the rooftop) to a tilt of around 70 degrees from the normal. This can decrease the SNR and the efficiency of the communication.

Some measurement campaigns show how the placement can affect the overall communication efficiency [13–17]. For example, in Ref. [16], we showed the relation of the K-factor and antenna height, as they related to the calculated channel capacity of MIMO systems. Ricean K-factor is defined as the power ratio of the LOS component to the diffused component. Generally, the higher values of K-factor imply that a direct path is stronger compared to the multipaths' power, while small K-factor values are associated with richer multipath environments and should provide higher MIMO capacities. In Ref. [16], it was shown that the capacity increases for lower K-factor values as long as the antenna height is fixed. However, lower K-factor values for lower heights did not show better capacity as compared to higher K-factor values for antenna elements at higher positions.

10.2 Directive Antennas for V2V MIMO Systems

The difficulty of having omnidirectional antennas on the vehicle body, as well as the advantages and control parameters that use of directional antenna patterns provides, motivates us to study the effect of directional patterns and beam steering in V2V MIMO systems. In the remaining part of this chapter the effect of antenna directive pattern main beam and orientation on the MIMO system performance is investigated using the Wireless InSite® ray tracing simulation program (from Remcom Inc. [18]). The study is done for a three dimensional (3D) environment and virtual cars, surrounded by 3D objects around the buildings. The study is performed based on a real site selected for simulations, a road next to the College of Engineering and Mines, the University of North Dakota, Grand Forks, ND, USA. The simulations were performed for both LOS and NLOS scenarios.

10.2.1 Antenna Positions

Fifteen antennas on the top, front, and rear of the cars at both transmitting and receiving ends were modeled. Simulations were performed by using either dipoles, as omnidirectional antennas, or directive antennas (horn), at transmitter (TX) or receiver (RX) ends. The horn antenna is only selected for the simplicity and availability in Wireless InSite simulation software, and it is understood that in reality other types of directive antennas are more suitable for vehicular systems. The main point of study here was to consider a directive beam; 225 cases were studied. Among them four different cases were examined in detail, since we found them to be the most revealing ones [16]. The front-to-back distance between receiver and transmitter cars was 10 m. To study the NLOS scenario, the obstacle vehicle was placed at the center between the transmitting and receiving cars. To simulate the dynamic scenario, the entire setup was moved to two different positions, +25 m and +40 m from the original position. In addition, the transmitting power in both scenarios was selected to be 0 dBm. Figure 10.1 shows the overview of the antennas' layout on the transmitter and receiver cars. Table 10.1 summarizes the spacing between each two antennas and center-to-center distances between each pair of routes. The spacing between elements in each route is uniform. Table 10.2 describes all the studied scenarios in terms of antenna types, antenna orientation, as well as their positions. Figure 10.2 [19] shows the antenna pattern for various scenarios as listed in Table 10.2. Antennas from routes 1, 3, 4, and 5 were selected for cases 1, 2, 3, and 4, respectively. This selection was done based on the previous results from Refs. [15–16,19]. Referring to Figure 10.1, where antennas are numbered, cases are defined as: case 1 TX(7,8)–RX(7,8), case 2 TX(7,9)–RX(7,9), case 3 TX(10,11)–RX(10,11), and case 4 TX(14,15)–RX(1,2). The frequency range opted for our simulation was from 5.855 GHz to 5.925 GHz, which is allocated to DSRC and wireless access in vehicular environment (WAVE) channels. The height of antenna element depends on the surface on

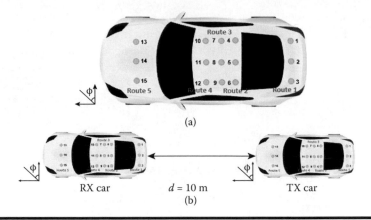

Figure 10.1 **(a) Layout of antenna positions and numbers on both transmitting and receiving vehicles, (b) transmitter (TX) and receiver (RX) cars. (Car models from http://www.clipartbest. com/clipart-jTxo5RLnc.)**

Table 10.1   Spacing between Antennas (Distances in Meters)

Antenna No./Route No.	Spacing (m)	Spacing (wavelength at 5.9 GHz)
1–2, 2–3 (Route 1)	0.50	9.43
4–5, 5–6 (Route 2)	0.35	6.60
7–8, 8–9 (Route 3)	0.35	6.60
10–11, 11–12 (Route 4)	0.35	6.60
13–14, 14–15 (Route 5)	0.50	9.43
Route 1 and Route 2	1.07	20.19
Route 2 and Route 3	0.40	7.55
Route 3 and Route 4	0.35	6.60
Route 4 and Route 5	1.60	30.19

Source: Mousavi, P., *IEEE Trans. Antenn. Propag.*, 59(8), 3123–3127, 2011.

which it is mounted. Antennas at the rear or front of the car have heights of 0.95 m, and those mounted on the roof are at 1.35 m.

10.3 Environment Modeling

The details of material selection for each scenario and assumed heights are summarized in Table 10.3 [16]. Google Maps was used to delineate the exact location of buildings. The image of each scenario was imported into Wireless InSite and the environmental setup was built. Figure 10.3a shows the map extracted from Google Maps. Figure 10.3b through d depicts its implementation in simulation setup and distances between the transmitter car and the closest interferers for different positions.

Table 10.2   Different Scenarios for Simulation

Scenario	Antenna Type	Antenna Orientation		Description
		TX Angle	Rx Angle	
1	Dipole	Omnidirectional	Omnidirectional	Position origin
2	Dipole	Omnidirectional	Omnidirectional	Position +25 m
3	Dipole	Omnidirectional	Omnidirectional	Position +40 m
4	Horn	0°	0°	Facing building, position origin
5	Horn	0°	0°	Facing building, Position +25 m
6	Horn	0°	0°	Facing building, Position +40 m
7	Horn	90°	270°	Facing each other, position origin
8	Horn	90°	270°	Facing each other, position +25 m
9	Horn	90°	270°	Facing each other, position +40 m
10	Horn	TX *i* (10°) TX *j* (170°)	RX *k* (10°) RX *l* (170°)	Not facing each other, position +40 m
11	Horn	TX *i* (45°) TX *j* (135°)	RX *k* (45°) RX *l* (135°)	Not facing each other, position +40 m
12	Horn	90°	90°	Not facing each other, position +40 m
13	Horn	TX *i* (10°) TX *j* (170°)	RX *k* (350°) RX *l* (190°)	Facing each other, position +40 m

10.4 MIMO Systems

We considered 2×2 MIMO systems (two antennas at both the transmitter and receiver sides) and compared them in terms of their capacity. The channel capacity of the MIMO system is calculated as [20]:

$$C = \log_2 \det\left(\mathbf{I} + \frac{\rho}{L_t} \mathbf{H}\mathbf{H}^T \right) \quad \text{bits/s/Hz} \tag{10.1}$$

where ρ is the receiver SNR, $\mathbf{H}$ is the channel matrix, $\mathbf{I}$ is the identity matrix, and $\mathbf{H}^T$ is the Hermitian or conjugate transpose of $\mathbf{H}$. All matrices have the size of $L_t \times L_r$, where L_t is the number of transmitter antennas and L_r is the number of receiver antennas ($L_t = L_r = 2$ for our cases).

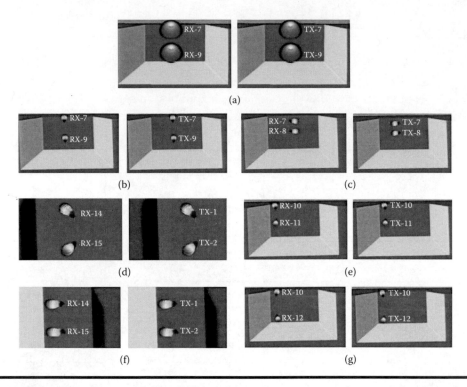

Figure 10.2 Antenna patterns for different scenarios: (a) scenarios 1–3, (b) scenarios 4–6, (c) scenarios 7–9, (d) scenario 10, (e) scenario 11, (f) scenario 12, and (g) scenario 13. (From Mirzaee, M., et al., Analysis of Static and Dynamic Scenarios of MIMO Systems for Physical Layer Modeling for Vehicular Communication, *National Wireless Research Collaboration Symposium (NWRCS)*, Idaho Falls, ID, pp. 1–10, 2014. With permission.)

TABLE 10.3   Material Selection for Simulation

Feature	Description	Type	Height (m)
Terrain	Wet earth	DHS	0.00
Babcock Hall	Brick	OLD	6.00
Harrington Hall	Brick	OLD	6.00
Gillette	Brick	OLD	6.00
Road	Asphalt	DHS	0.10
Cars	PVC, metal, glass	OLD, PEC, OLD	1.35

Source: Adhikari, N., and Noghanian S., Multiple antenna systems for vehicle to vehicle communication, *2013 International Conference on Electro/Information Technology*, Rapid City, SD, pp. 1–6, 2013.

Note: PVC, polyvinylchloride; DHS, dielectric half-space; OLD, one-layer dielectric; PEC, perfect electric conductor.

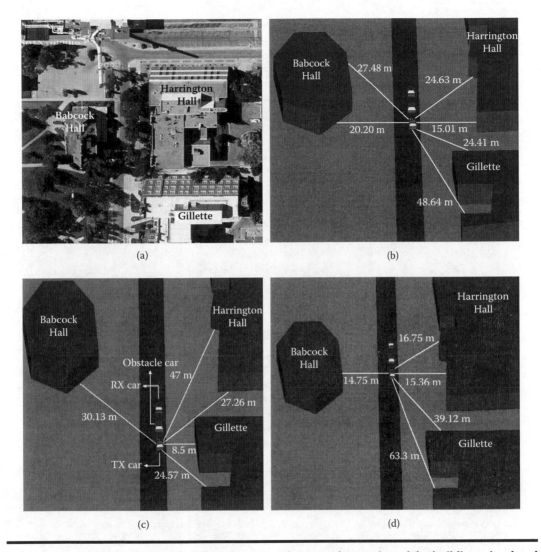

Figure 10.3 (a) Top view of the buildings from Google Maps, (b) top view of the buildings simulated in Wireless InSite positioned at origin, (c) 25 m away from origin and (d) 40 m away from origin.

10.4.1 Capacity of Directional and Omnidirectional MIMO

The system behavior is compared through the capacity of four different cases for each of 13 scenarios. Two setups were considered: the first setup was without considering a blocking car, and the second one was with considering a blocking car between transmitter and receiver vehicles. The car located at the back was considered to be the transmitting end and the front car was considered to be the receiving end. Multipath traveled by electromagnetic waves may be studied by ray tracing, which can be used to calculate the channel capacity and analyze it with different antenna configurations and placement. Figure 10.4 shows the multipath for various scenarios of case 2, without and with blocking obstacle car, when using dipole and horn antennas, respectively.

The MIMO system using horn antennas generates more multipath. It was observed that the received power in scenarios 4, 5, and 6 is lower than in scenarios 1, 2, and 3. The major reason for this behavior

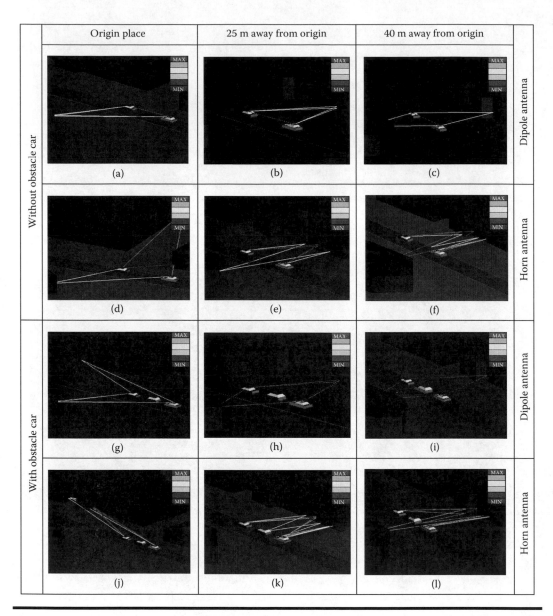

Figure 10.4 **(a) Paths for case 2 (with dipole antenna and without obstacle car) at the origin, (b) at +25 m, and (c) at +40 m; (d) paths for case 2 (with horn antenna and without obstacle car) at the origin, (e) at +25 m, and (f) at +40 m; (g) paths for case 2 (with dipole antenna and with obstacle car) at the origin, (h) at +25 m, and (i) at +40 m; (j) paths for case 2 (with horn antenna and with obstacle car) at the origin, (k) at +25 m, and (l) at +40 m.**

is that the transmitted signal for the MIMO system using directional antenna has a longer propagation path to reach the receiver car in comparison with the MIMO system using dipole antennas.

Figure 10.5a and b shows the variation of channel capacity for different cases. From Figure 10.5b, the channel capacity for case 1, offered by scenario 6, is greater than the other scenarios. For case 2, scenario 9 offers the greatest channel capacity. However, because of the long propagating path of the transmitted signal, the received power was around −125 dB. For cases 3

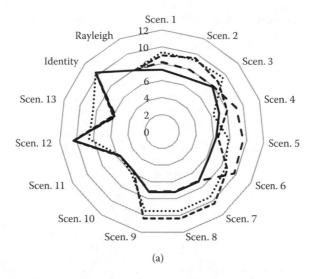

(a)

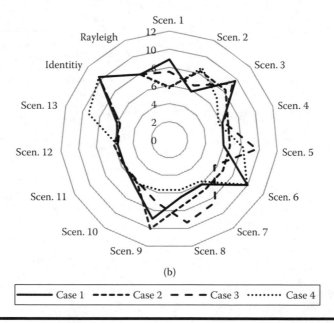

(b)

| —— Case 1 | ---- Case 2 | — — Case 3 | ········· Case 4 |

Figure 10.5 Channel capacity of cases 1–4 for scenarios 1–13 in comparison with identity and Rayleigh channel capacities: (a) without obstacle car between TX and RX cars and (b) with obstacle car between TX and RX cars.

and 4, scenarios 5 and 6 offer the greatest channel capacities, respectively. For all scenarios, except scenario 12 (due to the direction of antenna), the received power was relatively good, but scenario 6 demonstrated the highest channel capacity in most of the cases due to rich multipath caused by surrounding objects.

The effect of surrounding objects on the channel capacity can clearly be observed in this case. These capacity values can be compared with 2×2 identity MIMO capacity (ideal case) and 2×2 Rayleigh MIMO capacities (typical case). The capacity for 2×2 identity matrix channel

Table 10.4   Channel Capacity and K-factor for Different Scenarios without Obstacle Car

Scenario	Channel capacity (bits/s/Hz), mean K-factor (dB) (without obstacle car)			
	Case 1	Case 2	Case 3	Case 4
1	7.28, 23.87	9.00, 23.84	8.20, 26.56	9.38, 17.52
2	6.94, 22.09	9.49, 20.85	7.96, 19.66	9.24, 13.31
3	7.88, 21.16	8.76, 20.46	7.98, 21.85	9.53, 13.61
4	6.98, −42.22	8.01, −46.73	9.13, −51.58	6.26, −∞
5	6.34, −∞	6.53, −∞	9.58, −∞	7.82, −∞
6	6.24, −∞	8.83, −∞	9.77, −∞	8.54, −∞
7	7.22, 61.94	**10.36, 61.52**	7.14, 64.04	9.48, 74.73
8	7.22, 74.35	**10.36, 74.38**	7.14, 72.82	9.48, 74.15
9	7.22, 74.73	**10.36, 74.15**	7.14, 58.64	9.48, 74.40
10	5.69, −∞	5.65, −∞	5.70, −∞	6.08, −∞
11	5.67, −∞	5.73, −∞	5.67, −∞	5.67, −∞
12	**10.26, 8.76**	**10.36, 15.60**	**10.36, 28.00**	8.52, −∞
13	5.83, −∞	5.99, −∞	5.67, −∞	8.40, −∞

and the 2 × 2 Rayleigh channel are 10.3 bits/s/Hz and 7.9 bits/s/Hz, respectively, for 20 dB SNR. The mean K-factor and capacity for all cases for 20 dB SNR are summarized in Tables 10.4 and 10.5. The K-factor for all NLOS scenarios, when there was an obstacle car between the TX and RX, was −∞, as expected. In these scenarios the transmitted signal is completely blocked by the obstacle car.

It should be noted that the smaller K-factor means that there is more multipath and they can contribute to the higher channel capacity. There are also some other parameters that affect the channel capacity including the power and phase of received signals. For example, for scenarios 1 and 4 for case 2 (with TX-7 and RX-9) the power of received signal is −64.34 dB and −140.30 dB, respectively. In this case, the phase of received signal is 143.08° and 156.07°, respectively. The corresponding channel capacity for these scenarios without the obstacle car is 9.00 bits/s/Hz and 8.01 bits/s/Hz, respectively, while the K-factors are 23.84 dB and −46.73 dB, respectively. Here it can be seen how the power and phase of received signal can affect the channel capacity. Tables 10.6 and 10.7 show the calculated total received power. For scenarios without the obstacle car, the maximum power was realized in all the cases by a MIMO system using directive antennas, when the antennas were facing each other. It is worth mentioning that in all of the cases MIMO systems using directive antennas offer the highest channel capacity. Scenarios 7–9 for case 2 (without obstacle) offer the highest capacities. In addition, scenario 12, cases 1–3 offer high capacity, but in this scenario the received power is relatively low, especially for cases 1, 2, and 3. This is

Table 10.5 Channel Capacity for Different Scenarios with Obstacle Car

Scenario	Channel capacity (bits/s/Hz), (with obstacle car), K-factor = −∞			
	Case 1	Case 2	Case 3	Case 4
1	8.87	5.85	7.52	5.71
2	5.83	8.30	6.56	8.72
3	9.73	8.25	9.34	6.98
4	6.10	6.74	6.26	5.67
5	5.96	6.67	9.72	8.20
6	9.98	6.78	5.69	9.92
7	6.10	6.74	8.55	5.67
8	6.43	7.36	9.34	5.67
9	8.92	**10.05**	7.04	5.67
10	6.41	5.96	5.82	6.00
11	5.67	5.73	5.67	5.67
12	5.67	6.06	5.67	5.67
13	6.03	5.68	5.68	9.28

expected, since the main beam of RX is rotated away from the transmitter. Some measurement results using dipole antennas for omnidirectional antennas, and horn antennas for some cases are reported in Refs. [13,14,21].

10.5 Conclusion

In this chapter the importance of considering the antenna pattern in the performance of V2V communication was discussed. The use of MIMO systems in V2V communication, especially when the antenna pattern is directive, needs to be studied. The chapter is an effort to show the effect of antenna beam direction, orientation, and position on the MIMO system performance through site-specific numerical simulations. Directive antenna performance was studied and compared with omnidirectional antennas. This study was done through thirteen different scenarios for four different cases of MIMO antenna positions on the transmitter and receiver cars, in the presence and absence of the obstacle car.

It was observed that MIMO system performance for vehicles can be affected by the environment and antenna locations on the cars. While MIMO can be used to increase channel capacity, it should be noted that in different scenarios omnidirectional antennas might not provide the best available capacity. Having multiple directive antennas can provide an option to search and select the best possible channels.

Table 10.6 Received Power for Different Cases (Without Obstacle Car)

Case	TX	RX	Received Power (dBm)-Dipole			Received Power (dBm)-Horn (AFB)			Received Power (dBm)-Horn (AFEO)			Received Power (dBm)-Horn with Different Angles			
			Scen.1	Scen.2	Scen.3	Scen.4	Scen.5	Scen.6	Scen.7	Scen.8	Scen.9	Scen.10	Scen.11	Scen.12	Scen.13
1	7	7	−65.5	−64.2	−64.6	−145.5	−82.9	−87.6	−39.9	−39.9	−39.9	−94.3	−94.8	−155.3	−74.6
	7	8	−68.3	−63.8	−74.7	−140.6	−87.7	−83.0	−39.9	−39.9	−39.9	−86.7	−125.9	−116.0	−62.7
	8	7	−70.6	−66.6	−65.6	−144.3	−83.5	−76.6	−39.9	−39.9	−39.9	−69.0	−93.7	−115.2	−68.8
	8	8	−69.2	−68.2	−67.3	−142.2	−86.7	−77.1	−39.9	−39.9	−39.9	−93.5	−115.9	−140.0	−58.3
2	7	7	−65.5	−64.2	−64.6	−145.5	−82.9	−87.6	−39.9	−39.9	−39.9	−94.3	−94.8	−155.3	−74.6
	7	9	−64.3	−65.9	−66.2	−140.3	−82.1	−91.7	−40.1	−40.1	−40.1	−88.8	−103.9	−105.0	−65.6
	9	7	−65.9	−65.3	−70.7	−144.6	−84.0	−92.1	−40.1	−40.1	−40.1	−65.6	−101.5	−104.8	−60.4
	9	9	−69.2	−64.0	−67.2	−146.3	−78.6	−83.0	−39.9	−39.9	−39.9	−86.0	−116.9	−140.2	−58.1
3	10	10	−68.2	−74.4	−68.8	−145.1	−79.8	−76.5	−39.9	−39.9	−39.9	−69.5	−95.9	−150.1	−77.0
	10	11	−66.7	−69.0	−69.0	−138.2	−89.6	−80.8	−39.9	−39.9	−39.9	−90.8	−119.5	−116.3	−73.4
	11	10	−66.6	−68.8	−65.3	−137.2	−83.3	−79.5	−39.9	−39.9	−39.9	−71.1	−94.3	−116.1	−65.8
	11	11	−66.3	−66.6	−64.7	−140.3	−80.5	−76.8	−39.9	−39.9	−39.9	−84.3	−116.7	−137.7	−54.8
4	14	1	−63.6	−64.0	−63.5	−115.7	−69.5	−66.1	−37.8	−37.8	−37.8	−83.7	−74.5	−74.8	−61.3
	14	2	−64.1	−65.7	−64.0	−110.7	−68.2	−65.4	**−37.6**	**−37.6**	**−37.6**	−78.4	−94.7	−59.7	−64.3
	15	1	−61.6	−61.9	−62.5	−124.9	−66.5	−63.9	**−38.2**	**−38.2**	**−38.2**	−83.9	−89.4	−52.9	−67.1
	15	2	−62.3	−62.6	−61.3	−106.5	−69.5	−69.8	**−37.8**	**−37.8**	**−37.8**	−73.8	−113.5	−57.4	−57.1

Note: AFB, antenna facing building; AFEO, antenna facing each other. Bold values appearing in the table show the maximum received power.

Table 10.7 Received Power for Different Cases (With Obstacle Car)

Case	TX	RX	Received Power (dBm)-Dipole			Received Power (dBm)-Horn (AFB)			Received Power (dBm)-Horn (AFEO)			Received Power (dBm)-Horn with Different Angles			
			Scen.1	Scen.2	Scen.3	Scen.4	Scen.5	Scen.6	Scen.7	Scen.8	Scen.9	Scen.10	Scen.11	Scen.12	Scen.13
1	7	7	−78.5	−74.2	−84.6	−107.4	−77.9	−77.0	−107.4	−116.8	−121.7	−81.7	−94.8	−162.2	−79.1
	7	8	−77.8	−72.8	−69.9	−102.9	−91.7	−92.2	−102.9	−111.6	−120.3	−78.6	−139.0	−114.1	**−59.0**
	8	7	−74.8	−89.9	−72.1	−114.3	−84.8	−82.4	−114.3	−118.8	−121.2	−69.3	−93.7	−139.4	−65.6
	8	8	−77.0	−82.8	−86.8	−107.6	−89.9	−77.9	−107.6	−115.7	−129.4	−86.2	−116.7	−140.3	**−57.5**
2	7	7	−78.5	−74.2	−84.6	−107.4	−77.9	−77.0	−107.4	−116.8	−121.7	−81.7	−94.8	−162.2	−79.1
	7	9	−73.3	−82.2	−82.1	−100.2	−76.1	−82.7	−100.2	−113.1	−124.9	−77.0	−104.2	−167.9	**−59.8**
	9	7	−83.2	−75.2	−77.3	−108.7	−82.3	−79.4	−108.7	−114.8	−129.7	−65.2	−101.4	−157.8	−75.5
	9	9	−76.1	−75.2	−80.0	−108.9	−77.1	−83.2	−108.9	−114.9	−123.3	−86.0	−116.7	−162.2	**−58.0**
3	10	10	−79.8	−71.8	−84.2	−110.3	−79.2	−75.0	−110.3	−115.2	−121.1	−70.0	−98.4	−157.1	−92.2
	10	11	−91.0	−74.0	−72.62	−107.9	−88.7	−71.8	−107.9	−121.2	−126.3	−81.9	−121.5	−158.3	−65.9
	11	10	−90.9	−76.7	−73.90	−105.7	−83.8	−92.1	−105.7	−121.3	−120.3	−71.0	−94.3	−140.7	−67.1
	11	11	−86.3	−89.1	−75.86	−104.1	−79.9	−85.0	−104.1	−112.6	−126.0	−85.2	−118.3	−138.9	**−56.9**
4	14	1	−89.4	−81.2	−81.2	−69.49	−61.6	−71.0	−69.4	−69.5	−69.7	−79.21	−74.5	−120.9	**−58.1**
	14	2	−95.5	−72.6	−82.6	−92.93	−65.0	−66.8	−92.9	−94.9	−94.8	−71.63	−89.8	−115.4	−60.9
	15	1	−100.8	−75.3	−84.9	−99.16	−64.5	−65.0	−99.1	−101.4	−101.1	−83.95	−87.7	−136.8	−62.9
	15	2	−110.2	−77.1	−92.8	−113.6	−68.7	−74.1	−113.6	−93.1	−93.68	−68.76	−108.8	−136.9	**−57.2**

Note: AFB, antenna facing building; AFEO, antenna facing each other. Bold values appearing in the table show the maximum received power.

References

1. T. Mangel, O. Klemp, and H. Hartenstein, 5.9 GHz inter-vehicle communication at intersections: A validated non-line-of-sight path-loss and fading model, *EURASIP J. Wireless Commun. Netw.*, vol. 2011, pp. 1–11, 2011.
2. K.A. Hafeez, Z. Lian, B. Ma, and J.W. Mark, Performance analysis and enhancement of the DSRC for VANET's safety applications, *IEEE Trans. Veh. Technol.*, vol. 62, no. 7, pp. 3069–3083, September 2013.
3. M. Ali, G. Yang, H.-S. Hwang, and T. Sittironnarit, Design and analysis of an R-shaped dual-band planar inverted-F antenna for vehicular applications, *IEEE Trans. Veh. Technol.*, vol. 53, no. 1, pp. 29–37, January 2004.
4. L. Lizzi, F. Ferrero, J.-M. Ribero, and R. Staraj, Low-profile multimode antenna for vehicular applications, *IEEE Antennas and Propagation Society International Symposium (APSURSI)*, Orlando, FL, pp. 2069–2070, July 2013.
5. M.G.N. Alsath, and M. Kanagasabai, Planar pentaband antenna for vehicular communication application, *IEEE Antenn. Wireless Propag. Lett.*, vol. 13, pp. 110–113, December 2014.
6. P. Mousavi, Multiband multipolarization integrated monopole slots antenna for vehicular telematics applications, *IEEE Trans. Antenn. Propag.*, vol. 59, no. 8, pp. 3123–3127, August 2011.
7. E. Jedari, S. Noghanian, B. Shahrrava, and Z. Atalsbaf, Effects of antenna selection on vehicle to vehicle communication in highways, *2013 IEEE International Symposium on Antennas and Propagation, APS & CNC/USNC/URSI Symposium*, Orlando, FL, pp. 2109–2110, July 2013.
8. T. Abbas, J. Karedal, and F. Tufvesson, Measurement-Based analysis: The effect of complementary antennas and diversity on vehicle-to-Vehicle communication, *IEEE Antenn. Wireless Propag. Lett.*, vol. 12, pp. 309–312, March 2013.
9. D. Kornek, M. Schack, E. Slottke, O. Klemp, I. Rolfes, and T. Kurner, Effects of Antenna characteristics and placements on a vehicle-to-Vehicle channel scenario, *IEEE International Conference on Communications Workshops*, Capetown, pp. 1–5, May 23–27, 2010.
10. A.P. Subramanian, V. Navda, P. Deshpande, and S.R. Das, A measurement study of inter-vehicular communication using steerable beam directional antenna, *ACM VANET Workshop*, pp. 7–16, September 2008.
11. S. Kaul, K. Ramachandran, P. Shankar, S. Oh, M. Gruteser, I. Seskar, and T. Nadeem, Effect of antenna placement and diversity on vehicular network communications, *4th Annual IEEE Communications Society Conference on Sensor, Mesh and Ad Hoc Communications and Networks, SECON '07*, San Diego, CA, pp. 112–121, August 2007.
12. C.F. Mecklenbräuker, A.F. Molisch, J. Karedal, F. Tufvesson, A. Paier, L. Bernado, T. Zemen, O. Klemp, and N. Czink, Vehicular channel characterization and its implications forwireless system design and performance, *Proc. IEEE*, vol. 99, no. 7, pp. 1189–1212, July 2011.
13. N. Adhikari, A. Kumar, and S. Noghanian, A cost-effective test-bed for vehicular channel measurements, *IEEE Antenn. Wireless Propag. Lett.*, vol. 15, pp. 674–677, 2016, DOI: 10.1109/LAWP.2015.2468221.
14. N. Adhikari and S. Noghanian, Capacity measurement of multiple antenna systems for car to car communication, *2014 IEEE International Symposium on Antennas and Propagation, APS & CNC/USNC/URSI Symposium*, Memphis, TN, pp. 603–604, July 2014.
15. N. Adhikari and S. Noghanian, Antenna location effects on the capacity of MIMO DSRC channels, *USNC-URSI National Radio Science Meeting*, Boulder, CL, January 2014.
16. N. Adhikari and S. Noghanian, Multiple antenna systems for vehicle to vehicle communication, *2013 International Conference on Electro/Information Technology*, Rapid City, SD, Berlin, Germany, pp. 1–6, May 2013.
17. L. Reichardt, T. Fügen, and T. Zwick, B, Influence of antennas placement on car to car communications channel, *Proceedings of the European Conference on Antennas and Propagations*, pp. 630–634, 2009.
18. Remcom Inc. http://www.remcom.com/wireless-insite, visited February 2013.

19. M. Mirzaee, N. Adhikari, and S. Noghanian, Analysis of static and dynamic scenarios of MIMO systems for physical layer modeling for vehicular communication, *National Wireless Research Collaboration Symposium (NWRCS)*, Idaho Falls, ID, pp. 1–10, May 2014.
20. S. Sanayei and A. Nosratinia, Antenna selection in MIMO systems, *IEEE Comm. Mag.*, vol. 42, pp. 68–73, 2004.
21. A. Kumar and S. Noghanian, Wireless channel test-bed for DSRC applications using USRP software defined radio, *2013 IEEE International Symposium on Antennas and Propagation, APS & CNC/USNC/ URSI Symposium*, Orlando, FL, pp. 2015–2016, July 2013.

Chapter 11

The Fundamental Principles of Antenna Theory for V2I Deployments

Jonathan B. Walker*

Federal Highway Administration (FHWA)
Licensed Professor Engineer (PE)
Washington, DC

Contents

---

* The views expressed by the author do not necessarily represent the views of the Federal Highway Administration or the United States.

11.1 Introduction [1, 2]

In this chapter, we will introduce the propagation characteristics—the process of spreading to a larger area—of electromagnetic waves that are transmitted by an antenna used in a vehicle-to-infrastructure (V2I) deployment by a roadside unit (RSU). The intent of this chapter is to provide a simplistic explanation of antenna theory with a concentration on two-dimensional and three-dimensional pictorial representations of a V2I deployment. The fundamental principles of electromagnetic waves normally start with Maxwell's equations along with the study of electric and magnetic fields as a variable of time. However, the objective is placed on those with little or no knowledge of antenna theory but who are responsible for planning, managing, or owning a V2I deployment. The fundamental principles of electromagnetic waves and antenna theory are topics that can be found in most electrical engineering or radio communication textbooks. As a result, this chapter will summarize the basic concepts related to electromagnetic waves and focus on the application of antenna theory in a V2I deployment as the signal propagates from an RSU.

The antenna has become an effective means of distributing electrical charges and current in a specific direction for wireless communications. The electromagnetic waves can be transmitted over a short distance (*e.g.*, indoors or outdoors) or a long distance (*e.g.*, a satellite that is orbiting the Earth). An antenna's components will force the electric charges to oscillate when a changing power is applied to the wire(s) of an antenna. There are general concepts that a nonengineer can comprehend based on a few mathematical formulas and images. Similarly, the general concepts may prevent costly design changes, inaccurate procurement orders, and improper antenna placement.

With regard to V2I wireless communication, an antenna is designed to distribute the electromagnetic waves from the radio hardware (source of energy) to an end device (target) that may be running a safety application. There is a need to maintain a reliable wireless connection between the source and target so that the safety application can perform properly. For example, the red light violation warning (RLVW) is a V2I safety application that is defined in the US Department of Transportation (USDOT) connected vehicle reference implementation architecture (CVRIA). RLVW enables a connected vehicle to approach an instrumented signalized intersection to receive information from the infrastructure (*e.g.*, RSU) regarding the signal timing and the geometry of the intersection (*e.g.*, signal phase and timing or SPaT). The CVRIA states there is a mechanism to obtain the vehicle location and motion of the surrounding vehicles from the RSU. A poor wireless connection could obstruct the RLVW application from alerting the driver in time to avoid the traffic conflict.

In Figure 11.1, an RLVW scenario is depicted with electromagnetic waves emitting from the RSU, which are represented by the vertical and horizontal sinusoidal waves. The RSU is electrically wired to the signalized lights and the RSU has radio hardware that transits SPaT information within the electromagnetic waves. The two SUVs have an antenna that receives the electromagnetic waves from the RSUs. The two concentric patterns represent the reception coverage of electromagnetic waves by the radio hardware in the SUVs. Then, the V2I safety application utilizes the SPaT information to alert the driver of a traffic conflict. Likewise, the concentric patterns could represent the transmission coverage of electromagnetic waves to the RSU. The latter representation will be discussed in a section subtopic titled *concentric pattern and free-space path loss pattern*. In reality, the reception coverage is much greater in size than depicted by the concentric patterns in the figure. Nevertheless, the reception coverage was

Figure 11.1 **A red light violation warning scenario with two approaching vehicles that transmit and receive electromagnetic waves from/to the roadside units (RSU).**

Table 11.1 **Four Scenarios that Appear to be Reasonable Decisions when Installing, Procuring, and Specifying an Antenna in a V2I Deployment**

Scenario	Description
1	A project manager may find it reasonable to hire a well-known electrical company to design or install a V2I communication system.
2	A procurement officer may accept the lowest bid for a set of replacement antennas.
3	A traffic engineer may justify mounting the V2I antenna inside a roadside unit to prevent the equipment from being vandalized and rotating the antenna by 90 degrees to fit in the equipment cabinet.
4	A design engineer could write a technical specification requiring the white antenna housing and the associated mounting hardware coated with a non-glossy green paint to appease the planning commission.

reduced in size within the figure, to generate a clear image that was not overwhelmed with a concentric pattern.

In a V2I deployment using RSUs, there are several scenarios that require a basic understanding of antenna theory to ensure the system meets expectations. In Table 11.1, there are four scenarios that appear to be reasonable decisions when installing, procuring, and specifying an antenna in a V2I deployment. Nevertheless, each of these scenarios may have a significant impact on the V2I deployment in a negative manner that will be addressed after the general concepts are explained.

The major objective of each subtopic is to assist individuals such as design engineers, owner/operators, or closely related professions with realistic deployment expectation and the limitations of certain electromagnetic principles that affect the antenna. This will be achieved by explaining (a) the electromagnetic waves as the energy departs the antenna, (b) the antenna patterns,

(c) significance characteristics in antenna patterns, (d) common antenna patterns, (e) free-space antenna path loss, and (f) the antenna location on a vehicle and a roadside unit.

11.2 Antenna Patterns [3–7]

The antenna pattern is a graphical representation of electromagnetic waves (or energy) departing the antenna, which provides a means of estimating how the energy is distributed. Typically, the manufacturer will draw the antenna patterns in two dimensions while the academic communities have developed innovative ways to plot the antenna pattern in three dimensions for various analyses. Even though the antenna radiates in three-dimensional space, the manufacturer's two-dimensional drawings are represented in a *principal plane pattern* by rendering two slices through the three-dimensional pattern. The two patterns are frequently plotted in polar coordinates and called the *azimuth plane pattern* and the *elevation pattern*. The azimuth pattern represents the horizontal plane and the elevation pattern represents the vertical plane.

An elevation pattern and an azimuth pattern are represented in Figure 11.2a and b, respectively. In Figure 11.2a, the elevation pattern has electromagnetic radiation that propagates in the horizontal direction of 0 and 180 degrees, which is perpendicular to the antenna in the direction of 90 degrees. In Figure 11.2b, the azimuth plane pattern has the z-axis in the center of the graph and extends from the paper. Both figures have a hypothetical shape that does not exist in real life but the shape is used to characterize a pattern using a mathematical formula.

As an example, the graphical representation of the elevation plane pattern was plotted using a Lemniscate of Bernoulli (two-leaved rose, *i.e.*, the red curve):

$$r^2 = a^2 \cos 2\theta \text{ or } r = a + 2\cos 2\theta$$

whereby a is the size of the two-leaved rose along 90 and 270 degrees. When a is equal to 0, the two-leaved rose at 90 and 270 degrees will be of equal diameter to the leaved rose at 0 and 180 degrees. Then, the diameter will decrease to 0 as the value approaches the cosine multiplier (*i.e.*, the number 2). In Figure 11.2a, the elevation plane pattern is a function of $1 + 2 \cos (2\theta)$.

As we will discuss later, the signal strength of an electromagnetic wave is not uniform over distance. Nearly all electromagnetic energy has a limitation that is inversely proportional to the distance squared. This limitation must be accounted for in technical requirements when designing and deploying V2I systems. So, the antenna's principal plane pattern is essential for the traffic engineer or design engineer because the graph provides a means of estimating the area of effective (or ineffective) coverage for a V2I application.

11.2.1 Significance Characteristics in Antenna Patterns

There are four significance characteristics in antenna patterns that are described below and based on Figure 11.2a and b.

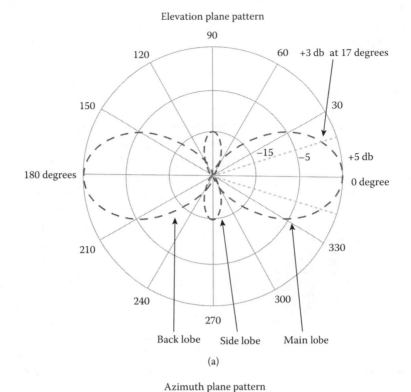

(a)

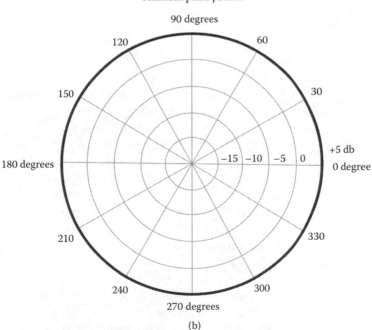

(b)

Figure 11.2 (a) The elevation plane pattern of a directional antenna that is plotted in polar coordinates as a function of 1 + 2 cos (2θ). (b) The azimuth plane pattern of a directional antenna that is plotted in polar coordinates as a function of one from 0 to 360 degrees.

11.2.1.1 Lobes

In Figure 11.2b, the azimuth plane is a hypothetical antenna pattern with a signal strength of +5 decibels relative to isotropic (or +5 dbi). The term isotropic has origins from "iso" meaning the same and "tropic" meaning direction. The isotropic antenna has a pattern that is equal in radiation all directions. In Figure 11.2a, the antenna pattern is characterized in terms of lobes: the main lobe, side lobe, and back lobe. The main lobe radiates in the direction of 0 degree if we continue to assume the antenna is orientated in the direction of 90 degrees, while the back lobe is radiating in the direction of 180 degrees. The maximum radiation is in the direction of 0 degree. The signal strength becomes weaker when moving in the direction of 30 and 330 degrees. In other words, there is a tremendous signal loss when a measurement is made outside of the two-leaved rose curve. In this hypothetical case, the back lobe is identical to the main lobe in signal strength and in the opposite direction. The side lobes radiate in the direction of 90 and 180 degrees with a signal strength at –15 dbi. The side lobes are a region of undesirable radiation and are several times lower in magnitude than the maximum radiation power of the main beam. In reality, the antenna pattern may be comprised of several side lobes but the manufacturer's technical specification may only list the first and nearest (or highest) side lobe to the main beam.

11.2.1.2 3 dB Beamwidth

The 3 dB beamwidth is defined as the angle between the points of the main lobe that are down from the maximum gain by 3 dB (or half-power). Normally, the decrease is experienced as the target moves away from the source; since, the signal strength is inversely proportional to the distance squared. In Figure 11.2a, the 3 dB beamwidth is located from 17 to 343 degrees for a total of 34 degrees (*i.e.*, the distance between the green lines).

11.2.1.3 Polarization

In the simplest form, an electromagnetic wave is comprised of two sinusoidal signals in the horizontal and vertical plane traveling through space, which are the electric and magnetic fields. A traveling wave is considered vertically (or linearly) polarized if the electrical field is moving vertically through space. An example of a vertical antenna is an antenna on an automobile, which is perpendicular to the Earth's surface. A polarized antenna is sensitive to the mounting position and provides the best transmission/reception when the electrical fields are on the same plane.

In Figure 11.3, the SUV antenna is linearly polarized; moreover, the antenna is mounted in the north direction with the electrical field traveling perpendicular (*i.e.*, the left signal moving up and down) to the Earth's surface. The RSU antenna is linearly polarized with a mounting position in the east direction but the electrical field is traveling parallel (*i.e.*, the right signal moving side to side) to the Earth's surface. Therefore, the electrical fields are not propagating on the same plane. In this scenario, a linearly polarized RSU antenna will receive the most energy from the SUV antenna when both antennas are physically mounted in the same direction and the electrical fields are traveling on the same plane (*i.e.*, perpendicular to the Earth's surface).

The polarization is directly related to the propagation direction (*e.g.*, horizontal or vertical) of the electrical fields that depart the antenna. An antenna can be designed with a sensitivity to the mounting direction, which implies the transmitting and receiving antennas must retain the same polarization for the best transmission/reception path. If the mounting direction is not observed during installation, the target will experience at least a 3 dB loss because the electrical fields are not propagating on the same plane. Ordinarily, the antenna manufacturer will provide a directional marking on the exterior body of the antenna such as an arrow to ensure the antennas are installed properly.

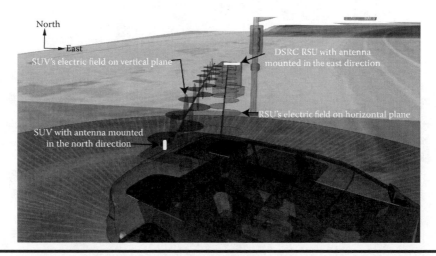

Figure 11.3 The SUV and RSU antennas are linearly polarized; however, the electromagnetic waves are not propagating on the same plane.

11.2.2 Hypothetical Antenna Patterns [2]

When studying antenna theory, there are hypothetical antenna patterns that capture the core of propagation waves. Moreover, the V2I application will determine the characteristic of the antenna or the orientation of a particular pattern. Below, there are four hypothetical antenna patterns with electromagnetic energy departing from an RSU, which are depicted in three dimensions. The antenna patterns are as following:

1. RSU 1 is an isotropic antenna based on a hypothetical lossless antenna that distributes the energy equally in all directions.
2. RSU 2 is a monopole antenna, manufactured as a single wire element, with a shape similar to half-a-donut, that is parallel to the Earth's surface.
3. RSU 3 is a dipole antenna, manufactured as two wire elements, that has a nondirectional (or circular pattern) in a given plane and a directional pattern in the perpendicular plane.
4. RSU 4 is a directional antenna, which is more effective in a specific direction and has a main lobe as well as minor lobe(s).

Figure 11.4a is a top view (or the three-dimensional azimuth plane pattern) as the electromagnetic energy departs from the RSU's antenna. The three-dimensional azimuth plane is a circular antenna pattern for RSU 1, RSU 2, and RSU 3, while RSU 4 has a concentrated antenna pattern in one direction.

Figure 11.4b and c are front viewpoints (or the three-dimensional elevation plane pattern) as the electromagnetic waves travel from the RSU's antenna along the Earth's horizon plane. Figure 11.4b is a close-up of RSU 1 and RSU 2 with the electromagnetic waves propagating outward along the horizon while Figure 11.4c is a close-up of RSU 3 and RSU 4. RSU 1 has a spherical radiation pattern and serves as a three-dimensional isotropic antenna. RSU 2 has radiation above the height of the physical antenna with no radiation below the RSU, ideally. RSU 3 has radiation in all directions along the Earth's horizon plane and has a shape similar to a donut. RSU 4 has a main lobe in front of the RSU with some radiation from the back and side lobes.

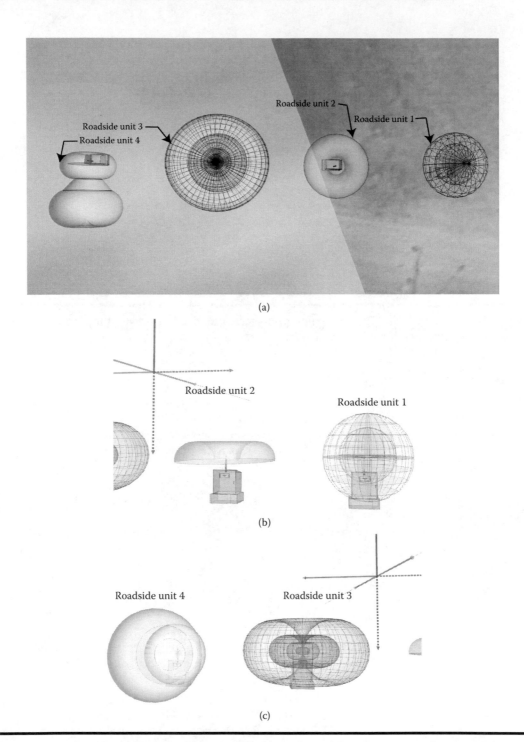

Figure 11.4 **(a) A three-dimensional azimuth plane pattern with electromagnetic waves depart-ing from RSU 1 (isotropic antenna), RSU 2 (monopole antenna), RSU 3 (dipole antenna), and RSU 4 (directional antenna). (b) A three-dimensional elevation plane pattern with electromagnetic waves along the Earth's horizon plane for RSU 1 and RSU 2. (c) A three-dimensional elevation plane pattern with electromagnetic waves along the Earth's horizon plane for RSU 3 and RSU 4.**

11.2.3 Concentric Pattern and Free-Space Path Loss Patterns [1,2,9,10]

At times, the azimuth plane pattern is represented by concentric circles with several annuli—ring-shaped regions that are bounded by two concentric circles. In Figure 11.5, the concentric circles represent two elements of radiation: (1) propagation of the electromagnetic waves along the horizontal plane from the source and (2) the free-space path loss which is the decrease in signal strength of an electromagnetic wave in the open air. The three-dimensional dipole sphere has several minispheres within the larger sphere. Furthermore, the larger sphere represents the outer boundary of an azimuth plane pattern. For example, the larger sphere could represent +5 dB, similar to the azimuth plane pattern in Figure 11.2b. Accordingly, the minispheres represent inner circles within the polar coordinate graph such as 0, –5, and –10 dB. As stated earlier, the signal strength (or reception coverage) was reduced in size, within the figure, to produce a clear image that was not inundated with a concentric pattern.

Also, the sphere and concentric circle depict the free-space path loss (FSPL) which is "not due to dissipation, but rather to the fact that the power flux density decreases with the square of the separation distance" as quoted from the IEEE standard 145–1993. The equation for FSPL is derived below in terms of dB:

$$FSPL = \left(\frac{4\pi df}{c} \right)$$

$$FSPL(dB) = 20\,Log_{10}\left(\frac{4\pi df}{c} \right)$$

$$FSPL(dB) = 20\,Log_{10}(d) + 20\,Log_{10}(f) + 20\,Log_{10}\left(\frac{4\pi}{c} \right)$$

$$FSPL(dB) = 20\,Log_{10}(d) + 20\,Log_{10}(f) + 92.45$$

where c is the speed of light (3×10^8 m/s), d is the distance from the transmitter (in km), and f is the signal frequency (in GHz).

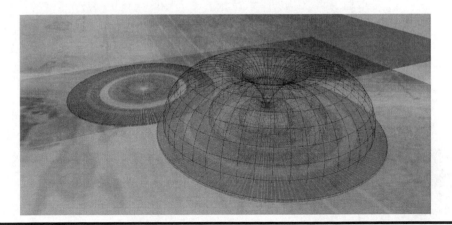

Figure 11.5 An azimuth plane pattern is represented as a three-dimensional dipole sphere and as a two-dimensional concentric circle.

Normally, a manufacturer will produce an antenna with gain in reference to an isotropic antenna. The antenna gain for the radio hardware transmitter (source) and receiver (target) must be accounted for in the free-space path loss such that

$$FSPL(dB) = 20\,Log_{10}(d) + 20\,Log_{10}(f) + 92.45 - GTx - GRx$$

whereby d is the distance from the transmitter (in km), f is the signal frequency (in GHz), GTx is the transmitter gain (in dBi), and GRx is the receiver gain (in dBi). The antenna gain for the transmitter and receiver can be different in value but the designer of the V2I system will make the final decision, which is often based on the V2I application or surrounding environment.

In Figure 11.6, there are four graphs depicted using the free-space path loss formula to show the effect of frequency and antenna gain. The top graph (large dots) has a frequency of 5.9 GHz with no antenna gain for the transmitter or receiver. Just below, the next graph (dashed line) has a frequency of 5.9 GHz with an antenna gain of +5 dB for the radio transmitter and 0 for the receiver. The third graph from the top (solid line) has a frequency of 2.4 GHz with no antenna gain for the transmitter or receiver. Just below, the next graph (dotted line) has a frequency of 2.4 GHz with an antenna gain of +5 dB for the transmitter and 0 for the receiver.

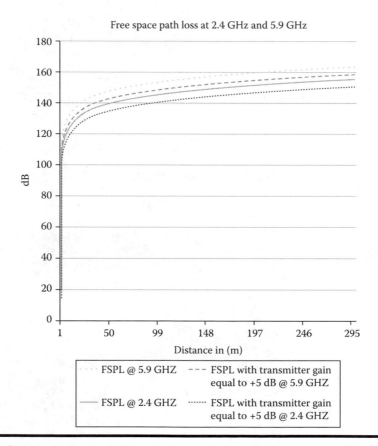

Figure 11.6 The free-space path loss (FSPL) to show the effect of frequency (2.4 GHz vs. 5.9 GHz) and antenna gain (0 dB vs. +5 dB).

This scenario shows how the path loss will increase as the frequency increases; consequently, the frequency will have an impact on the reception and transmission coverage of a V2I deployment. A designer can estimate the signal strengths that may be expected in a given V2I deployment. In addition, the free-space path loss has a major impact on the number of RSUs and the spacing of RSUs, so a V2I application requires contiguous wireless coverage.

11.2.4 Antenna Location on a Vehicle and RSU

The antenna placement is a critical aspect to the reception and transmission coverage of electromagnetic energy since the signal strength of an electromagnetic wave is not equal (or infinite) over distance. In a V2I deployment, the location of the vehicle and roadside antennas are critical for the initial and continuous wireless communication over time.

With regard to vehicle installation, the car manufacturers have developed techniques to mount an antenna on the rooftop with minimum signal strength loss. When an antenna is mounted inside the vehicle, the signal strength is attenuated (or reduced) because the electromagnetic energy must propagate through a layer of glass, metal, or fiberglass. Usually, the mounting location on the rooftop is positioned in the rear-center but the front or center rooftop positions are sufficient if there are no vehicle accessories (*e.g.*, ski racks, luggage racks, or a taxi advertisement sign).

When considering a roadside installation, the antenna is typically mounted on (or near) (a) the RSU, (b) a traffic light signal pole, or (c) a highway gantry. In Figure 11.7a, the antenna is mounted on the RSU and the concentric pattern represents the transmission coverage of electromagnetic energy (*i.e.*, the source) with several annuli—ring-shaped regions that are bounded by two concentric circles. The RSU is located in the center of the concentric pattern and each annulus represents the free-space path loss as the signal strength decreases in the open air. Each SUV is approximately 110 m from the street's intersection or a total of 200 m apart. On a similar note, one SUV is 88 m from the RSU and the second is 132 m from the RSU because the radio hardware is located several meters from the street intersection. The closest SUV will receive a stronger signal strength and potentially a better wireless connection if there is a clear line of sight. In other words, the electromagnetic waves will travel a direct path from the source (RSU) to the target (SUV) with no (or very little) obstruction.

In Figure 11.7b, the antenna for the RSU is located on the traffic light signal pole. When the antenna is located on a signal pole or highway gantry, the electromagnetic energy is radiated more homogeneously along the street intersection and the transmission coverage is aligned with both driving directions. In turn, the signal strength will be similar as well as the wireless communication data rate. Conversely, the polarization is a vital component when mounting the antenna on a signal pole or highway gantry (source) because the antenna polarization of source must match the vehicle antenna (target). Because the vehicle is moving, there will be a misalignment between the source and target at various points on the road.

In Figure 11.7b, the electric field of the RSU's antenna is angled slightly downward to align with the antenna polarization of SUV 1. The purpose of mounting the RSU's antenna at a slight angle is to maximum the amount of electric field energy received by the SUV's antenna from the RSU. SUV 1 has a concentric pattern that represents the reception coverage of electromagnetic energy (*i.e.*, the target), which depicts the signal strength of wireless communication coverage. SUV 2 has a concentric pattern as well to represent the wireless reception coverage. Once again, the signal strength (or reception coverage) was reduced in size within the figure, to produce a clear image that was not inundated with a concentric pattern. SUV 1 is within wireless communication range of the RSU's antenna while SUV 2 is out of wireless communication range with the RSU's electromagnetic field. The relative polarization will change as each vehicle approaches and departs

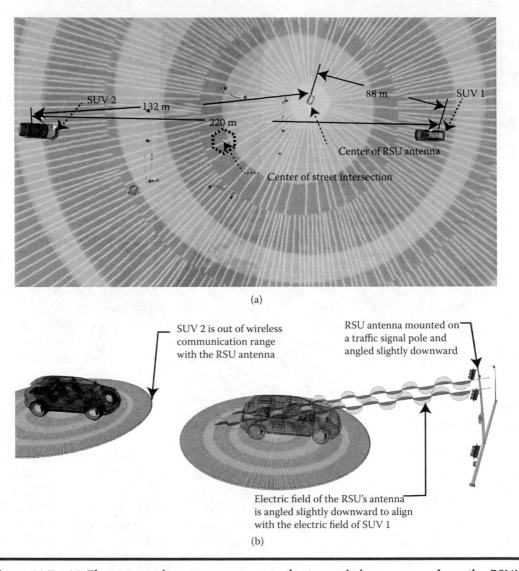

(a)

(b)

Figure 11.7 (a) The concentric pattern represents the transmission coverage from the RSU's antenna, which is closer to SUV 1 even though SUV 1 and SUV 2 are equal distance from the center of the street intersection. (b) To maximize electromagnetic energy, the RSU antenna is mounted on the traffic light signal pole and angled slightly downward to match the SUV's antenna polarization.

from the signal pole, which will impact the total signal strength. The source antenna must be mounted in a tilted fashion to initiate and maintain wireless communication before the vehicle is underneath the light signal pole. In a V2I application such as RLVW, there is a need to initiate and maintain wireless communication before the SUV is underneath the signal light to provide ample warning to the driver. The speed of the SUV will determine if there is ample time to initiate or maintain wireless communication. For example, a vehicle traveling at twice the posted speed limit will have less time to initiate and maintain wireless communication compared to a vehicle moving at one-half the posted speed limit.

11.3 Conclusion

There are numerous subtopic topics related to antenna theory that are not discussed in this chapter. The intent of this chapter was to provide a simplistic explanation of fundamental principles on antenna theory and utilize two-dimensional and three-dimensional image representations of a V2I deployment using a roadside unit. The antenna of a roadside unit has become an effective means of distributing electrical energy and maintaining a reliable wireless connection if the equipment is mounted properly. Thus, there are subtle mistakes, which may lead to an ineffective V2I deployment. In Table 11.2, the four scenarios are used to close out the chapter by applying our knowledge of antenna theory in real-world mistakes that could significantly impact the V2I deployment.

Table 11.2 Four Scenarios that Appear to be Reasonable Decisions but there is a Significant Impact to Each Scenario, which may Impact the Wireless Communication of the V2I Deployment

Scenario	Description	Significant Impact on the V2I Deployment
1	A project manager may find it reasonable to hire a well-known electrical company to design or install a V2I communication system.	The average electrical company may not have the experience or knowledge of antenna theory. Hence, the electrical company could produce costly delays such as (a) purchasing the wrong antenna for the application, (b) installing an insufficient number of antennas to maintain a reliable wireless communication, or (c) purchasing a 2.4 GHz antenna when the application requires a 5.9 GHz antenna.
2	A procurement officer may accept the lowest bid for a set of replacement antennas.	The replacement antenna may be a directional antenna whereby the application requires a dipole antenna. Consequently, the procurement officer's decision may impact the reception and transmission coverage of a V2I deployment.
3	A traffic engineer may justify mounting the V2I antenna inside a roadside unit to prevent the equipment from being vandalized and rotating the antenna by 90 degrees to fit in the equipment cabinet.	A vertically polarized antenna is sensitive to the mounting position and rotating the antenna 90 degrees will impact the maximum amount of electric field energy received by the SUV's antenna from the RSU. When the antenna is mounted in the equipment cabinet, the electromagnetic energy will experience a signal strength loss because the electromagnetic waves must penetrate the metal cabinet before traveling in open air.
4	A design engineer could write a technical specification requiring the white antenna housing and the associated mounting hardware coated with a non-glossy green paint to appease the planning commission.	The antenna's manufacturer discourages painting the antenna housing because the signal strength will be attenuated (or reduced) because the electromagnetic waves must propagate through a layer of paint. Also, there are some paints that contain lead and the lead may affect the antenna pattern.

References

1. Cheng D. K., *Field and Wave Electromagnetics No Title*, Second Edition. Reading, MA: The Addison-Wesley, 1989.
2. Hayt, Jr. W. H., *Engineering Electromagnetics,* Fifth Edition. New York: McGraw-Hill, 1989.
3. Samuel P. D. S. D. and Selby M., Ed., *Standard Mathematical Tables,* Seventeenth Edition. Cleveland: The Chemical Rubber, 1969.
4. Mirsa I. S., *Wireless Communications and Networks 3G and Beyond,* New Delhi: Tata McGraw-Hill Education Private, 2009.
5. Cisco Systems Inc., *Antenna Patterns and Their Meaning (White Paper),* San Jose, CA: Cisco Systems, Inc., 2007.
6. Link Technologies, Inc., *The Link Budget and Fade Margin (Application Notes)*, Logan, UT: Link Technologies, Inc., 2016.
7. Shivaldova V., Winkelbauer A., and Mecklenbrauker C. F., Vehicular link performance: From real-world experiments to reliability models and performance analysis, *IEEE Veh. Technol. Mag.*, vol. 8, no. 4, pp. 35–44, 2013.
8. Shulman M. and Deering R. K.R.K., *Third Annual Report of the Crash Avoidance Metrics Partnership,* Washington, DC: U.S. Department of Transportation, 2005.
9. Meneses R., Montes L., and Linares R., The RFID Radio channel performance in the vehicular control, *J. Vectorial Relativ.*, vol. 4, pp. 78–84, 2009.
10. Cidronali G. M., Alessandro C., Maddio S., and Passafiume M., Car talk, *IEEE Microw. Mag.*, vol. 17, no. 11, pp. 40–60.

Chapter 12

Radio Channel Measurements and Modeling for V2V Communications

Ruisi He

Beijing Jiaotong University
Beijing, China

Contents

12.1 Introduction

Recently, vehicle-to-vehicle (V2V) communications have gained much interest among governments, industries, and academia because of their wide application in the field of intelligent transportation systems (ITS). In the future, ITS will be such that each vehicle is able to acquire information through onboard sensors and exchange it with neighboring vehicles through wireless vehicular networks. This will improve vehicular safety, as each vehicle has detailed knowledge of the surrounding traffic situation. However, such networks require reliable communication links that meet strict packet delay deadlines. Several international standards have been proposed to support the applications of V2V communications, such as the wireless access in vehicular environments (WAVE) [1] and its European counterpart, ITS G5 [2].

For the design, implementation, and analysis of V2V wireless communication system, it is essential to understand the propagation channel [3,4]. In the past, much research effort has been devoted to V2V channel measurements and modeling [5,6]. Despite some progress, many open topics of V2V channels still remain, and the current amount of available V2V channel measurements and modeling do not allow the formulation of statistically significant statements about real-world V2V channels. Therefore, the goal of this chapter is twofold: (1) survey the recent progress of V2V channel characterizations and (2) provide some thoughts on the future topics of V2V channel measurements and modeling.

12.2 Vehicular Environment

Propagation channel is largely affected by the environment in which the system operates. Before proposing a channel model, a series of detailed propagation scenarios should be defined. The typical V2V communication link is influenced by the surrounding cars and the traffic. However, classification of different traffic environments is somewhat arbitrary, and there is generally no strict scientific principle to gauge whether one environment is "significantly" different from another. It is noteworthy that with the application of millimeter wave (mmWave) in V2V, the environment-dependent characteristics would be more pronounced, as mmWave bands have increased reflectivity and scattering, resulting in poor diffraction and penetration capabilities. The following are some of the typical categories of V2V channels [6–12]:

- *Highway*: It usually has two to six lanes in each direction with few houses around. Viaduct is widely used in highways. Speeds vary from 20 to 50 m/s according to different countries. Traffic density depends on time and is usually high in urban areas and low in rural areas.
- *Rural*: It usually has one to two lanes with few buildings on both sides. Traffic density is usually light and speed can be 20–30 m/s.
- *Suburban*: It usually has one to three lanes with some buildings on both sides (some of them could be very close to the road). Traffic density is usually light and speed can be 10–20 m/s.
- *Urban*: It usually has two to four lanes with many buildings around. Traffic density is usually high and speed can be 10–20 m/s. Note that mostly urban streets with light traffic and with heavy traffic are considered separately, as a large number of cars in heavy traffic usually lead to non-line-of-sight (NLOS) propagation, which significantly affects propagation characteristics.
- *Vehicle obstruction*: It happens when there are large vehicles (such as trucks and buses) on the road. These large vehicles usually lead to shadowing and extra delay spread. Note that the relatively low heights of vehicular antennas (mostly below 2 m) imply that the line-of-sight (LOS) can be easily blocked by such obstructions.

- *Road crossing*: In this scenario, the LOS path may be obstructed for long durations due to buildings, and other propagation paths, such as reflections from nearby buildings, play an important role.
- *Merge lanes*: It is similar to the road crossing and the LOS path is usually obstructed. It usually occurs when the Rx car is driving on a highway, whereas the Tx car is entering it from an entrance ramp. It is an important safety-related ITS scenario.
- *Traffic congestion*: Even though traffic density has been considered in some of the above scenarios, traffic congestion is sometimes considered individually, as the communication links may be largely obstructed by the surrounding cars.
- *Tunnel*: It usually has one to two lanes. Traffic density is usually light and speed can be 10–20 m/s. It leads to rich scatterings and a more dense impulse response.
- *Parking garage*: It is usually a multi-floored structure, probably fully enclosed, underground or aboveground, or may have sides that are open to the outside. The propagation characteristics are hence distinct from most indoor environments.

12.3 Radio Channel Measurements and Characterization

12.3.1 Path Loss

Path loss is the single most important quantity of any wireless system design [4]. According to V2V measurements that exist in the literature, it has been found that the path loss depends on the type of environment [13–15]. The two-ray model has been found to fit to V2V measurements in some open areas such as the rural scenario [16] and even in urban streets [8] when LOS path exists. However, modeling path loss in nonrural scenarios with the two-ray model is found to produce nonmeaningful results. Therefore, the power law path loss model is widely used [16], given by

$$PL[d] = PL_0 + 10n\log_{10}(d/d_0) + X_{\sigma_2} + \zeta PL_c, \quad d > d_0$$

where n is the path loss exponent, PL_0 is the path loss at a reference distance d_0, and X_{σ_2} is a zero-mean normally distributed random variable with standard deviation σ_2. PL_c is a correction term that accounts for the offset between forward and reverse path loss, and ζ is defined as

$$\zeta = \begin{cases} 1 & : \text{ for reverse path loss} \\ -1 & : \text{ for forward path loss} \\ 0 & : \text{ for convoy path loss.} \end{cases}$$

The typical parameters for all models can be found in Ref. [16].

As mentioned before, the path loss model of V2V channel is affected by traffic density. For example, the light traffic highway scenario has a path loss exponent around 1.8–1.9 [17–19], and the crowded highways have more severe path loss and experience larger variations [6]. Moreover, the breakpoint path loss model is also widely used for V2V channels [20]. For mmWave band, it is found that the effects of road surface such as surface roughness and curvature significantly affect path loss [21]. In Table 12.1, we summarize some typical path loss exponents based on V2V measurements.

Table 12.1 Path Loss Exponent for V2V Channels

Scenario	Highway	Rural	Suburban	Urban
Path loss exponent	n = 1.77 [16] n = 1.8 [17] n = 1.85 [18] n = 1.6–1.9 [22]	Two-ray [16] n = 1.79 [18] n = 1.65–1.89 [22]	n = 1.59 [16] n = 2.32–2.75 [20] n = 1.86 [23]	n = 1.68 [16] n = 1.61 [18]

Table 12.2 Standard Deviation of Shadow Fading for V2V Channels

Scenario		Highway	Rural	Suburban	Urban
Standard deviation of shadow fading	LOS	3.2 dB [18] 4–7.9 dB [22] 4.49 dB [26] 3.95 dB [27]	3.3 dB [18] 3.9–8.4 dB [22]	5.5–7.1 dB [20] 3.99 dB [26]	2.2–2.62 dB [8] 3.4 dB [18] 3.95 dB [26] 4.15 dB [27]
	NLOS	6.5–17 dB [22] 6.12 dB [27]	5.1–13.9 dB [22]	–	2.9–3.5 dB [8] 6.67 dB [27]

12.3.2 Shadow Fading

As in the cellular communications, the large-scale fading of V2V channels is modeled by using the log-normal distribution [4,15,24,25], which means that the dB-valued shadow fading component has a Gaussian distribution with a standard deviation of σ. The standard deviation of shadowing strongly influences V2V communication performance, as it determines the instantaneous signal-to-noise-and-interference ratio. Many measurements have been conducted to address this issue, and some typical results are summarized in Table 12.2.

12.3.3 Small-Scale Fading

The narrowband small-scale fading of V2V channels has been mostly modeled with the Rice, Nakagami, and Rayleigh distributions [15,28–30]. Reference [20] indicates that the Nakagami m-factor can be quite high in V2V channel (up to 4) if the distance between TX and RX is less than 5 m. However, when the distance is over 70 m, the m-factor is less than 1. In Ref. [8], the Akaike information criterion is used to show that the Nakagami distribution has the best fit, and in Ref. [26], the K–S test is used to show that the Rice distribution has the best fit. Reference [31] finds that both the Nakagami and Rice distributions offer good agreement of measured data.

The analysis of small-scale fading in wideband measurements usually considers a discretized channel impulse response (CIR), given by [6]

$$h(t,\tau) = \sum_{i=0}^{N} c_i(t)\delta(\tau - i/B),$$

where B is the system bandwidth and $c_i(t)$ are the (complex) amplitudes of the resolvable delay bins. It is thus meaningful to also analyze the fading statistics of each bin. In Ref. [32], the Weibull distribution is suggested to model the fading of each tap. In Ref. [33], a tapped delay line (TDL) model is proposed for V2V channels, and each tap is modeled with the Rice distribution. In general [6], for V2V LOS scenarios, the first delay bin shows "better than Rayleigh" fading, whereas bins with longer delays have distributions ranging from Rice to Rayleigh fading. In Table 12.3, we summarize some typical small-scale distributions observed from V2V measurements.

12.3.4 Doppler Spread

The Doppler spread of channel measures how fast the channel changes, and it can be used to estimate the coherence time of channel, given by [5]

$$T_{coh} \approx \frac{1}{2\pi f_D}$$

where f_D is the Doppler spread. Due to the high mobility of TX/RX and surrounding scatterers, the V2V channel generally has high Doppler spreads, which can be characterized by the mean value. It is found that the mean Doppler spread of V2V channel ranges from 100 to 300 Hz and also has large variations [40]. The maximum Doppler frequency in a V2V scenario can be up to four times higher than in a cellular scenario. The distribution of Doppler spread is found to depend on scenarios and TX/RX velocities [17,18,20]. It is noteworthy that the instantaneous Doppler spectrum of V2V channel may change rapidly, which implies that the V2V channel is not a wide-sense stationary. In Table 12.4, we summarize some typical measurements of Doppler spread for V2V channels.

Table 12.3 Small-Scale Fading Distributions for V2V Channels

Scenario	Highway	Rural	Suburban	Urban
Small-scale fading distribution	Rice [26] Weibull [32] Rice [33] Weibull [34] Nakagami [35]	Rice [33] Rice [36] Rice [37]	Nakagami [10] Nakagami [20] Rice [26] Rice [33] Rice [38] Rice [37] Nakagami [39] Weibull [34]	Nakagami [8] Rice [26] Weibull [32] Rice [33] Rice [38] Rice [37] Nakagami [39] Weibull [34]

Table 12.4 Mean Value of Doppler Spread for V2V Channels

Scenario		Highway	Rural	Suburban	Urban
Mean value of Doppler spread	Same direction	92 Hz [18] 297.8 Hz [41] 120 Hz [31] 15.5–31 Hz [42]	108 Hz [18] 19.4–28.6 Hz [42]	20–120 Hz [20]	33 Hz [18] 73.8 Hz [41] 85.6 Hz [31] 23.6–364 Hz [42]
	Opposite directions	761–978 Hz [40] 252.4 Hz [41]	782 Hz [40] 239.7 Hz [41] 26.2–196 Hz [42]	65.1 Hz [41]	263–341 Hz [40] 100.9 Hz [41] 42.4–107 Hz [42]

12.3.5 Delay Spread

The delay spread is widely used to characterize the delay dispersion/frequency selectivity of channel [15,43] and is thus of interest for digital communication applications. The delay spread also determines the coherence bandwidth of the channel. The root-mean-square (rms) delay spread can be expressed as [44]

$$\sigma_\tau = \left(\frac{\int_0^\infty (\tau - \mu_\tau)^2 \, \psi_h(\tau) d\tau}{\int_0^\infty \psi_h(\tau) d\tau} \right)^{1/2}$$

where $\psi_h(\tau)$ is the power delay profile of the measured received power as a function of delay τ. Parameter μ_τ is the average delay of the profile. A large number of measurements have been conducted to characterize the rms delay spreads of V2V channels. It is found that the mean rms delay spreads generally range from 100 to 200 ns [6]. Furthermore, the urban scenario, with the most scatterers, has the highest delay spreads, whereas for the rural and suburban scenarios, the delay spreads are found to be low. The NLOS propagation also leads to increased delay spreads. In addition, the distribution of the rms delay spread is often modeled by a log-normal distribution [44]. In Table 12.5, we summarize some typical mean value of rms delay spread in V2V channel measurements.

12.3.6 Non-Stationary Characterization

Due to the rapidly changing environment when both link ends move, V2V radio channels have been widely considered to be non-stationary, and the wide-sense stationarity (WSS) assumption of conventional cellular channels cannot be used anymore. This means that V2V channel statistics are valid only for a short period of time. A visual inspection of the measured time-varying power delay profile (PDP) validates the non-WSS of V2V channels [6]. Non-stationarity of V2V channels should be carefully considered as using the WSS assumption could lead to (erroneous) optimistic BER simulation results for V2V systems [46].

The first step of non-stationary channel characterization is to divide the time-variant channels into a series of smaller segmentations. Within each segmentation, the channel is WSS so

Table 12.5 Mean Value of rms Delay Spread for V2V Channels

Scenario	Highway	Rural	Suburban	Urban
rms delay spread (mean)	247 ns [17] 41 ns [18] 141–398 ns [40] 7.8–10.8 ns [42] 165 ns [26] 53–127 ns [32] 53.4–149.6 ns [45]	52 ns [18] 22 ns [40] 8.5–21.6 ns [42] 160 ns [32] 98.9–174.6 ns [45]	35.8–111.5 ns [10] 104 ns [26]	59.8–134.6 ns [8] 47 ns [18] 158–321 ns [40] 373 ns [26] 126 ns [32] 15.2–62.6 ns [42] 138.4–294.9 ns [45]

that the small-scale fading parameters can be rigorously evaluated and the channel modeling becomes physically meaningful, and therefore, the birth/death process of MPCs can be characterized over different segmentations to lead to a non-stationary channel model. As suggested by Refs. [23,47], such local region is called *quasi-stationarity* region, which means that the channel statistics within the quasi-stationarity region are similar enough compared to the statistics of the neighboring region such that the statistics can be approximately considered to be stationary. Several measures of the similarity between channel statistics are proposed to estimate the size of quasi-stationarity region. In Ref. [48], the correlation matrix distance (CMD) is proposed to evaluate the changes in the spatial structure of the channel and has been used for V2V channels [49,50]. In Ref. [51], the spectral divergence (SD) [52], which measures the distance between two spectral densities, is applied to the PDPs to analyze the non-WSS of V2V channels. Reference [23] compares the estimated size of quasi-stationarity regions by using both CMD and SD, based on V2V measurements. Figure 12.1 shows some example plots of the estimated CMD and SD in the V2V channels in Ref. [23], where two quasi-stationarity regions (corresponding to the regions of 10–25 s and 70–95 s) can be generally observed for both algorithms. It is found that both CMD and SD lead to similar results under some conditions; however, different thresholds should be used. In Ref. [53], the shadowing correlation is used to characterize quasi-stationarity region, and it is found that the decorrelation time/ distance of shadow fading is close to the estimated quasi-stationarity region by using CMD. Finally, based on a large body of V2V channel measurements, in Ref. [23] it is found that the size of the quasi-stationarity region ranges from 3 to 80 m in different V2V scenarios, and it is suggested to rely on SD and shadowing metrics when measurements are conducted with small electrical aperture size of arrays and to use CMD metric if arrays with large electrical apertures are employed. However, further characterization of the size of the quasi-stationarity region is needed, especially for some safety-related V2V scenarios, such as NLOS scenario with vehicle obstructions, road crossing, merge lanes, and tunnel. Since the quasi-stationarity also depends on carrier frequency, V2V non-stationary channel characterizations at mmWave bands are required for future applications.

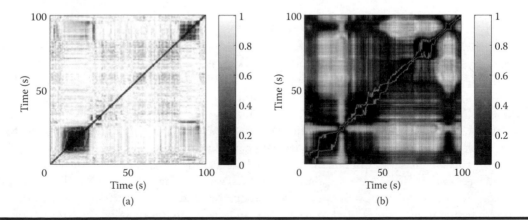

Figure 12.1 **Example plots of the estimated CMD and SD in the V2V measurements. (a) CMD and (b) SD. The corresponding time-variant quasi-stationarity intervals are marked using grey curves. (He, R., et al.,** *IEEE Transactions on Antennas and Propagation***, vol. 63, no. 5, pp. 2237–2251, © 2015, IEEE.)**

For the non-stationary channel modeling, several approaches have been used. In Ref. [32], a birth/death process is used to account for the appearance and disappearance of taps for wideband V2V TDL model; however, it does not account for the movements of scatterers into a different delay bin. In Refs. [34,54], the geometry-based stochastic channel models (GSCMs) are used, which take nonstationarities into account automatically; however, they generally consider the channel statistics of MPC to be constant within the lifetime and do not analyze the angular domain variations. In Refs. [39,55], a dynamic wideband directional model is proposed for time-variant V2V channels, where the birth/death process is incorporated into the directional CIRs. The dynamic directional CIR is expressed as [39]

$$
h\big(t_i, \tau(t_i), \phi(t_i)\big) = \sum_{j=1}^{i-1} \sum_{l_{j\to i}=1}^{N(t_{j\to i})} \left[a_{l_{j\to i}}(t_i) e^{j\varphi_{l_{j\to i}}} \delta(\tau - \tau_{l_{j\to i}}(t_i)) \delta(\phi - \phi_{l_{j\to i}}(t_i)) \right]
$$

$$
+ \sum_{l_{i\to i}=1}^{N(t_{i\to i})} \left[a_{l_{i\to i}}(t_i) e^{j\varphi_{l_{i\to i}}} \delta(\tau - \tau_{l_{i\to i}}(t_i)) \delta(\phi - \phi_{l_{i\to i}}(t_i)) \right]
$$

where t is time; a, τ, ϕ are the amplitude, delay, and angle of MPC, respectively; φ is the phase and is assumed to be uniformly distributed over the range of 0–360 degrees; δ is the Dirac delta function; i and j are time indices; N is MPC number; $i \to i$ indicates that the MPCs are first observed at time t_i; and $j \to i$ indicates that the MPCs are first observed at time t_j and exist at time t_i. The first term on the right side of the above equation represents the MPCs that have been observed before time t_i, that is, the *old* MPCs, and the second term of the above equation represents the MPCs that are first observed at time t_i, that is, the *new* MPCs. And we thus have

$$
N(t_i) = N(t_{i\to i}) + \sum_{j=1}^{i-1} N(t_{j\to i})
$$

This approach is able to incorporate the "birth" and "death" evolutions of MPCs at any time instant. By modeling the dynamic evolutions of new and old MPCs, the model can generate the dynamic CIRs at any time. An MPC distance-based tracking algorithm is further used to identify the "birth" and "death" of such paths over different quasi-stationarity regions. Figure 12.2 shows some example plots of MPC tracking in suburban measurements in Ref. [39], where the results of MPC tracking are marked with linear least square regression curves. It is found that the evolutions of MPCs on angular and delay domains are independent, and the track of evolution (i.e., the slope of the least square fit curves) is generally independent of angle and delay. This is because that the scatterers are randomly distributed in the dynamic V2V channels. The PDFs of the number of MPCs, lifetime, and position are fitted to some theoretical distributions in the dynamic model of Ref. [39]. A delay-based model is used to describe the MPC's power, and a linear polynomial function is used to model the variations of the MPC's behaviors within its lifetime. By incorporating both angular and delay domain properties and dynamic behaviors of the MPC's evolutions, the model is capable of handling time-variant V2V scenarios with dynamic scatterers.

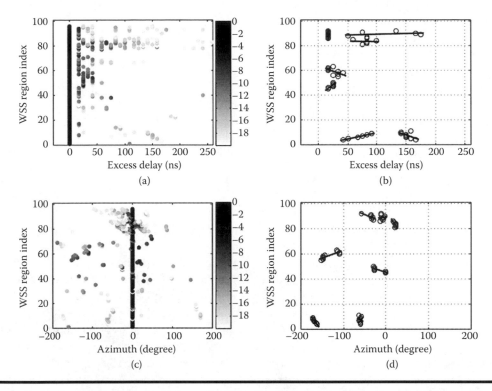

Figure 12.2 Example plots. (a) Detected MPCs (in dB) on delay domain, (b) MPC tracking on delay domain, (c) detected MPCs (in dB) on azimuth domain, and (d) MPC tracking on azimuth domain. (He, R., et al., *IEEE Transactions on Industrial Electronics*, vol. 62, no. 12, pp. 7870–7882, © 2015, IEEE.)

12.4 Modeling Approach

12.4.1 Statistical Model

Statistical channel models reproduce the statistics of the channel parameters (such as the path loss, shadowing, and the small-scale fading) based on the realistic measurements and have been widely used in the V2V scenarios. The narrowband statistical models mainly characterize the fading statistics together with the Doppler spectrum and do not characterize the frequency selectivity. In Section 17.3, the statistical characterizations for the typical V2V channel parameters have been summarized. In the wideband statistical model, the TDL model is widely used for V2V channels [6,32,33,56], where each tap represents signals received from several paths with a different delay and different type of Doppler spectrum. The power of taps is usually assumed to decay exponentially in delay domain, and each tap may have different fading distributions. It is noteworthy that the V2V channel may have correlated fading for different taps [5].

12.4.2 Deterministic Model

The deterministic model requires detailed information of the environment, for example, geometry of the environment, locations of the scatterers, and electromagnetic properties of objects, and solves Maxwell's equations (or an approximation thereof) under the boundary conditions imposed by a specific environment [8]. In Refs. [31,57], ray tracing for V2V environments is conducted, which provides a site-specific simulation of the propagation channel. It is found that by appropriately modeling the environment, agreement between measured and simulated receive powers can be brought within 3 dB standard deviation [58,59]. In Ref. [60], the ray tracing is used to develop a simulator for the 802.11p protocol, and the measurements in both LOS and NLOS scenarios are used for validations. However, the complexity of the shape of realistic cars often makes the ray tracing oversimplified, and in many cases, the influence of a normal traffic environment is neglected [8]. Another drawback of ray tracing lies in its high computational demands.

12.4.3 Geometry-Based Stochastic Model

The GSCMs [61] have been widely used for theoretical analysis of channel statistics and performance evaluation of MIMO communication systems. For V2V applications, two approaches exist: irregular-shaped GSCMs with scatterers in realistic positions and regular-shaped GSCMs with scatterers on regular shapes. For the former, a realistic placement of scatterers is used, where the locations as well as properties of scatterers are closely adapted to measurements so that the model better reproduces the physical reality. In such a model, the signal contributions of the scatterers are determined by using a simplified ray tracing, and the total signal is summed up at RX to obtain the complex impulse response. In Ref. [54], a GSCM is proposed for V2V channels, which makes a distinction between discrete and diffuse scattering contributions. An example of V2V scenario in Refs. [5,54] is shown in Figure 12.3. In the model, a roadside TX and a vehicular RX are deterministically placed in the simulated highway area. The mobile discrete scattering components, static discrete scattering components, and diffuse scattering components, which all correspond to the certain objects in V2V environments, are generated randomly according to specified statistical distributions. The speed of the scatterers can be stochastically assigned from a specified distribution to reflect a real highway scenario. The model is then parameterized from measurements and supplies an implementation recipe for simulations. In Ref. [34], a similar approach is used for V2V channel modeling, where the fading is further parameterized, and the diffuse scattering components are modeled in a purely stochastic manner. The irregular-shaped GSCMs can easily handle the non-stationarity of V2V channels by

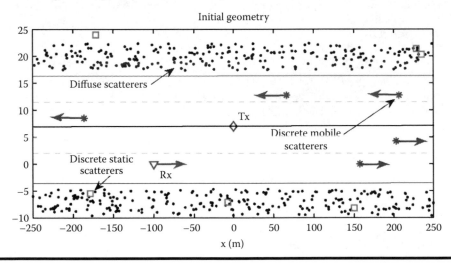

Figure 12.3 An example plot of the GSCM with scatterers in realistic positions for V2V channels. (Mecklenbräuker, C. F., et al., *Proceedings of IEEE*, vol. 99, no. 7, pp. 1189–1212, © 2011, IEEE and Karedal, J., et al., *IEEE Transactions on Wireless Communication*, vol. 8, no. 7, pp. 3646–3657, © 2009, IEEE.)

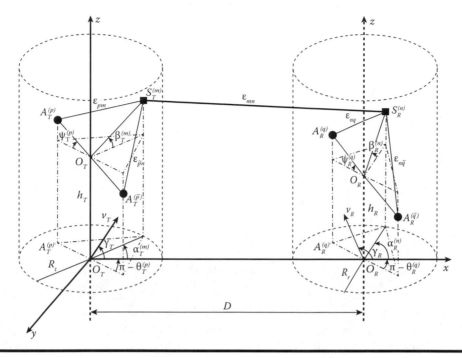

Figure 12.4   An example plot of the three-dimensional two-cylinder model for V2V channels. (Zajic, A. G., and Stuber, G. L., *IEEE Transactions on Vehicular Technology*, vol. 57, pp. 2042–2054, © 2008, IEEE.)

including the motions of the TX/RX and scatterers. It is also noteworthy that generally only single-bounced rays are considered in the irregular-shaped GSCMs.

The regular-shaped GSCMs have also been widely used for V2V channel modeling [62], which are able to preserve the mathematical tractability by assuming that all the effective scatterers are located on regular shapes. The simplest regular shape geometrical model is the two-ring model [63–66], where the two rings of scatterers are placed around the TX and RX. This model can be used to derive the joint space–time correlation function and the Doppler power spectral density. Several generalizations of the two-ring model exist: in Ref. [67], an LOS component is added; in Ref. [68], additional scatterers are added on an ellipse to depict the stationary roadside environments; the two-cylinder model [69–72] (as shown in Figure 12.4) and the two-sphere model [73,74] are proposed to extend the model to the three-dimensional case so that scatterers are placed in both the horizontal and vertical planes; in Ref. [75], a two-sphere model and an elliptic-cylinder model are combined to study the impact of the vehicular traffic density on channel statistics and jointly consider the azimuth and elevation angles; and in Ref. [76], a two-sphere model and multiple confocal elliptic-cylinder models are combined to extend the model for wideband applications.

12.5 Open Topic

12.5.1 Vehicle Obstruction

Although the impact of vehicles between the TX and RX significantly affects the performance of V2V networks, it has not been explored much. The vehicle obstruction leads to shadowing of the desired signals and larger delay spreads, and also causes changes in the angular distributions of MPCs. In Refs. [8,77], analysis and modeling of V2V channels when the LOS between the TX and RX is blocked by a large school bus are presented, based on realistic measurements. The impact of vehicle obstruction on path loss, shadow fading, small-scale fading, delay spread, and correlation is analyzed. However, due to the limitation of the measurement setup, the bus is kept stationary during the measurements, and the surrounding traffic, for example, moving cars and other large vehicles, is not considered. In Refs. [22,27], the impact of vehicle obstruction is measured with moving scatterers; however, only path loss, shadow fading, and delay spread are analyzed, which is not enough to develop a double-directional channel model. Due to the huge difficulty of measuring the dynamic V2V channels with large vehicle obstruction, there still lacks an accurate and complete model for this scenario. This results in that almost all VANET simulators neglect the impact of vehicles as obstacles on propagation, due to the lack of a model to incorporate the effect of vehicles [78]. It is necessary and important to explore the impact of large vehicles between the TX and RX in V2V communications.

12.5.2 Vehicular Channels at LTE and mmWave Bands

Even though many vehicular standards are developed in the 5.9 GHz ITS band, extra frequency bands have been considered to improve and enhance vehicular communications [79]. Several bands of long-term evolution (LTE) and mmWave have been considered [7,80,81]; however, the propagation channels in those bands have been less explored for V2V environments. The research on V2V communications has considered mmWave bands [82] and the ultrahigh frequency band [83]. Some measurements have also been made in the 2.1–2.4 GHz band [84]. In Ref. [84], the 900 MHz RF wireless V2V measurements are conducted. The path loss is modeled, and the rms delay spread ranges between 8.9 and 20.8 ns. The Rician K factor mostly ranges from 5.0 to 11.0. It is concluded that the 900 MHz V2V channel consists of two dominant components—one is the LOS component and the other is the strong reflection from the roadway. In Ref. [86], a path loss model is proposed for planning of V2V wireless systems at 900 MHz. In Ref. [87], the statistical channel models for small-scale fading are presented for a frequency selective V2V channels in an expressway environment, based on the measurements at 2.45 GHz. The proposed model implies a non-separable channel with a persistent Rician behavior across multiple model taps. In Ref. [88], the Doppler spectra are measured at 2.45 GHz in V2V environments. It is found that different environments produced quite different spectra. In Refs. [21,89–92], mmWave V2V channel measurements are conducted; however, most of them only focus on path loss model. It can be seen that the V2V channels at LTE and mmWave bands have been less explored. The current measurements can only characterize a few channel parameters and are not able to develop a complete channel model. The study of V2V channels at LTE and mmWave bands should be useful for future V2V applications.

12.5.3 3D Directional Channel Modeling

To further use the potential of spatial multiplexing, the three-dimensional (3D) channel needs to be modeled. In V2V scenarios, since the vehicular antennas are usually lower than most of the scatterers, a large number of MPCs with high elevations are expected, and a 3D directional channel modeling is required. However, most of the V2V measurements and channel models only cover the azimuth domain. Even though the 3D regular-shaped GSCMs include the scatterers on elevation domain, the parameterizations of the distributions on elevation domain usually lack measurement-based validations. Figure 12.5 shows an example plot of the azimuth and elevation of DOA estimates in V2V NLOS scenarios in Ref. [93], based on measurements. It can be seen

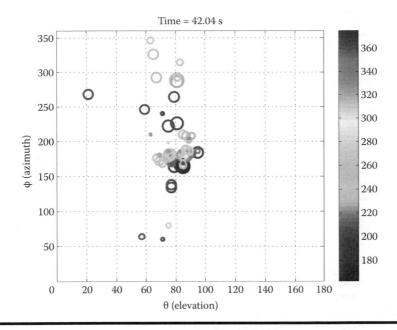

Figure 12.5   An example plot of the azimuth and elevation of DOA estimates in V2V scenarios. (Reprinted from Elsevier Measurement, vol. 95, Garcia Sanchez, M., *Millimeter wave radio channel characterization for 5G vehicle-to-vehicle communications*, pp. 223–229, Copyright (2017).)

that the V2V NLOS propagation leads to large spreads in both azimuth and elevation domains. Therefore, a 3D directional channel model is required for V2V communications.

12.5.4 Non-Stationary Channel Modeling

It has been widely accepted that V2V channel is non-stationary, and a few models have been proposed to incorporate non-stationarity [32,34,39,54]. However, they all use different approaches, and none of them is widely accepted to fully solve the problem of non-stationary channel modeling: the current irregular-shaped GSCMs generally cannot reflect the complete dynamic behaviors of MPCs in double-directional channels; even though the regular-shaped GSCMs lead to the time-variant CIRs, there lacks validations that the resulting quasi-stationary region size fits to the measurements; the birth/dead-based TDL model does not include angular information of MPCs and needs to be modified to provide time-varying tap locations and statistics; the ray-tracing model has high computational demands to include dynamic traffic environments. In general, further efforts are still required to provide a generic non-stationary channel model for V2V communications.

12.6 Summary

In this chapter, the current researches in V2V radio channel measurements and modeling are reviewed. First, we present the typical vehicular environments in ITS, which are useful for channel measurements. Then, the typical channel parameters extracted based on measurements are summarized to show the insights of V2V channels, and the non-stationary channel characterization, which is of great importance for V2V channels, is discussed. Next, three typical channel modeling approaches for V2V channels are introduced and compared. Finally, some key insights and future research topics are suggested. Overall, the V2V radio channels will remain a vibrant research area and continue to provide worthwhile challenges in the next years.

Acknowledgment

This work is supported by the National Natural Science Foundation of China under Grant 61501020 and Grant 61771037, the State Key Laboratory of Rail Traffic Control and Safety, Beijing Jiaotong University, China, and the Huawei Corporation.

References

1. IEEE Draft Standard for Information Technology-Telecommunications and Information Exchange between Systems—Local and Metropolitan Area Networks—Specific Requirements—Part 11: Wireless LAN Medium Access Control (MAC) and Physical Layer (PHY) Specifications Amendment 7: Wireless Access in Vehicular Environmesnts, IEEE Unapproved Draft Std P802.11p /D7.0, May 2009.
2. Intelligent Transport Systems (ITS), European profile standard on the physical and medium access layer of 5 GHz ITS, *Draft ETSI ES*, vol. 202, no. 663, p. V1.1.0, 2009.
3. R. Chen, Z. Zhong, C.-Y. Chang, B. Ai, and R. He, Performance analysis on network connectivity for vehicular ad hoc networks, *International Journal of Ad Hoc and Ubiquitous Computing*, vol. 20, no. 2, pp. 67–77, 2015.
4. A. F. Molisch, *Wireless Communications*, 2nd ed. IEEE-Wiley, NJ, 2011.
5. C. F. Mecklenbrauker, A. F. Molisch, J. Karedal, F. Tufvesson, A. Paier, L. Bernado, T. Zemen, O. Klemp, and N. Czink, Vehicular channel characterization and its implications for wireless system design and performance, *Proceedings of IEEE*, vol. 99, no. 7, pp. 1189–1212, 2011.
6. A. F. Molisch, F. Tufvesson, J. Karedal, and C. Mecklenbrauker, A survey on vehicle-to-vehicle propagation channels, *IEEE Wireless Communications*, vol. 16, no. 6, pp. 12–22, 2009.
7. V. Va, T. Shimizu, G. Bansal, and R. W. Heath, Millimeter wave vehicular communications: A survey, *Foundations and Trends in Networking*, vol. 10, no. 1, pp. 1–113, 2015.
8. R. He, A.F. Molisch, F. Tufvesson, Z. Zhong, B. Ai, and T. Zhang, Vehicle-to-vehicle propagation models with large vehicle obstructions, *IEEE Transactions on Intelligent Transportation Systems*, vol. 15, no. 5, pp. 2237–2248, 2014.
9. A. Paier, L. Bernado, J. Karedal, O. Klemp and A. Kwoczek, Overview of vehicle-to-vehicle radio channel measurements for collision avoidance applications, in *Proceedings of IEEE Vehicular Technology Conference*, Taipei, pp. 1–5, 2010.
10. R. He, O. Renaudin, V.-M. Kolmonen, K. Haneda, Z. Zhong, B. Ai, S. Hubert, and C. Oestges, Vehicle-to-vehicle radio channel characterization in cross-road scenarios, *IEEE Transactions on Vehicular Technology*, vol. 65, no. 8, pp. 5850–5861, 2016.
11. T. Abbas, L. Bernadó, A. Thiel, C. Mecklenbrauker, and F. Tufvesson, Radio channel properties for vehicular communication: Merging lanes versus urban intersections, *IEEE Vehicular Technology Magazine*, vol. 8, no. 4, pp. 27–34, 2013.
12. D. W. Matolak, V2V communication channels: State of knowledge, new results, and what's next, *International Workshop on Communication Technologies for Vehicles*, pp. 1–21, 2013.
13. R. He, Z. Zhong, B. Ai, G. Wang, J. Ding, and A. F. Molisch, Measurements and analysis of propagation channels in high-speed railway viaducts, *IEEE Transactions on Wireless Communications*, vol. 12, no. 2, pp. 794–805, 2013.
14. R. He, Z. Zhong, B. Ai, and J. Ding, An empirical path loss model and fading analysis for high-speed railway viaduct scenarios, *IEEE Antennas and Wireless Propagation Letters*, vol. 10, pp. 808–812, 2011.

15. T. S. Rappaport, *Wireless Communications Principles and Practice*, 2nd ed. Prentice Hall, 2001.
16. J. Karedal, N. Czink, A. Paier, F. Tufvesson, and A. F. Molisch, Path loss modeling for vehicle-to-vehicle communications, *IEEE Transactions on Vehicular Technology*, vol. 60, no. 1, pp. 323–328, 2011.
17. A. Paier, J. Karedal, N. Czink, H. Hofstetter, C. Dumard, T. Zemen, F. Tufvesson, A.F. Molisch, and C.F. Mecklenbrauker, Car-to-car radio channel measurements at 5 GHz: Pathloss, power-delay profile, and delay-Doppler spectrum, *Proceedings of International Symposium on Wireless Communication Systems*, Trondheim, Norway, pp. 224–228, 2007.
18. J. Kunisch and J. Pamp, Wideband car-to-car radio channel measurements and model at 5.9 GHz, in Proc. IEEE VTC 2008 Fall, Calgary, BC, Canada, 2008.
19. L. Heng, B. E. Henty, F. Bai, and D. D, Stancil, Highway and rural propagation channel modeling for vehicle-to-vehicle communications at 5.9 GHz, *Proceedings of IEEE Antennas Propagation Society International Symposium*, San Diego, CA, pp. 1–4, 2008.
20. L. Cheng, B.E. Henty, D. D. Stancil, F. Bai, and P. Mudalige, Mobile vehicle-to-vehicle narrow-band channel measurement and characterization of the 5.9 GHz Dedicated Short Range Communication (DSRC) frequency band, *IEEE JSAC*, vol. 25, pp. 1501–1516, 2007.
21. R. Schneider, D. Didascalou, and W. Wiesbeck, Impact of road surfaces on millimeter-wave propagation, *IEEE Transactions on Vehicular Technology*, vol. 49, no. 4, pp. 1314–1320, 2000.
22. D. Vlastaras, T. Abbas, M. Nilsson, R. Whiton, M. Olback, and F. Tufvesson, Impact of a truck as an obstacle on vehicle-to-vehicle communications in rural and highway scenarios, in *Proceedings of the IEEE 6th International Symposium on Wireless Vehicular Communications*, Vancouver, Canada, pp. 1–6, September 2014.
23. R. He, O. Renaudin, V. -M. Kolmonen, K. Haneda, Z. Zhong, B. Ai, and C. Oestges, Characterization of quasi-stationarity regions for vehicle-to-vehicle radio channels, *IEEE Transactions on Antennas and Propagation*, vol. 63, no. 5, pp. 2237–2251, 2015.
24. R. He, Z. Zhong, B. Ai, and C. Oestges, Shadow fading correlation in high-speed railway environments, *IEEE Transactions on Vehicular Technology*, vol. 64, no. 7, pp. 2762–2772, 2015.
25. G. L. Stüber, *Principles of Mobile Communication*, 2nd ed. Norwell, MA: Kluwer, 2001.
26. O. Renaudin, V. M. Kolmonen, P. Vainikainen, and C. Oestges, Wideband MIMO car-to-car radio channel measurements at 5.3 GHz, in *IEEE 68th Vehicular Technology Conference, VTC 2008-Fall*, Calgary, BC, pp. 1–5, 2008.
27. T. Abbas, F. Tufvesson, K. Sjöberg, and J. Karedal, Measurement based shadow fading model for vehicle-to-vehicle network simulations, *International Journal of Antennas and Propagation*, vol. 2015, pp. 1–13, 2015.
28. R. He, Z. Zhong, B. Ai, J. Ding, Y. Yang, and A. F. Molisch, Short-term fading behavior in high-speed railway cutting scenario: Measurements, analysis, and statistical models, *IEEE Transactions on Antennas and Propagation*, vol. 61, no. 4, pp. 2209–2222, 2013.
29. P. Marinier, G. Y. Delisle, and C. L. Despins, Temporal variations of the indoor wireless millimeter-wave channel, *IEEE Transactions on Antennas and Propagation*, vol. 46, no. 6, pp. 928–934, 1998.
30. H. Hashemi, M. McGuire, T. Vlasschaert, and D. Tholl, Measurements and modeling of temporal variations of the indoor radio propagation channel, *IEEE Transactions on Vehicular Technology*, vol. 43, no. 3, pp. 733–737, 1994.
31. J. Maurer, T. Fugen, and W. Wiesbeck, Narrow-Band measurement and analysis of the inter-vehicle transmission channel at 5.2 GHz, in *Proceedings of IEEE VTC 2002 Spring*, Birmingham, AL, pp. 1274–1278, 2002.
32. I. Sen and D. W. Matolak, Vehicle-vehicle channel models for the 5-GHz band, *IEEE Transactions on Intelligent Transportation System*, vol. 9, no. 2, pp. 235–245, 2008.
33. G. Acosta-Marum and M. Ingram, Six time-and frequency-selective empirical channel models for vehicular wireless LANs, *IEEE Vehicular Technology Magazine*, vol. 2, no. 4, pp. 4–11, 2007.
34. O. Renaudin, V. -M. Kolmonen, P. Vainikainen, and C. Oestges, Wideband measurement-based modeling of inter-vehicle channels in the 5-GHz band, *IEEE Transactions on Vehicular Technology*, vol. 62, no. 8, pp. 3531–3540, 2013.
35. J. Yin, G. Holland, T. Elbatt, F. Bai, and H. Krishnan, DSRC channel fading analysis from empirical measurement, in *Proceedings of IEEE ChinaCom*, Beijing, China, pp. 1–5, 2006.

36. L. Bernadó, T. Zemen, F. Tufvesson, A. F. Molisch, and C. F. Mecklenbräuker, Time- and Frequency-Varying K-Factor of non-stationary vehicular channels for safety-relevant scenarios, *IEEE Transactions on Intelligent Transportation on Systems*, vol. 16, no. 2, pp. 1007–1017, 2015.

37. S. Zhu, T.S. Ghazaany, S.MR. Jones, R.A. Abd-Alhameed, J.M. Noras, T. Van Buren, J. Wilson, T. Suggett, and S. Marker, Probability distribution of Rician K-Factor in Urban, Suburban and rural areas using Real-World captured data, *IEEE Transactions on Antennas and Propagation*, vol. 62, no. 7, pp. 3835–3839, 2014.

38. X. Wang, E. Anderson, P. Steenkiste, and F. Bai, Improving the accuracy of environment-specific vehicular channel modeling, in *Proceedings of the 7th ACM International Workshop Wireless Network Testbeds, Experimental Evaluation Characterization*, New York, pp. 43–50, 2012.

39. R. He, O. Renaudin, V.-M. Kolmonen, K. Haneda, Z. Zhong, B. Ai, and C. Oestges, A dynamic wideband directional channel model for vehicle-to-vehicle communications, *IEEE Transactions on Industrial Electronics*, vol. 62, no. 12, pp. 7870–7882, 2015.

40. I. Tan, W. Tang, K. Laberteaux, and A. Bahai, Measurement and analysis of wireless channel impairments in DSRC vehicular communications, in *Proceedings of IEEE ICC*, Beijing, China, pp. 4882–4888, 2008.

41. Paschalidisp, M. Wisotzki, A. Kortke, et al. A wideband channel sounder for car-to-car radio channel measurements at 5.7 GHz and results for an urban scenario, in *VTC 2008: Proceedings of the IEEE 68th Vehicular Technology Conference*, Calgary, 2010.

42. T. Abbas, J. Karedal, and F. Tufvesson, Measurement-Based analysis: The effect of complementary antennas and diversity on vehicle-to-vehicle communication, *IEEE Antennas and Wireless Propagation Letters*, vol. 12, pp. 309–312, 2013.

43. S. Sangodoyin, V. Kristem, A. F. Molisch, R. He, F. Tufvesson, and H. M. Behairy, Statistical modeling of ultrawideband MIMO propagation channel in a warehouse environment, *IEEE Transactions on Antennas and Propagation*, vol. 64, no. 9, pp. 4049–4063, 2016.

44. D.W. Matolak, I. Sen, W. Xiong, and N. T. Yaskoff, T 5 GHz wireless channel characterization for vehicle to vehicle communications, in *Proceedings of IEEE MILCOM*, Atlantic City, NJ, pp. 1–7, 2005.

45. W. Qiong, D. W. Matolak, and I. Sen, 5-GHz-Band vehicle-to-vehicle channels: Models for multiple values of channel bandwidth, *IEEE Transactions on Vehicular Technology*, vol. 59, pp. 2620–2625, 2010.

46. D. W. Matolak, Channel modeling for vehicle-to-vehicle communications, *IEEE Communications Magazine*, vol. 46, no. 5, pp. 76–83, 2008.

47. A. Ispas, G. Ascheid, C. Schneider, and R. Thomä, Analysis of local quasi-stationarity regions in an urban macrocell scenario, in *Proceedings IEEE 71st Vehicular Technology Conference (VTC'10)*, Taipei, pp. 1–5, 2010.

48. M. Herdin, N. Czink, H. Özcelik, and E. Bonek, Correlation matrix distance, a meaningful measure for evaluation of non-stationary MIMO channels, in *Proceedings of the IEEE 61st Vehicular Technology Conference (VTC'05)*, Stockholm, Sweden, vol. 1, pp. 136–140, 2005.

49. L. Bernadó, Non-stationarity in vehicular wireless channels, PhD dissertation, Technische Universität Wien, Vienna, Austria, 2012.

50. O. Renaudin, V. -M. Kolmonen, P. Vainikainen, and C. Oestges, Non-stationary narrowband MIMO inter-vehicle channel characterization in the 5-GHz band, *IEEE Transactions on Vehicular Technology*, vol. 59, no. 4, pp. 2007–2015, 2010.

51. L. Bernadó, and T. Zemen, A. Paier, G. Matz, J. Karedal, N. Czink, C. Dumard, F. Tufvesson, M. Hagenauer, A. F. Molisch, F. Andreas, et al., Non-WSSUS vehicular channel characterization at 5.2 GHz-spectral divergence and time-variant coherence parameters, in *Proceedings of 29th URSI General Assembly*, Chicago, Illinois, pp. 9–16, 2008.

52. T. T. Georgiou, Distances and Riemannian metrics for spectral density functions, *IEEE Transactions on Signal Process*, vol. 55, no. 8, pp. 3995–4003, 2007.

53. R. He, O. Renaudin, V. -M. Kolmonen, K. Haneda, Z. Zhong, B. Ai, and C. Oestges, Non-stationarity characterization for vehicle-to-vehicle channels using correlation matrix distance and shadow fading correlation, in *Proceedings of 35th Progress in Electromagnetics Research Symposium*, Guangzhou, China, pp. 1–5, 2014.

54. J. Karedal, F. Tufvesson, N. Czink, A. Paier, C. Dumard, T. Zemen, C.F. Mecklenbrauker, and A.F. Molisch, A geometry-based stochastic MIMO model for vehicle-to-vehicle communications, *IEEE Transactions on Wireless Communication*, vol. 8, no. 7, pp. 3646–3657, 2009.

55. R. He, O. Renaudin, V.-M. Kolmonen, K. Haneda, Z. Zhong, B. Ai, and C. Oestges, Statistical characterization of dynamic multi-path components for vehicle-to-vehicle radio channels, in *Proceedings of IEEE VTC*, Glasgow, UK, pp. 1–6, 2015.

56. W. Viriyasitavat, M. Boban, H. M. Tsai, and A. Vasilakos, Vehicular communications: Survey and challenges of channel and propagation models, *IEEE Vehicular Technology Magazine*, vol. 10, no. 2, pp. 55–66, 2015.

57. J. Maurer, T. Schafer, and W. Wiesbeck, A realistic description of the environment for inter-vehicle wave propagation modeling, in *Proceedings of IEEE VTC 2001 Fall*, Atlantic City, NJ, pp. 1437–1441, 2001.

58. W. Wiesbeck and S. Knorzer, Characteristics of the mobile channel for high velocities, in *International Conference Electromagnetic in Advanced Applications*, Torino, Italy, pp. 116–20, 2007.

59. J. Maurer, T. Fugen, T. Schafer, and W. Wiesbeck, A new Inter-Vehicle Communications (IVC) channel model, in *Proceedings of IEEE VTC. 2004 Fall*, Los Angeles, CA, pp. 9–13, 2004.

60. S. Biddlestone, K. Redmill, R. Miucic, and Ü. Özgüner, An integrated 802.11 p WAVE DSRC and vehicle traffic simulator with experimentally validated urban (LOS and NLOS) propagation models, in *IEEE Transactions on Intelligent Transportation Systems*, vol. 13, no. 4, pp. 1792–1802, 2012.

61. J. Fuhl, A. F. Molisch, and E. Bonek, Unified channel model for mobile radio systems with smart antennas, in *IEEE Proceedings-Radar, Sonar and Navigation*, vol. 145, no. 1. pp. 32–41, 1998.

62. Y. Li, R. He, S. Lin, et al., Cluster-based non-stationary channel modeling for vehicle-to-vehicle communications, *IEEE Antennas and Wireless Propagation Letters*, vol. 16, no. 1, pp. 408–411, 2017.

63. G. J. Byers and F. Takawira, Spatially and temporally correlated MIMO channels: Modeling and capacity analysis, *IEEE Transactions on Vehicular Technology*, vol. 53, no. 3, pp. 634–643, 2004.

64. M. Pätzold, B. O. Hogstad, N. Youssef, and D. Kim, A MIMO mobile-to-mobile channel model: Part I-The reference model, in *Proceedings of IEEE PIMRC*, Berlin, Germany, vol. 1, pp. 573–578, September 2005.

65. A. G. Zajic and G. L. Stüber, Space-time correlated MIMO mobile-to-mobile channels, in *Proceedings of IEEE PIMRC*, Helsinki, Finland, pp. 1–5, September 2006.

66. A. S. Akki and F. Haber, A statistical model for mobile-to-mobile land communication channel, *IEEE Transactions on Vehicular Technology*, vol. 35, no. 1, pp. 2–10, 1986.

67. L. C. Wang, W. C. Liu, and Y. H. Cheng, Statistical analysis of a mobile-to-mobile Rician fading channel model, *IEEE Transactions on Vehicular Technology*, vol. 58, no. 1, pp. 32–38, 2009.

68. X. Cheng, C.-X. Wang, and D.I. Laurenson, S. Salous, and A.V. Vasilakos, An adaptive geometry-based stochastic model for non-isotropic MIMO mobile-to-mobile channels, *IEEE Transactions on Wireless Communications*, vol. 8, no. 9, pp. 4824–4835, 2009.

69. A. G. Zajic and G. L. Stuber, Three-dimensional modeling, simulation, and capacity analysis of space-time correlated mobile-to-mobile channels, *IEEE Transactions on Vehicular Technology*, vol. 57, pp. 2042–2054, 2008.

70. A. G. Zajic and G. L. Stuber, Three-dimensional modeling and simulation of wideband MIMO mobile-to-mobile channels, *IEEE Transactions on Wireless Communications*, vol. 8, no. 3, pp. 1260–1275, 2009.

71. A. Zajic, G. L. Stüber, T. Pratt, and S. Nguyen, Wide-band MIMO mobile-to-mobile channels: Statistical modeling with experimental verification, *IEEE Transactions on Vehicular Technology*, vol. 56, no. 2, pp. 517–534, 2009.

72. A. Zajic and G. L. Stüber, Space-time correlated mobile-to-mobile channels: Modelling and simulation, *IEEE Transactions on Vehicular Technology*, vol. 57, no. 2, pp. 715–726, 2008.

73. P. T. Samarasinghe, T. A. Lamahewa, T. D. Abhayapala, and R. A. Kennedy, 3D mobile-to-mobile wireless channel model, in *Proceedings of 2010 AusCTW*, Canberra, ACT, Australia, pp. 30–34, 2010.

74. T. -M. Wu and T. -H. Tsai, Novel 3-D mobile-to-mobile wideband channel model, in *Proceedings of 2010 IEEE APSURSI*, Toronto, ON, Canada, pp. 1–4, 2010.

75. Y. Yuan, C. -X. Wang, X. Cheng, B. Ai, and D. I. Laurenson, Novel 3D geometry-based stochastic models for non-isotropic MIMO vehicle-to-vehicle channels, *IEEE Transactions on Wireless Communications*, vol. 14, no. 1, pp. 298–309, 2014.

76. Y. Yuan, C. X. Wang, Y. He, et al. 3D wideband non-stationary geometry-based stochastic models for non-isotropic MIMO vehicle-to-vehicle channels, *IEEE Transactions on Wireless Communications*, vol. 14. pp. 6883–6895, 2015.

77. R. He, A. F. Molisch, F. Tufvesson, Z. Zhong, B. Ai, and T. Zhang, Vehicle-to-vehicle channel models with large vehicle obstructions, in *Proceedings of IEEE International Conference on Communications*, Sydney, NSW, Australia, pp. 1–6, 2014.

78. R. Meireles, M. Boban, P. Steenkiste, O. Tonguz, and J. Barros, Experimental study on the impact of vehicular obstructions in VANETs, in *Proceedings of IEEE Vehicular Networking Conference*, Jersey City, NJ, pp. 338–345, 2010.

79. R. He, B. Ai, G. Wang, et al., High-speed railway communications: From GSM-R to LTE-R, *IEEE Vehicular Technology Magazine*, vol. 11, no. 3, pp. 49–58, 2016.

80. D. Matolak, Q. Wu, J. Sanchez-Sanchez, D. Morales-Jimenez, and M. Aguayo-Torres, Performance of LTE in vehicle-to-vehicle channels, in *Proceedings of IEEE VTC Fall*, San Francisco, CA, pp. 1–4, September 2011.

81. Ruisi He. Bo Ai, Gordon L. Stuber, Gongpu Wang, and Zhangdui Zhong, "Geometrical based modeling for millimeter wave MIMO mobile-to-mobile channels," *IEEE Transactions on Vehicular Technology*, to appear, 2018.

82. T. Wada, M. Maeda, M. Okada, K. Tsukamoto, and S. Komaki, Theoretical analysis of propagation characteristics in millimeter-wave inter-vehicle communication system, *IEICE Transaction on Communication*, vol. 83, no. 11, pp. 1116–1125, 1998.

83. S. Sai, E. Niwa, K. Mase, M. Nishibori, J. Inoue, M. Obuchi, T. Harada, H. Ito, K. Mizutani, and M. Kizu, Field evaluation of UHF radio propagation for an ITS safety system in an urban environment, *IEEE Communication Magazine*, vol. 47, no. 11, pp. 120–127, 2009.

84. K. Konstantinou, S. Kang, and C. Tzaras, A measurement based model for mobile-to-mobile UMTS links, *Proceedings of IEEE Vehicular Technology Conference*, Singapore, pp. 529–533, 11–14 May, 2008.

85. J. S. Davis and J. P. M. G. Linnartz, Measurements of vehicle-to-vehicle propagation, *Proceedings of Asilomar Conference*, Monterey, CA, October 31–November 1, 1994.

86. J. Turkka and M. Renfors, Path loss measurements for a non-line-of-sight mobile-to-mobile environment, *2008 8th International Conference on ITS Telecommunications*, Phuket, pp. 274–278, 2008.

87. G. Acosta and M. A. Ingram, Model development for the wideband expressway vehicle-to-vehicle 2.4 GHz channel, in *IEEE Wireless Communications and Networking Conference, WCNC 2006*, Las Vegas, NV, pp. 1283–1288, 2006.

88. G. Acosta, K. Tokuda, and M. A. Ingram, Measured joint Doppler-delay power profiles for vehicle-to-vehicle communications at 2.4 GHz, *Global Telecommunications Conference, GLOBECOM '04. IEEE*, Dallas, TX, pp. 3813–3817, 2004.

89. K. Sato and M. Fujise, Propagation measurements for inter-vehicle communication in 76-GHz band, in *ITS Telecommunications Proceedings*, Chengdu, China, pp. 408–411, 2006.

90. A. Yamamoto, K. Ogawa, T. Horimatsu, K. Sato, and M. Fujise, Effect of road undulation on the propagation characteristics of inter-vehicle communications in the 60 GHz band, in *IEEE WCACE*, Honolulu, HI, pp. 843–846, 2005.

91. S. Takahashi, A. Kato, K. Sato, and M. Fujise, Distance dependence of path loss for millimeter wave inter-vehicle communications, in *IEEE VTC-Fall*, Orlando, FL, pp. 26–30, 2003.

92. M. Garcia Sanchez, M. P. Taboas, and E. L. Cid, Millimeter wave radio channel characterization for 5G vehicle-to-vehicle communications, *Elsevier Measurement*, vol. 95, pp. 223–229, 2017.

93. R. He, O. Renaudin, V.-M. Kolmonen, K. Haneda, Z. Zhong, S. Hubert, and C. Oestges, Angular dispersion characterization of vehicle-to-vehicle channel in cross-road scenarios, in *Proceedings of EuCAP*, Lisbon, Portugal, pp. 1–4, 2015.

PHYSICAL LAYER
TECHNOLOGIES

Chapter 13

Design Methodologies Comparison of an OFDM System and Implementation Issues Using FPGA Technology for Vehicular Communications

George Kiokes
Hellenic Air Force Academy
Dekeleia, Greece

Erietta Zountouridou
I-SENSE Group, Institute of Communication and Computer Systems (ICCS)
Athens, Greece

Contents

13.1 Introduction

The IEEE 802.11p standard, based on IEEE 802.11a, has been endorsed by the American Society for Testing and Materials (ASTM) as the platform of the PHY and medium access control (MAC) layers for dedicated short-range communications (DSRC) [1,2]. It defines an international standard for wireless access in vehicular environments and has been proposed to face situations in cellular communications where high data transfer rates are required in circumstances where minimizing latency in the communication link is important. This includes data exchange between moving vehicles and between the vehicles and roadside units in the licensed ITS band of 5.9 GHz. Orthogonal frequency-division multiplexing (OFDM) [3] is a multicarrier transmission scheme where all subcarriers are orthogonal to each other. The multicarrier transmission idea occurred recently so as to efficiently use the entire spectrum available. OFDM is a combination of configuration and multiplexing allowing effective and reliable digital data transmission. It is a frequency division system, in which a large number of subcarriers transfer a small segment of information in parallel within the transmission channel. Using this technique, the information remains unaltered and spectrum economy is achieved, given that the subcarriers transmitted in parallel are rectangles transmitted one adjacent to the other. OFDM is based on the orthogonality of signals ensuring that at the points of the subchannels showing peaks, the adjacent subchannels show zero. In order for the subcarriers to be orthogonal, all subcarrier frequencies must be integer multiples of the same frequency.

Digital signal processing (DSP) [4,5], as the name implies, is the processing of signals in digital format. The digital form of the signal is created through sampling of the voltage level at frequent intervals and converting the voltage levels of a specific time period to a digital number relevant to the voltage.

13.2 OFDM Mathematical Model

The overall OFDM signal to be transmitted through the wireless channel is represented by the complex enclosure of the combination of all subcarriers for a symbol period T and is given by the equation in Ref. [6]:

$$\tilde{s}(t) = \sum_{k=-\infty}^{\infty} \tilde{s}_n(t) \tag{13.1}$$

where $\tilde{s}_n(t)$ n-in OFDM symbol. Thus, an OFDM symbol can be expressed as:

$$\tilde{s}_n(t) = \sum_{n=0}^{n-1} d_{i,n} \exp\left(j2\pi f_n t\right) \quad \text{for } (i-1)T \le t < iT,$$

$$0, \qquad\qquad\qquad\qquad \text{else}$$

where d_i is the complex value of QAM symbol and for each different time iT there is selection of a different symbol, $d_{i,n}$ [7]. If A_i and φ_i are the width and phase respectively for i-in the subcarrier, then d_i is defined as:

$$d_i = A_i \exp\left(j\varphi_i\right) = I_i + jQ_i$$

where $I_i = A_i \cos(j\varphi_i)$ and $Q_i = A_i \sin(j\varphi_i)$ in the phase and square part respectively.

The incoming data pulse sequence has M-QAM modulation and is demultiplexed to n independent data pulse sequences. Following the configuration of the individual subcarriers, the individual data pulse sequences are again immediately combined. The overall complex signal of the base zone is then converted to zone-pass signal and is represented as follows:

$$s(t) = \text{Re}\left[\tilde{s}(t)\exp\left(j2\pi f_c t\right)\right]$$

where $\tilde{s}(t)$ is the complex enclosure. Using this multiple carriers configuration method, the number of bits transmitted per symbol period, T, is the number of the subcarriers multiplied by the number of bits per symbol configured. Then, through inverse fast Fourier transformation, a group of samples in the frequency field is converted to the time field. Inverse Fourier transformation is mathematically described by equation:

$$D_m = \frac{1}{M}\sum_{n=0}^{M-1} d_n \exp(j2\pi mn/M), \quad \text{for} \quad n = 0,1,2,...,M-1$$

where D_m and d_n are the samples in the time field and the frequency field respectively. In the context of transformation, subcarrier frequencies are selected so that:

$$f_n = \frac{n}{T} \quad \text{for} \quad n = 0,1,2,...N$$

And each subcarrier and the output are sampled M times per symbol period, that is, for #0 $\leq t \leq T$, the samples are given by the equation:

$$t = \frac{m}{M}T, \quad \text{for} \quad m = 0,1,2,...,M-1$$

Thus, for n - in the symbol period, the following M samples of configured waveforms are received:

$$\tilde{s}\left(\frac{mT}{M}\right) = \sum_{n=0}^{n-1} d_{n,i} \quad \exp\left(j2\pi mn/M\right), \quad \text{for} \quad m = 0,1,2,...,M-1$$

For the inverse discrete Fourier transformation equation (IDFT), there should be $M = n$. However, IDFT can be effectively implemented as inverse fast Fourier transformation (IFFT) if M is a power of 2. This ensures the data subcarriers, a number of additional subcarriers used for

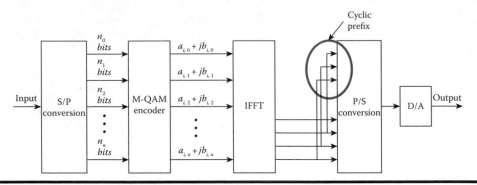

Figure 13.1 OFDM transmitter.

synchronization reasons on the receiver, and a number of null carriers for the assurance of a buffer zone for protection against adjacent channel interference.

13.2.1 OFDM Transmitter

The standard implementation of OFDM transmitter with its standard constituting parts is depicted in Figure 13.1.

- Serial-to-parallel format converter
- QAM configuration
- Inverse Fourier conversion (IFFT)
- Parallel-to-serial format converter
- Digital to analog conversion

Initially the incoming binary data pulse sequence is converted from serial to parallel, where the bits are divided into n subgroups consisting of M bits each. These n subgroups will be mapped as per the selected constellation using Gray encoding, and this way, a $a_i + jb_i$ value will be received at the modulator constellation. QAM modulator converts input data to complex values and depicts them on points in accordance with the given constellation—QPSK, 16-QAM, 32-QAM, etc. The number of data transmitted by each subcarrier depends on the constellation, that is, QPSK configuration transmits two data bits, 16-QAM transmits four data bits per subcarrier. The configuration to be used depends on the quality of the communication channel. On a channel with great interference, it is optimal to use a small configuration scheme such as BPSK, given that the signal to noise ratio (SNR) required at the receiver will be low, while on a channel without interference it is preferred to use a larger constellation due to the higher rate of binary digits.

The IFFT converts the signals from the frequency field to the time field. An IFFT converts a group of data whose length is a power of 2, within the same number of data but in the time field. The subcarriers number determines the subzones into which the available spectrum is divided. The cyclic prefix (CP) having length v, greater than length L of the discrete, equivalent, channel impulse response (CIR), is a copy of the last n samples of the output of IFFT and is placed at the beginning of the OFDM frame. It is converted from a discrete to an analog signal, which is then transmitted. At the receiver, following the analog-to-digital conversion of the signal, the cyclical prefix is removed and the signal is submitted to the inverse process using FFT of n points. This is used for the effective treatment of intersymbol interference (ISI) [8] and for reasons of synchronization with the receiver.

The receiver must "know" with great accuracy the time at which the transmitted signal reaches it, so as to initiate reception and further processing thereof. In order to avoid intersymbolic interference, the duration of the cyclical prefix must be at least equal to the time dispersal of the channel. Therefore, it is important to choose the necessary CP, so as to maximize system performance.

13.2.2 OFDM Receiver

The standard simplification of OFDM receiver with its key components is depicted in Figure 13.1 [9].

- Analog-to-digital format conversion
- Serial-to-parallel format conversion
- Cyclical prefix removal
- Fast Fourier transformation (FFT)
- M-QAM demodulation
- Parallel-to-serial format conversion

The symbol received is in the time field and due to its extension with the cyclic prefix the result may be altered. Thus, the signal received is converted from parallel to serial and then the cyclic prefix is removed. Following removal of the cyclic prefix, the signals are converted from the time field to the frequency field through the FFT of n points. FFT transformation is followed by the demodulation of symbols so that they are decoded as per the respective symbol of the constellation with which they were transmitted. The constellation point of the transmitted symbol may have changed due to the additional noise on the communication channel, an erroneous adaptation of the sampling time with the receiver, or for various other reasons. Therefore, it is necessary to define a threshold for the reception of decisions on the receiver's constellation. This operation is performed by the M-QAM decoder.

13.3 OFDM Implementation

IEEE 802.11p standard specifies an OFDM physical layer that employs 64 subcarriers; 52 out of the 64 subcarriers are used for actual transmission consisting of 48 data subcarriers and 4 pilot subcarriers to provide transmission of data at rates of 3, 4.5, 6, 9, 12, 18, 24, or 27 Mbps. The basic

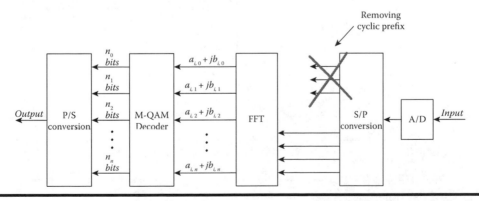

Figure 13.2 OFDM receiver. (From *Digital Modulation in FPGAs Xilinx using system generator (ASK, BPSK, FSK, OOK, QPSK)'* Diego Barragán, http://www.mathworks.com/matlabcentral/fileexchange/14650-digital-modulation-in-fpgas-xilinxusing-system-generator--ask--bpsk--fsk--ook--qpsk. With permission.)

principal of operation is to divide a high-speed binary signal to be transmitted into a number of lower data rate subcarriers. The system uses pilot subcarriers as a reference to disregard frequency or phase shifts of the signal during transmission. OFDM is a promising technique for achieving high data rate and combating multipath fading in wireless communications. Standard 802.11p specifies seven 10 MHz channels consisting of one control channel and six service channels. Four complex modulation methods (BPSK, QPSK, 16-QAM, and 64-QAM) are employed, depending on the data rate that can be supported by channel conditions. IEEE 802.11p main characteristics values are depicted in Table 13.1 [8,10].

13.4 OFDM Design Using Handwritten VHDL

The first design comprises handwritten VHDL code for the OFDM scheme. OFDM modulation is exploiting the fact that multiple data signals can be carried out from the same carry signal. The time space between the data signals which are carried from the carry signal is defined from the ratio $1/f_0$, where f_0 is the transmission duration of an OFDM symbol (symbol = wavelet). During the OFDM data modulation, every data sequence is considered to be available for transmission at any time and they are just separately distributed in the phase spectrum. Then, data spectrums are carried from the frequency to the time field through the inverse Fourier transform and they are transmitted in a wavelet sequence. During the OFDM demodulation, each wavelet is collected

Table 13.1 Main Characteristics Values of IEEE 802.11p

Parameter	Value	Parameter	Value
N_{SD} : Number of data subcarriers	48	T_{SHORT} : Short training sequence duration	16 µs ($10 \times T_{FFT}$ /4)
N_{SP} : Number of pilot subcarriers	4	T_{LONG} : Long training sequence duration	32 µs ($T_{GI2} + 2 \times T_{FFT}$)
N_{ST} : Number of subcarriers, total	52 ($N_{SD} + N_{SP}$)	Data rate	3, 4.5, 6, 9, 12, 18, 24, 27
IFFT/FFT size	64	Modulation	BPSK, QPSK, 16QAM, 64QAM
T_{FFT} : IFFT/FFT period	6.4 µs(1/ΔF)	Convolutional code	K = 7 (64 states)
$T_{PREAMBLE}$: PLCP preamble duration	32 µs ($T_{SHORT} + T_{LONG}$)	Coding rates	1/2, 2/3, 3/4
T_{GI} : GI duration	1.6 µs (T_{FFT} /4)	Channel spacing (MHz)	10
T_{GI2} : Training symbol duration	3.2 µs (T_{FFT} /2)	Signal bandwidth (MHz)	8.3
T_{SYM} : Symbol period	8.0 µs ($T_{GI} + T_{FFT}$)	Subcarrier spacing (KHz)	156.2

in serial sequence and after their transformation to parallel form, they are carried out into the frequency field through the Fourier transformation. A representative diagram of the above description is given in Figure 13.3. The crucial contribution in the modulation and demodulation procedure is given by the serial-to-parallel and parallel-to-serial components. These components are implemented in parallel.vhd and serial.vhd files respectively. The parallel component is converting serial data to parallel accepting as input sequential bits and producing bit arrays as output. Analogically, the parallel-to-serial component is using the same technique in order to accomplish the bit array to sequential bit transformation.

Obviously, in order to implement the different digital data transmission through the same analog transmission signal, a transformation must occur so that the transmitted digital data could be safely carried through the carry signal. In this project, the QPSK modulation is selected mostly for simplicity and implementation reasons. The QPSK modulation and demodulation are implemented from the qam and qamdecoder modules respectively. During modulation, data are modulated only according to phase, so the modulation could produce four states of data signals: sinc, simple, logical, and operations. The main component of the project is the modem component because it has the role of data receiver-transmitter but also the role of OFDM modulator-demodulator. The receiver-transmitter property is implemented with the coordination of three subcomponents: the txmodem, trx, and the rxmodem modules. The txmodem component is responsible for modulating data through the OFDM modulation and transmitting them. The rxmodem component is responsible for receiving data and demodulation them. Finally, the txrx component is responsible for the data passing through the QAM modulator-demodulator and also for the communication with system's inputs and outputs.

The txmodem implementation is achieved with the presence of two submodules: the input component and the OFDM component. The input component is accepting the signal through QAM modulation data and passing them to the OFDM component. It also checks the memory status and it locates the memory addresses where the modulated data are stored.

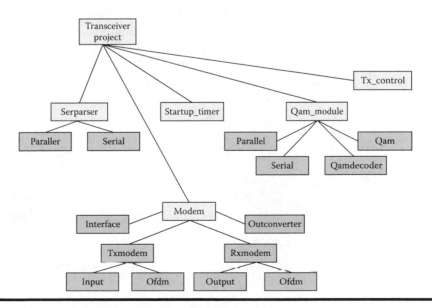

Figure 13.3 Transceiver schematic algorithm diagram.

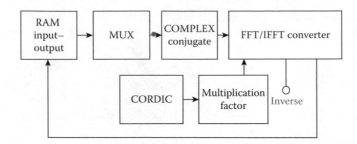

Figure 13.4   CFFT subcomponents.

The OFDM component accepts the data coming from the input component as data bit arrays (12 bit length) and modulate the incoming data through OFDM modulation. The OFDM modulation consists of two submodules: the cfft_control and the cfft component. Both OFDM and FFT modules have normal and reversed operation. In order to achieve such behavior, addition data flow control modules are required in order to obtain proper sequence conversion component activation. This is the reason for the cfft_control component presence which sets the behavior of the cfft component. This means that through the cfft_control component, the inverse or normal fast Fourier conversion could be accomplished. The cfft component consists of the following submodules (Figure 13.4):

- RAM input–output
- MUX
- COMPLEX conjugate
- FFT–IFFT converter
- CORDIC
- Multiplication factor

The fast Fourier subcomponent is using the CORDIC [11] method of transformation. The fast Fourier algorithm is using N samples and thus N angles in trigonometric angle, so with proper CORDIC rotations we can modulate the samples in such a way that we can achieve only simple operations (for example shift bit and add/subtract operations). The rxmodem module is operating in an analogical way with the txmodem and it is driven by the OFDM and the output subsystem. The OFDM subsystem is operating as a demodulator and thus the flag Tx_nRx is set to 0. The demodulated data are driven in the output subcomponent. The output subcomponent is responsible for the proper transfer of the OFDM demodulated data to the QAM demodulator. The serparser component is responsible for converting data from serial to parallel data and reverse. The presence of this component is dealing with system's timing as the multiple conversion data is causing time delays which are useful for time management. The startup timer component is responsible for the system initiation and timing. Finally, the outconverter component implements useful functions that are memory related. The functions are mainly converters of the arithmetic value of the counter to memory addresses.

In hardware programming, XILINX ISE® [12] software was used to perform the implementation [13,14]. All the designs were targeted in two XtremeDSP Developments boards with Xilinx Virtex-4 XC4VSX35-10FF668 FPGA.

The XtremeDSP Development Kit-IV from Nallatech serves as an ideal development platform for the Virtex-4 FPGA technology and provides an entry into the scalable DIME-II systems.

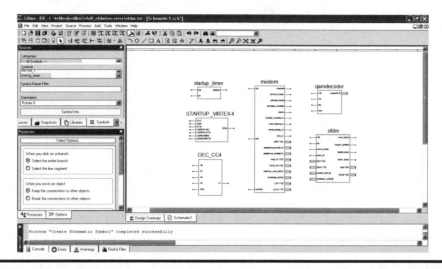

Figure 13.5   Xilinx ISE project navigator OFDM handwritten code receiver schematic symbols.

Its dual channel high-performance ADCs and DACs, as well as the user programmable Virtex-4 device are ideal to implement high-performance signal processing applications such as software-defined radio, 3G wireless, and networking [15]. Nallatech's FUSE System Software [16], is a Java-based application that provides configuration, control, and communications functionality between host systems and Nallatech FPGA computing hardware. This enables developers to design complex processing systems, with seamless integration between software, hardware, and FPGA applications. FUSE provides several interfaces, including the FUSE Probe Tool which is ideal for configuration of the FPGA. The XtremeDSP Development Kit-IV features three Xilinx FPGAs—a Virtex-4 User FPGA, a Virtex-II FPGA for clock management, and a Spartan-II Interface FPGA. The Virtex-4 device is available exclusively for user designs while the Spartan-II is supplied pre-configured with firmware for peripheral component interconnect (PCI)/universal serial bus (USB) interfacing. The Virtex-4 XC4VSX35-10FF668 device is intended to be used for the main part of a user's design; the Virtex-II XC2V80-4CS144 is intended to be used as a clock configuration device in a design. To connect the development platform at the computer, three options exists: the Joint Test Action Group (JTAG) interface, the USB, and finally the PCI interface can be used. At the beginning the design ran through synthesis, place, and route stages in ISE software, and then the programming bit file and the schematic symbols were generated. After that with FUSE software we located the XtremeDSP Motherboard through the USB interface and finally we assigned two bit files to the devices for the Virtex-4 and Virtex-2 configuration (Figure 13.5).

13.5 OFDM Design Using XILINX System Generator

The Xilinx System Generator is a tool for the design of DSP systems using the Simulink® environment of Mathworks for the design in FPGA [17]. The drawings are formed through modeling in the Simulink using the blocks availed of by the System Generator libraries. The System Generator allows full simulation of the designed system, control, evaluation of the results, and HDL code generation. All stages of implementation in FPGA are performed automatically, including the compilation, placement and routing, and generating an FPGA programming file.

The transmitter of the OFDM configuration system consists of the following parts:

- Signal generator
- TX controller
- QPSK mapping
- Pilot generator
- Subcarriers union
- Preamble
- IFFT and CP

The receiver of the basic OFDM configuration system consists of the following:

- RX controller
- Remove CP
- FFT
- Subcarriers disunion
- Pilot extractor
- QPSK demapping

The transmitter of an OFDM system consists of the blocks depicted in the following Figure 13.6. The first block of the OFDM system is the random binary digits generator that creates the data through which the entire system will operate. Data are generated serially and converted by the serial/parallel converter to parallel format. Then, there is configuration of the binary data by QPSK mapping, while at the same time there is importation of the zero subcarriers and the pilot generator generates the pilot subcarriers.

All subcarriers are integrated in the subcarriers union in order to be imported to the IFFT and CP block. The IFFT and CP block converts the symbols to the frequency field and adds the cyclical prefix—this creates the OFDM symbols. Preamble generates the short training symbols and long training symbols. Two MUXs blocks, MRe and Mim, sometimes allow the transmission of the preamble and at other times, the transmission of OFDM symbols. The data transmission rate is $\dfrac{1}{4\ \mu s} \times 96$ bits $= 24$ Mbps. The Tx controller controls the proper operation of the transmitter, organizing all the above.

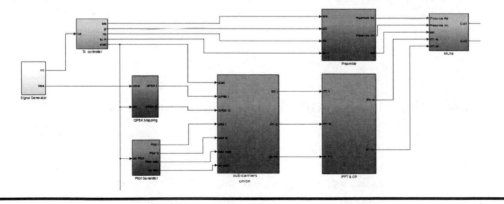

Figure 13.6 OFDM transmitter.

13.5.1 Transmitter Controller

The controller used on the Tx controller transmitter is implemented as a Moore machine which generates signals to guide the other blocks of the transmitter. The controller consists of an m-code and a counter. It is activated by an *init* signal and is guided by the counter, so as to control the stages of the Tx controller and thus activate the relevant signals each time. When *sts* is activated, that is, it changes from 0 to 1, then there is activation of the short training symbols. Upon activation of the *gi*, there is generation of the guard interval and the activation of the *lt* initiates the generation of long training symbols. As long as *sym* is 0, there is transmission of preamble symbols, but when *sym* turns to 1, then it allows the transmission of the OFDM symbols created. The controller status diagram is presented Figure 13.7.

13.5.2 Preamble

The preamble is the first part of each OFDM frame and the primary reason it is used is the synchronization of the transmitter with the receiver. Preamble accepts as an input signals *sts*, *gi*, *lts*, and *sym*. The four inputs of the block are activated in accordance with the controller. The outputs of the preamble are preamble *re* and preamble *im*, which export the real and imaginary components of each symbol respectively. The preamble is presented in more detail in Figure 13.8. For its implementation counters, ROMs and MUXs blocks were used. For the generation of the *sts* symbols, there was usage of two ROMs for the storage of real and imaginary values and a counter was used to read them. Moreover, the controller is used for the activation of the MUXs. The MUX3 and MUX5 are responsible for the selection of the appropriate symbol for transmission, while MUX4 is responsible for the selection of the counter of the symbol to be read.

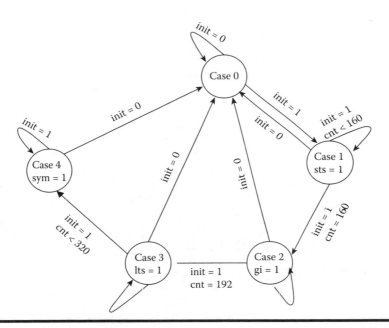

Figure 13.7   Transmitter controller status diagram.

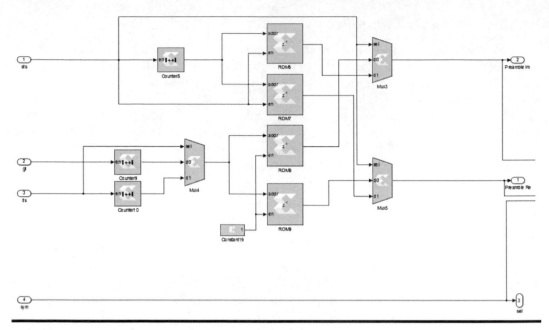

Figure 13.8 Preamble block implementation.

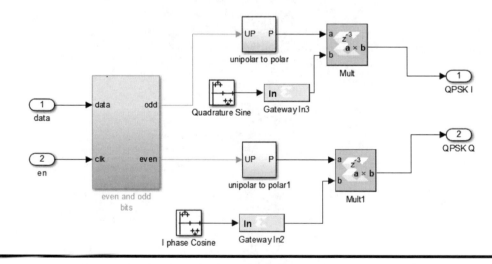

Figure 13.9 QPSK mapping implementation. (From Zhou, H., *Design and FPGA Implementation of OFDM System with Channel Estimation and Synchronization' Master thesis Montréal*, Québec, Canada, 2013, http://spectrum.library.concordia.ca/977626/1/Zhou_MASc_F2013.pdf)

13.5.3 QPSK Configuration

In QPSK mapping there is importation of the binary data to be configured. Initially the data are converted from serial to parallel format. Then the bits are converted to polar format and finally they are multiplied by a fixed value so as to be mapped as per the QPSK constellation. Figure 13.9 [18] presents the implementation of a QPSK configurator. Figure 13.10 shows the QPSK mapping waveforms.

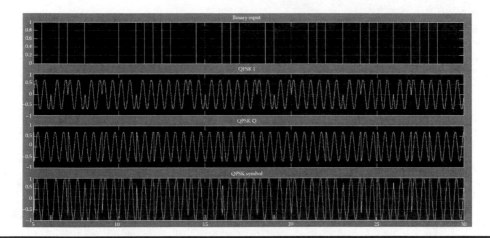

Figure 13.10 QPSK simulation.

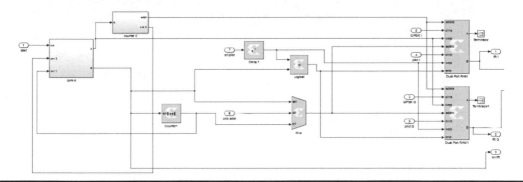

Figure 13.11 Subcarriers union implementation.

13.5.4 Pilot

The pilot generator block is used for importing pilot subcarriers. Pilot subcarriers can be used for the synchronization of frames, frequency synchronization, channel approach, and transmission function verification, while they can also be used for following the noise phase.

The pilot generator is implemented using three ROMs and a counter. The ROMs store the addresses where the pilot subcarriers must be placed, as well as the values (real and imaginary) to be generated. The counter is used for reading them. Pilot subcarriers generation starts when the pilot is activated by the controller.

13.5.5 Subcarriers Distribution

In the subcarriers union block there will be unification of the data, the pilot, and zero subcarriers. Then, they will be distributed on the IFFT input. The primary elements contained in the subcarriers union block are two dual port RAMs, two counters, and a controller. The dina ports of the RAMs are used for writing the subcarriers data, while addra ports are used for reading data through counter 0. The dinb ports are used for writing the pilot subcarriers, while addrb ports are used for importing their locations. Moreover, addrb ports are used for reading the

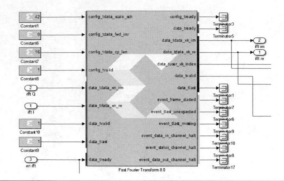

Figure 13.12   Fast Fourier 8.0.

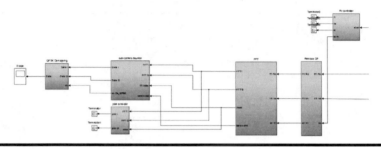

Figure 13.13   OFDM receiver.

64 subcarriers through counter 1. The zero subcarriers are placed concurrently to the reading of the data of the subcarriers and their placement depends on the determination of the content achieved by counter 0. The implementation is depicted in Figure 13.11.

13.5.6 IFFT and CP Addition

Using IFFT and CP blocks, there is conversion of the entire spectrum required to a temporal signal using IFFT. IFFT transformation is performed quite effectively and ensures the orthogonality of all generated signals of the subcarriers. For the IFFT transformation of the input data, there was usage of the fast Fourier transform 8.0 block which is available in the Xilinx blockset library. FFT 8.0 effectively implements the IFFT transformation of 64 points at the input data. Figure 13.12 shows the fast Fourier 8.0 block used for the transformation of the signal. In order to add the cyclic prefix to the OFDM, it is necessary to select *cyclic prefix insertion* in the block parameters of FFT 8.0.

13.6 Receiver Implementation

The receiver performs the inverse procedure of that of the transmitter. Initially, from the OFDM symbol of 80 samples received, there is deduction of the first 16 samples of the cyclical prefix. The remaining 64 samples will be placed at the input of FFT 8.0 in order to be converted from the time field to the frequency field. Then a subcarriers disunion block will remove zero subcarriers and separate pilot subcarriers from data carriers. Finally, data carriers will be demodulated in the QPSK demapping block. In order to check the proper receiver function, an Rx controller is used (Figure 13.13).

13.6.1 Receiver Controller

The receiver controller—Rx controller—is similar to that of the transmitter. It is implemented as a Moore machine and generates signals to guide the remaining blocks of the receiver. The receiver consists of an m-code and two counters. It is activated by a start signal and is guided by the two counters. At each of the controller stages, the necessary output signal is activated in order to achieve efficient receiver control.

13.6.2 FFT and CP Deduction

The real and imaginary symbols received by the receiver will be imported to the FFT 8.0 block for the decoding of the signal, using Fourier transformation (FFT). The transformation must be performed solely on the 64 desired samples of the OFDM symbol. Therefore, out of the 80 samples received, there will be deduction of the cyclical prefix of 16 samples. This is achieved using two registers before the FFT 8.0 block inputs where the real and imaginary OFDM symbols are to be placed. The two registers are activated according to the *sym* signal of the controller (Figure 13.14).

13.6.3 Subcarriers Redistribution

Following the decoding of OFDM symbols by FFT, there is redistribution thereof by the subcarriers disunion block. The subcarriers disunion exports the data containing the information separating them from the pilot subcarriers and deducting the zero ones. For its implementation, there was usage of RAMs, two counters, and a controller.

13.6.4 QPSK Demodulation

During the demodulation procedure, there is demodulation of the configuration symbols. At this stage there is performance of the reverse process from that of the modulation, in accordance with the QPSK constellation used on the transmitter, that is, it converts the complex information to bits. For the conversion to binary data, values I and Q are compared with threshold 0. Then, they are converted from parallel to serial format.

13.7 System Simulation Results

Figure 13.15 presents the System Generator model implemented for the simulation process.

The System Generator symbol allows selecting the FPGA implementation device, as well as the type of translation to machine language, depending on the programming file generated. At the end of the process, System Generator creates the files (netlist) necessary for the implementation of the programming of the default device materialized through ISE.

The use of the Xilinx ISE tool achieves the generation of the bitstream file from the VHDL code of the system. The FPGA development kit achieves a top frequency of 105 MHz which means tclk = 9.5 ns. Implementation results are shown in Table 13.2 which lists speed and resource usage (in terms of the building blocks of an FPGA): (a) period, the critical path of the circuit in ns, (b) LUT, the number of 4-input look-up table function generators required, (c) FF, the number of 1-bit flip-flops required, and (d) slice, the number of FPGA slices required. The transmitter implementation with VHDL code occupies a total of 1113 slices which means that the device utilization is 7%, 5% of the slice flip-flops, and 6% of the dedicated LUTs. The implementation of the

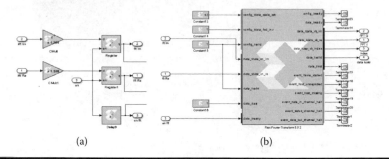

| (a) | (b) |

Figure 13.14 **(a) Remove CP block and (b) FFT block.**

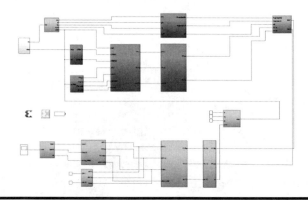

Figure 13.15 **Screenshot of the entire model.**

Table 13.2 Utilization Summary of the Entire System

	Period	LUT	FF	Slices
OFDM VHDL TRANSMITTER	2824 ns	1817	1668	1113
OFDM VHDL RECEIVER	2939 ns	1946	1735	1281
OFDM S.G. TRANSMITTER	3145 ns	2345	2034	1854
OFDM S.G. RECEIVER	3.256 ns	2454	2249	688

receiver is a lot more complex than that of the transmitter since it requires more computations, use of larger arrays of data and the storing of these data to more memory blocks. However, the device utilization is kept at reasonable rates; this is a crucial part of area-restricted applications, such as the implementation on an embedded system. The receiver occupies a total of 1.281 slices available on the FPGA, meaning that the device utilization is at 9% in terms of occupied slices, 6% of the slice flip-flops, and 7% of the dedicated LUTs. System Generator transmitter and receiver implementation occupies many more slices, flip-flops, and LUT than the handwritten design. In System Generator the main advantage is that hardware cosimulation entails automatic generation of an FPGA bitstream from Simulink, as well as its incorporation back into Simulink itself. This allows the user to exploit FPGAs to significantly accelerate simulation, while also providing the ability to validate a working design in hardware, all without necessarily having to invoke a traditional FPGA tool explicitly. The main drawback is the increase of the sources in the FPGA—something really important to the DSP design industry.

13.8 CONCLUSIONS

In the context of the present chapter, there was theoretical analysis, design, and implementation in FPGA of an OFDM system with QPSK configuration based on IEEE 802.11p PHY. On the system transmitter there was application of IFFT, while on the receiver there was application of FFT. The selection of IFFT/FFT instead of the discrete Fourier transformation (DTF) is attributed to the improved speed offered, supported by the lower computing complexity.

In order to assess the correctness of the results of the system designed, the model was initially implemented through handwritten VHDL code and afterward in the System Generator environment, where, through appropriate simulation, its behavior was observed. Then the OFDM transceiver was implemented on Virtex-4 XC4VSX35-10FF668 where the above results were confirmed. During the implementation process there was an assessment of system requirements in terms of surface and speed. From the results, it was derived that the overall system occupied 6% of the registers, 7% of the LUTs, and 9% of the slices of the FPGA in the first design (receiver) and increased values in the System Generator design.

Taking everything into consideration, the results ensure that the hardware design was correctly working when implemented in the FPGA; the handwritten design tends to outperform the other design in terms of sources but it needs much more working hours.

ACKNOWLEDGMENTS

The author would like to thank Xilinx and their university program representatives for their contribution by providing us the Xilinx OFDM Library v1.0 FEC Blockset and WIMAX 802.16-2004 demonstration design and collateral.

References

1. ASTM E2213-03 Standard Specification for Telecommunications and Information Exchange Between Roadside and Vehicle Systems—5 GHz Band Dedicated Short Range Communications (DSRC) Medium Access Control (MAC) and Physical Layer (PHY) Specifications. https://www.astm.org/Standards/E2213.htm.
2. IEEE 802.11p/D3.0, "Draft Amendment to Standard for Information Technology-Telecommunications and Information Exchange between Systems—Local and Metropolitan Area Networks Specific Requirements—Part 11: Wireless LAN Medium Access Control (MAC) and Physical Layer Specifications—Amendment 7: Wireless Access in Vehicular Environment" 2007. https://www.ietf.org/mail-archive/web/its/current/pdfqf992dHy9x.pdf
3. Berkeley Wireless Research Center "OFDM Introduction" http://bwrc.eecs.berkeley.edu/classes/ee225c/Lectures/Lec16_OFDM.pdf, May 2014.
4. http://www.radio-electronics.com/info/rf-technology-design/digital-signal-processing/dsp-basics-tutorial.php, Radio-Electronics.com is operated and owned by Adrio Communications Ltd.
5. F. Taylor, *Digital Filters: Principles and Applications with MATLAB* (Introduction to Digital Signal Processing), Wiley-IEEE Press. The Atrium Southern Gate, Chichester.

6. H. Zhou, 'Design and FPGA Implementation of OFDM System with Channel Estimation and Synchronization', MSc Thesis in The Department of Electrical and Computer Engineering, Concordia University.

7. S. Haykin and M. Moher, *Communication Systems*. Wiley, United States.

8. T.S. Rappaport, 'Inter-carrier interference cancellation for OFDM systems', EE 381K-11: Wireless Communications. The University of Texas Department of Electrical and Computer Engineering, May 6, 2013, https://www.ece.utexas.edu/graduate/courses/ee-381k-11, (Accessed date May 2013).

9. Hongyan Zhou 'Design and FPGA Implementation of OFDM System with Channel Estimation and Synchronization' Master thesis Montréal, Québec, Canada June 2013, http://spectrum.library.concordia.ca/977626/1/Zhou_MASc_F2013.pdf

10. G. C. Kiokes, A. Amditis, N. K. Uzunoglu, Performance evaluation of OFDM - 802.11p system for vehicular communications, *Journal IET Intelligent Transport Systems,* 3(4), 429–436, December 2009.

11. S. Eichler, Performance Evaluation of the IEEE 802.11p WAVE Communication Standard, Institute of Communication Networks, *Technische Universität München, Vehicular Technology Conference,* 2007. VTC-2007 Fall. IEEE, pp. 2199–2203, 66th September 30–October 3, 2007.

12. C-Y. Yu, S-G. Chen and J-C. Chih, Efficient CORDIC Designs for Multi-Mode OFDM FFT, *31st IEEE International Conference on Acoustics, Speech, and Signal Processing (ICASSP)*-Toulouse, France, pp. 1036–1039, May 14–19, 2006.

13. http://www.xilinx.com/products/design-tools/ise-design-suite.html, San Jose, CA.

14. University of Pennsylvania Digital Design Laboratory, Introduction to Xilinx ISE 8.2i http://www.seas.upenn.edu/~ese201/ise/ISEIntroduction.pdf, June 2008.

15. Nallatech XtremeDSP Development kit-iv user guide http://www.xilinx.com/support/documentation/boards_and_kits/ug_xtremedsp_devkitIV.pdf, June 2009.

16. FUSE System Software User Guide, http://www.nallatech.com/indexx.php/product- briefs.html

17. Xilinx System Generator, http://www.xilinx.com, June 2009.

18. 'Digital Modulation in FPGAs Xilinx using system generator (ASK, BPSK, FSK, OOK, QPSK)' Diego Barragán (Accessed date July 2008) http://www.mathworks.com/matlabcentral/fileexchange/14650-digital-modulation-in-fpgas-xilinxusing-system-generator--ask--bpsk--fsk--ook--qpsk

Chapter 14

A Novel Comparative Study of Different Coding Algorithms and Implementation Issues through FPGA Technology for Wireless Vehicular Applications

George Kiokes

Hellenic Air Force Academy
Dekeleia, Greece

Contents

14.1 Introduction

It is known that communication systems performance is limited by the available signal power, the unavoidable noise levels, and the need to reduce bandwidth. The significance of the information rate is that it introduces us to a theorem attributed to Shannon, which is deemed fundamental in communications theory. This theorem refers to the information transmission rate over communication channel. Although the term *communication channel* has been used in various circumstances, it includes all attributes and constituent parts of a transmission system, which introduce noise, or reduce bandwidth. Shannon's theorem states that it is possible to devise a means through which a system can transmit information with an arbitrarily low error probability, given that information rate R is less than, or equal to C rate, which is called channel capacity. The technique used to approach this threshold is called channel coding.

In order to optimally express the above statements, the following is provided:

Assume a source M of equiprobable messages, where $M \gg 1$, which generates information at rate R. Assume a channel with capacity C. Then, if

$$R \leq C$$

then, there is a channel coding technique such that the source output can be transmitted through the channel with an arbitrarily low error probability in the receiver's message [1].

Channel coding achieves its objective through the intentional introduction of the concept of redundancy in the messages (Figure 14.1). The deliberately introduced redundancy allows us to identify the occurrence of an error and even correct it. However, the introduction of redundancy cannot be deemed a guarantee that an error would be either detectable or correctable, because errors are caused by an unpredictable, random process called noise. As a consequence, though

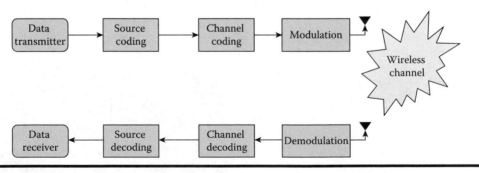

Figure 14.1 Introduction of redundant information in a "structured" manner so as to treat more efficiently any distortions introduced by the channel.

noise level may be fairly low, so that it is impossible to encounter more than one error, there is always a finite chance for two errors to occur. In such a case, we would know that an error has already occurred but we would have the tendency to read 1 as 0 and 0 as 1. Even more, there is always the possibility, albeit low, for three errors to occur. In this case, we would read erroneously a digit and, moreover, we wouldn't even suspect that an error has occurred. Therefore, it is concluded that though coding allows us to assume detection and correction to a great extent, it is normally not able to detect, or correct all errors.

There is a correspondence between the redundancy intentionally introduced in the coded message prior to its transmission through a channel, and the redundancy that is part of the language.

Channel coding (Figure 14.2) expediency lies in the fact that it allows us to increase the rate at which information can be transmitted through a channel, while retaining the error rate at a predefined value. Alternatively, coding allows us to reduce the erroneous information transmission rate, while retaining a constant, specific transmission rate. More generally, coding allows the design of a communications system in which both the data transmission rate and the error rate are independent and arbitrarily set, but also subject to bandwidth limitations. This requires less transmission power to achieve the same error probability. This power reduction (in dB) is called coding gain. The "price" paid for our desire to approach Shannon threshold to the greatest extent possible, is the increased complexity of the hardware both on the transmitter, where encoding takes place, and the receiver, where decoding is performed. Theoretically, smart enough coding and unlimited complexity would allow us to reach Shannon threshold, which means that we would be able to transmit, within channel capacity at an error rate that could be as low as desirable. A measure of the effectiveness of a code is precisely the extent to which it allows us to approach Shannon threshold.

14.2 Error Correction Codes

Error correction codes are distinguished in two large categories: block codes and convolutional codes. This differentiation is related to the mechanism used for the realization of the encoding process.

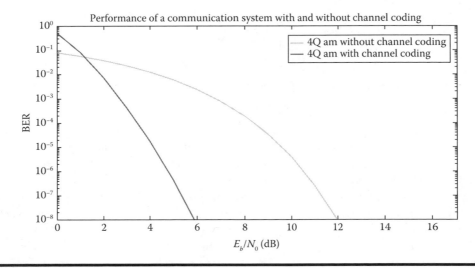

Figure 14.2 Assessment of the performance of a communication system with and without channel coding.

14.2.1 Block Codes

Assume that a source can generate M equipossible messages. Then, each message is initially represented with κ bits as $2^\kappa = M$. These k bits are those carrying the information. Then, to each k bit message we add r, redundancy bits (parity check bit). Hence, each message is expanded to a coded word of n bits, where

$$n = k + r \tag{14.1}$$

with the total number of possible coded n-bit words being 2^n while the total number of messages possible is 2^k. Therefore, there are $2^n - 2^k$ possible words of n bits, which do not represent possible messages.

The codes formulated through the reception of a block of k information bits and adding $r = n - k$ redundant bits to form a coded word are called block codes and indicated as (n,k) codes. When a coded word of n bits consists of k information bits and r redundant bits, then the code is called systemic. A nonsystemic code presents n bits of coded word and k information bits which are not precisely represented in the coded word.

If the coded word of n bits is intended to be transmitted in a time period that does not exceed that required for the transmission of the k information bits, and if T_b and T_c are the durations of the bits both in the noncoded and the coded word, then the following must apply

$$nT_c = kT_b \tag{14.2}$$

the code rate is set to

$$R_c = k/n \tag{14.3}$$

According to $f_b = 1/T_b$ and $f_c = 1/T_c$, we get

$$\frac{f_c}{f_b} = \frac{T_b}{T_c} = \frac{n}{k} = \frac{1}{R_c} \tag{14.4}$$

14.2.2 Low-Density Parity Check Codes

The low-density parity check codes (LDPC) [2], constitute a category of *linear block codes*. These codes, as well as the relevant repetitive coding algorithm, were proposed by R. G. Gallager in 1960 in his thesis, yet, they were not exploited until the beginning of the 1990s. The reason for which LDPC codes were rather set aside was the exceptionally large, for the time, computing costs involved, as the computers of that time were not able to cope with the complexity of the algorithm on which LDPC codes were based. An exception to the above is Layer's work in 1981, who introduced LDPC codes representation using graphs, called *Tanner graphs* or, otherwise, *bipartite graphs*.

In general, any code used for channel coding is a function or, similarly, a one-to-one representation of elements from a set A to a set B. Set A consists of the elements called information words and is denoted as u_i where $i = 1,2,...,M$, while set B consists of the elements called code words and is denoted as c_i. Without prejudice to generality, it is assumed that all information words have a fixed length of k digits. The possible number of words is $M = 2^k$ (it is assumed that the words are binary, that is, digits can assume value of 0 or 1). Given that the code was defined as a one-to-one representation of elements, it arises that the number of code words is also M. In general, however, code words do not have the same length as the information words, but they are longer, say n, as there has been addition of some redundancy bits. Therefore, only one part (M to be precise) of the 2^n possible words $2^n > 2^k$, considering that $n > k$ are code words, are called valid code words. Linear block codes are called the codes for which the sum of any valid code words is also a code word.

The channel encoder is fed by the source encoder output, which generates symbols at a fixed rate. Hence, at the source encoder output there is a constant data flow. However, for the function of a linear block encoder, this flow must be segmented in packets or fixed-length words, and then must be processed by the encoder, while for other encoder types this precondition is not required.

It is assumed that the length of the input word at the encoder is k bits while the length of the output word is n bits, where $k < n$.

The conversion of the initial k bits word (information word) to an n bits word (code word) is the encoding process. This is materialized with the help of a $k \times n$ binary matrix G, which is called a code generator matrix. The code word is generated through the multiplication of the information word with the generator matrix, which is expressed using the following formula:

$$c_i = u_i \otimes G$$

Each code word, c_i, due to its method of generation described above, has some special characteristics. These can be characterized as the satisfaction of a series of equations which equivalently determine which of the 2^n possible words of n length is a valid code word. During composition of all these equations which must be satisfied by the code words, there is creation of a matrix called a parity check matrix, denoted by H. The validity check of a code word is now performed using the equation:

$$c_i \otimes H^T = 0$$

The repetitive process to find a code word meeting the above equation is the decoding process, which requires much greater computing complexity.

The representation of these LDCP codes can be performed in two ways, either through the check matrix or using certain graphs representing message transmission and information exchange among the processing units of the code. The two representations are described below in detail.

In the code representation using a parity check matrix, which can determine precisely the specific code, there is a binary matrix of size $((n - k) \times n)$ characterized by the fact that it contains very few elements that are equal to one. That is, it is sparse. An example of such a matrix is depicted in the following matrix:

$$H = \begin{bmatrix} 1 & 1 & 1 & 0 & 0 & 0 & 0 & 0 & 0 \\ 1 & 1 & 0 & 1 & 0 & 0 & 0 & 0 & 0 \\ 0 & 0 & 1 & 1 & 1 & 0 & 0 & 0 & 0 \\ 0 & 0 & 0 & 0 & 1 & 1 & 0 & 1 & 0 \\ 0 & 0 & 0 & 0 & 0 & 1 & 1 & 0 & 1 \\ 0 & 0 & 0 & 0 & 0 & 0 & 1 & 1 & 1 \end{bmatrix}$$

Each line of matrix *H* corresponds to a parity check equation, and the unit, on position (i, j) of matrix *H*, indicates that *j* at the data symbol participates in the *i*-th parity check equation.

Two important sizes that should be considered in a parity check matrix are the number of nonzero elements on each line of the matrix, w_r, and the number of nonzero elements in each column of the matrix, w_c. These sizes are called line score and column score, respectively. In order to be able to characterize a matrix as low density, conditions $w_c \ll n$ and $w_r \ll m$ must be met. An LDPC whose parameters w_c, w_r are constant for all matrix columns and lines is called regular. However, there are codes for which this does not apply. These belong to the category of irregular LDPC codes.

The second way of representing LDPC codes was proposed by Tanner [3], who introduced Tanner graphs which consists of a graphical representation. Tanner graphs, in addition to the fact that they provide a full description of the code, also aid in the explanation of the encoding algorithm used. Tanner graphs belong to the category of bipartite graphs. A graph is called bipartite when its nodes are divided into two groups and groups of different nodes can be joined together. The two types of nodes existent in a Tanner graph, for LDPC codes, are called variable nodes, usually referred to as VPUs, and check nodes, usually referred to as CPUs.

The category of decoding algorithms used to decode LDPC codes is called message-passing algorithms as their function can be described by the successive exchange of messages between the two node types of the decoder, the check and variable nodes. This way, having an initial estimate of the information bits from the channel and performing certain calculations on these nodes, the decoder improves its estimate having always as a criterion the satisfaction of all parity check equations.

An LDPC decoder (Figure 14.3) can be implemented in series, in fully parallel mode, or partially parallel mode. The number and connections between the processing units determine the parallelism of the eventual architecture, which constitutes a regulatory factor in the trade-off between the required area, decoding speed (throughput), energy consumption, and flexibility [4]. One of the great advantages of LDPC decoders, which are deemed exceptionally adequate for hardware implementations, is the ability to select the parallelism degree depending on the requirements of the eventual application.

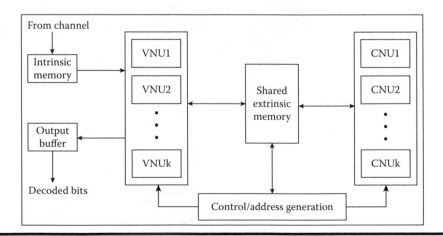

Figure 14.3   General decoding process diagram.

14.2.3 Convolutional Codes in Parallel Connection

Turbo codes form a special category of chain codes containing a degree of coupling placed between two parallel or serially connected encoders. Using these codes, it is possible to approach Shannon threshold even up to 0.7 dB for small signal-to-noise ratios (SNR). They were first discovered by Berrou, Glavieux, and Thitimajshima in 1993 [5]. A classic turbo encoder layout consists of two individual, convolutional encoders of 1/2 rate, which are distinguished between them by an N-bit interleaver. Figure 14.4 describes such a coding layout. The interleaver (Π) is a function which receives a structure of N data, rearranges it on the basis of a matrix and feeds it to the second recursive systematic code (RSC). The two key requirements for the appropriate and effective operation of an interleaver is, first, to rearrange a large number of data and, second, such rearrangement to take place as randomly as possible. Data number, N, is called information data length. The interleaver protects the telecommunications system from burst errors caused by the channel, because, in essence, it decorrelates the input data. Hence, it provides the ability to create (with high probability) different sequences of parity bits and the code word originates from two distinct paths in each of the two RSC decoders.

The code consists of two constituent codes divided by an N-long coupler. These constituent codes are normally recursive, systemic, convolutional codes of 1/2 rate, and usually the same code is used on both constituent codes. Recursive, convolutional codes differ from the nonrecursive ones in the existence of feedback in their implementation using shift registers. Consequently, contrary to nonrecursive convolutional codes, which are made as finite impulse response digital filters, the recursive codes are infinite impulse response filters. Through the introduction of RSC codes to turbo codes, we can increase the maximum minimum code distance, thus achieving enhanced correction ability.

In the encoder structure of Figure 14.4, I information bits enter the first constituent encoder. The same information bits are coupled and coupled again in the second encoder. Given that the encoders are systemic, each one generates the I information bits applied at the input, followed by C parity check bits. Following encoding, the I information bits and the C_1 and C_2 parity check bits enter a multiplexer and in the end, a total of $IC_1 C_2$ bits are transmitted through the channel. The total rate is $K = 1/3$. Through puncturing, a rate of 1/2 can be achieved [6].

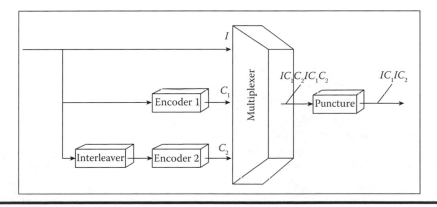

Figure 14.4 Block diagram of a turbo encoder with puncturing.

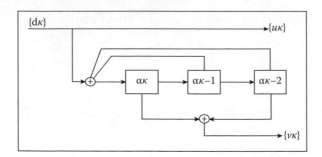

Figure 14.5 **Recursive systematic convolutional encoder. (From Sklar, B.,** *Fundamentals of Turbo Codes,* **http://ptgmedia.pearsoncmg.com/images/art_sklar3_turboc odes/elementLinks/ art_sklar3_turbocodes.pdf.)**

14.2.4 Turbo Encoder Implementation

In the general case scenario, the output of a nonsystemic convolutional encoder with rate ½ and control length *K* can be described using the following expressions:

$$u_k = \sum_{i=0}^{K-1} g_{1i}d_{k-i} \mod 2 \tag{14.5}$$

$$v_k = \sum_{i=0}^{K-1} g_{2i}d_{k-i} \mod 2 \tag{14.6}$$

where $G_1 = \{g_{1i}\}$ and $G_2 = \{g_{2i}\}$ the connection polynomials for the code. Based on this code, it is possible to construct the RSC as the coded information recursively enters the encoder input continuously (Figure 14.5) [7].

The recursive code shall be described with the expression

$$a_k = d_k + \sum_{i=0}^{K-1} \gamma_i \alpha_{k-1} \mod 2 \tag{14.7}$$

Where $\gamma_i = g_{1i}$ if $u_k = d_k$ and $\gamma_i = g_{2i}$ if $v_k = d_k$

14.2.5 Turbo Decoder

The turbo decoding mechanism is based on the structure of repetitive decoding. The point in which turbo codes differ compared to all other code types is their repetitive decoding method. This means that, while the decoder produced before a final hard-decision in its output for each symbol, in the case of turbo codes the information retrieved during one repetition of the algorithm must be used by the next repetition so as to improve the reliability of the final decision. Hence, the decoding of turbo codes requires an algorithm generating soft decisions. Such an algorithm is BCJR [8]. The repetitive decoding includes partial decoding and feeding of the results to the next tier, creating a form of feedback and this process continues repetitively. This reduces complexity but increases decoding delay. Hence, the use of information extracted from one stage by the next one improves the decoding quality and the result is stabilized very close to the optimum decision after the lapse of a specific number of repetitions. Figure 14.6 depicts a block model of

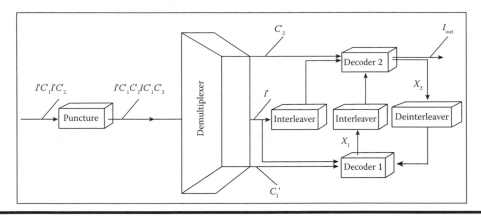

Figure 14.6 Systematic turbo decoder.

a turbo decoder. The serial data, after crossing the reverse puncturing path, are routed to the demultiplexer and then through the demultiplexing tiers we get the initial data following a finite number of repetitions.

Initially, the systematic information flow enters the first decoder and with the help of the first parity wave there is execution of algorithm BCJR, which generates at its output the probabilities of both the systematic information bits and the parity bits probabilities. The latter constitute the extrinsic information, which is used as a priori information for the second decoding. During the second decoding, the systematic information flow enters the decoder again, along with the parity bits of the second decoder.

The performance limits of turbo codes were developed assuming trellis codes, the trellis length of which is nonfinite. In a system using systematic convolutional encoders and then the symbols flow enters a BPSK modulator, the following shall apply on the decoder output

$$v_t = \left[u_t^{(0)}, u_t^{(0)} \right] = \left[x_t, u_t^{(1)} \right] \tag{14.8}$$

The modulated signals shall be

$$a^{(p,q)} = \left[a_t^{(0,p,q)}, a_t^{(1,p,q)} \right] =$$
$$= \sqrt{E_C} \left[2x_t^{(p,q)} - 1, 2u^{(1,p,q)} - 1 \right] \tag{14.9}$$

The signal received at the receiver shall be

$$r_t = \left[r_t^{(0)}, r_t^{(1)} \right] = \left[a_t^{(0)} + n_t^{(0)}, a_t^{(1)} + n_t^{(1)} \right] \tag{14.10}$$

The probability of the systematic bit shall be

$$P(x_t = x) = N_x p\left(r_t^{(0)} \mid x_t = x \right) P(x_t = x) \bullet$$

$$\bullet \left[\sum_{(p,q) \in S_x} a_t(p) p\left(r_t^{(1)} \mid a_t^{(1)} = a^{(1,p,q)} \right) b_{t+1}(q) \right] = \tag{14.11}$$

$$= N_x P_{s,t}(x) P_{p,t}(x) P_{e,t}(x)$$

where

$$P_{s,t}(x) = p\left(r_t^{(0)} \mid x_t = x\right)$$

denotes the systematic probability,

$$P_{p,t}(x) = P(x_t = x)$$

indicates the a priori systematic probability, and

$$P_{e,t}(x) = \sum_{(p,q)\in S_x} a_t(p)\, p\left(r_t^{(1)} \mid a_t^{(1)} = a^{(1,p,q)}\right)\beta_{t+1}(q)$$

indicates the extrinsic probability.

A noteworthy point of turbo codes that should be mentioned is the region in which the value of bit error rate (BER) is saturated in terms of SNR. Indeed, while there is an initial abrupt slope of the characteristic curve of the BER rate for small SNR values, in a demonstration of the exceptional performance of the code, through the increase of the SNR value, this characteristic curve follows in an asymptotic manner the characteristic curve of the asymptotic free distance and there is observation of saturation in the BER value achieved. This region is called the error floor-flare region, and code performance is limited by its free distance.

14.3 Design of the V2V System

This chapter describes the steps followed for the design of the V2V system, that is, the transmitter and the receiver. System implementation was performed in a simulation environment, always in accordance with the requirements and specifications introduced by standard 802.11p [9–12]. It should be noted that the system is able to function in eight different modes, depending on the modulation type and, as a consequence, depending on the coding assigned (code and modulation). The key characteristics of the standard are presented in Table 14.1 [13].

14.3.1 IEEE 802.11p Transmitter Implementation

This section describes all successive steps performed by the transmitter before the commencement of data transmission to the wireless channel.

14.3.2 Information Source—Bus Coding

As described in the standard, the information bits must be randomized prior to their transmission. In this model, instead of performing this randomization process, there is usage of a ready block available in the software library. Data are randomly generated by the block of the random Bernoulli Binary Generator. There is usage of 480 samples per time unit for a data transmission rate of 3.0 Mbps and for a number of 20 OFDM symbols. The OFDM symbol time is 8.0 μs as derived from the standard, while the sample time value is 3.3333×10^{-7} sec in the case of QPSK modulation. This way there is parallel generation of the data to be transmitted and they are also

Table 14.1   802.11p Standard Main Characteristics Values

Parameter	Value
N_{SD} : Number of data subcarriers	48
N_{SP} : Number of pilot subcarriers	4
N_{ST} : Number of subcarriers, total	52 ($N_{SD} + N_{SP}$)
IFFT/FFT size	64
T_{FFT} : IFFT/FFT period	6.4 μs(1/ΔF)
$T_{PREAMBLE}$: PLCP preamble duration	32 μs ($T_{SHORT} + T_{LONG}$)
T_{GI} : GI duration	1.6 μs (T_{FFT}/4)
T_{GI2} : Training symbol duration	3.2 μs (T_{FFT}/2)
T_{SYM} : Symbol period	8.0 μs ($T_{GI} + T_{FFT}$)
T_{SHORT} : Short training sequence duration	16 μs (10 × T_{FFT}/4)
T_{LONG} : Long training sequence duration	32 μs (T_{GI2} + 2 × T_{FFT})
Data rate	3, 4.5, 6, 9, 12, 18, 24, 27
Modulation	BPSK, QPSK, 16QAM, 64 QAM
Convolutional code	K = 7 (64 states)
Coding rates	1/2, 2/3, 3/4
Channel spacing (MHz)	10
Signal bandwidth (MHz)	8.3
Subcarrier spacing (KHz)	156.2

randomized already. Following the generation of the data by the generator, the data are imported in a bus encoding subsystem. There, they will be initially encoded, sometimes by a turbo encoder and sometimes by an LDPC. This encoder has a fixed coding rate equal to ½ for all code and modulation cases. After the encoder the puncturing process is performed. This process aims at increasing the fixed encoding rate so as to achieve the desired total encoding rate of ½. Data interleaving is used to scatter the error bursts and thus increase the effectiveness of the forward error correction (FEC) previously used.

14.3.3 Symbol Modulation

The stage following channel coding is symbol modulation. All wireless communication systems use a modulation technique to match the encoded bits to a structure easily transmitted over a communications channel. Hence, bits are matched to amplitude and phase carrier represented by a complex in-phase and quadrature-phase vector. During modulation the binary bits sequence is converted to a sequence of complex symbols ($I + jQ$). In particular, the input bits are divided in groups of N_BPSC (1,2,4,or 6) bits and are converted to complex numbers representing points of the BPSK, QPSK, 16-QAM, or 64-QAM schemes. In this comparison, QPSK was used.

14.3.4 Creation of the OFDM Symbol

This complex subsystem generates the OFDM symbols which are transmitted. It consists of many individual blocks which aim at creating an OFDM symbol the best way possible, as per the standard. The physical OFDM level of the standard defines that data transmission is to be performed using 64 carriers. This total number of carriers is defined by the number of points required to execute the IFFT algorithm. In order to successfully generate the OFDM, the relation between all carriers must be carefully examined in order to maintain their orthogonality. Therefore, the required spectrum derived from modulation is then transformed to a temporal signal using an inverse Fourier transformation. IFFT executes this transformation quite effectively and ensures that the signals of the carriers generated are orthogonal. The final stage prior to the completion of the OFDM symbol creation is the addition of the cyclic prefix. In case of addition of a cyclic prefix an initial part of each OFDM symbol is repeated at its end, extending thus its length. Finally, parallel data are then converted to serial again and these are the transmitter output, ready to be transmitted via the wireless bus.

14.3.5 IEEE 802.11p Receiver Implementation

The receiver follows the opposite logic of the transmitter. The signal received is the sum of the copies of the initial signal with a temporal delay due to the multiple paths, along with the AWG noise. Initially, after the serial data become parallel, the cyclic prefix that was inserted to the transmitter is removed and thus intrasymbol interference is treated.

The system receiver undertakes the task to extract from the signal received, the signal transmitted from the transmitter. Then, at the OFDM symbol received, the FFT is applied to acquire the symbols to be introduced to the M-QAM decoder. Before the decoder and prior to the transformation the signals are converted to frequency signals through adaptive modulation, which corrects and estimates the signal as close as possible to the transmitted signal. The adaptive modulation of the channel does not include an estimate of the noise introduced to the signal. The M-QAM decoder that follows transforms the complex information to bits. The bits are rearranged through the deinterleaver and are decoded through a Viterbi decoder.

Finally, the bits (and packets) transmitted and the bits (and packets) received are compared to deduce the number or erroneous bits (and packets) and the respective BER and packet error rate. The latter are a particularly useful tool in the assessment of the performance of the telecommunications system, for many of its modulations (number of subcarriers, cyclic prefix length, encoding rate, etc.).

14.4 IEEE 802.11p System Simulation

In order to examine the characteristics and performance of the systems, there was performance of many simulations in the MATLAB® environment using the BER tool. DSRC technology refers to communications between vehicles and among vehicles and units at the side of the road. The radio channel comprises LOS and NLOS paths. The term LOS path refers to the situation where we have visual line of sight contact between the transmitting antenna and the receiving antenna. NLOS path is a term used to describe radio transmission across a path that is partially obstructed and there is no eye contact between the vehicles. For the NLOS path, it is found that the amplitude of

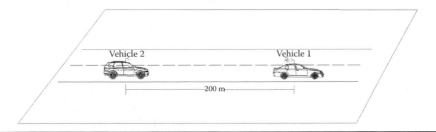

Figure 14.7   Simulation scenario.

the received signal due to fading obeys a Rayleigh fading distribution and the delay spreads were several times greater than that for LOS [14–16].

The modulation and coding rate used were in accordance with ASTM E2213-03 standard [17]. The total signal bandwidth was 10 MHz divided to 64 sub-buses. Each of the 64 subcarriers is assumed to carry information. The sizes examined during the simulations were the BER, the number of errors, the number of bits and lastly the E_b/N_0 ratio. This ratio is more appropriate than the SNR, as it provides a measure for the comparison of systems using different modulation schemes. In order for the results of the sizes examined to be reliable, there was performance of successive repetitions of data packet transmissions until certain statistical sizes converged.

The coupling characteristics are described below:

- Tx power: 18 dBm
- Carrier frequency: 5.9 GHz
- Antenna type: omnidirectional
- Antenna location: on the roof of the vehicle
- Polarization: vertical
- Antenna height from ground: 1.8 and 1.50 m
- Vehicle 1 speed: 120 km/h
- Vehicle 2 speed: 70 km/h

During simulation, there was no consideration of the effect of other vehicles, or obstacles Figure 13.7. In order to examine the performance of turbo code in the standard, a recursive systematic, convolutional code was used in the standard; this had a control length of 3 and connection polynomials 5_{OCT} and 7_{OCT}. Given that the parallel concatenation of two such encoders in a turbo coding schema generates a code with rate 1/3, and in order to compare it with codes having rate 1/2, there was application of puncturing. The simulations were performed over an AWGN channel as well as on NLOS while the size of the information for transmission was 1024 bits. The comparison was implemented for QPSK modulation and the transmitter–receiver distance was 200 m. Doppler dispersion was found to be 400 ns, while coding repetitions reached 10. A log-domain algorithm [18] was used for LDPC decoding. Figure 14.8 shows a block diagram illustrating the flow chart of all steps required for LDPC decoding.

The simulation results (Figure 14.9) for the AWGN environment showed that LDPC coding provides excellent performance from the first values of the ratio, slightly better than turbo coding. In particular, the performance of the scheme even reaches 10^{-5} dB. In the case of NLOS (Figure 14.10), the results showed that the performance of the LDPC is better for 1 dB at the tenth iteration up to the point where $E_b/N_0 = 3.5$ dB BER and then they perform the same. The effect of iterations is presented in Figure 14.11 where the optimization of the performance of the iterative decoding is obvious.

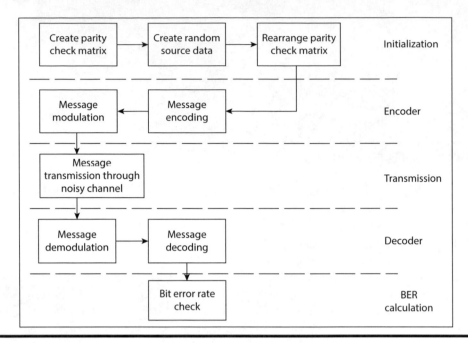

Figure 14.8   MATLAB® flow chart for LDPC transceiver.

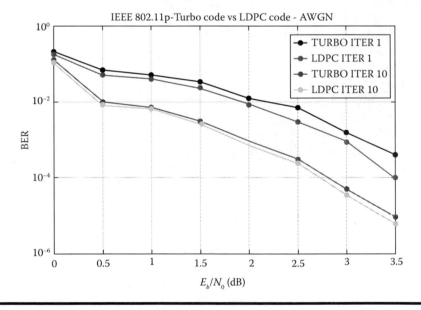

Figure 14.9   Simulation results for the first and tenth iteration in an AWGN environment for LDPC and turbo coding comparison.

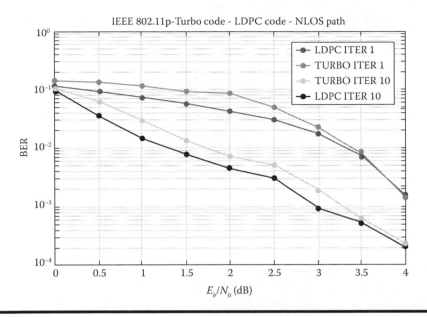

Figure 14.10 Simulation results for the first and tenth iteration in a Rayleigh fading environment for LDPC and turbo coding comparison.

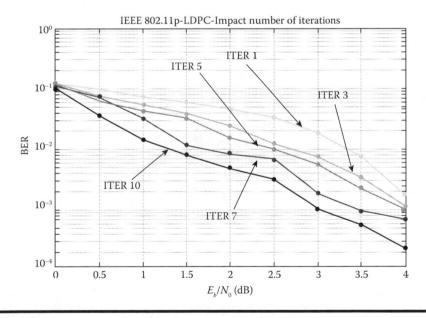

Figure 14.11 Simulation results for different iterations in a Rayleigh fading environment for LDPC coding.

14.5 Implementation in FPGA

Logic circuits are used for the development of circuits that constitute the basis of computers, named with the international term hardware, while they also form the basis of other products. All these products can be called digital systems [19]. The reasoning behind the use of term "digital" is deduced from the method of representation of the information on the computers, considering that the electronic signals correspond to information bits. The development of integrated circuits rendered possible the installation of multiple transistors and hence of an entire circuit in a piece called an integrated circuit, or chip.

In most digital products, it is also necessary to design the manufacture of some logic circuits from the beginning. In order to implement such circuits, there can be usage of three forms of integrated circuits: standard integrated circuits, programmable logic arrays, and dedicated integrated circuits. Contrary to the standard integrated circuits performing specific functions, it is possible to develop integrated circuits containing user-organizable circuits, so as to implement various logic circuits. These circuits have a very generic structure and include a set of programmable logic switches, which allow the internal circuits of the integrated circuit to be organized in various ways. The designer may implement any functions desired for the performance of a specific application, by means of selecting the appropriate switches' layout. The switches are programmable by the end-user and not the manufacturing firm, at the time of construction of the integrated circuit. These integrated circuits are called programmable logic devices (PLDs). Most PLD forms can be programmed many times over.

One of the most developed versions of the PLDs are the field programmable gate arrays (FPGA). The integrated circuit consists of a large number of small logic circuit elements that can be connected using programmable switches. The elements of the logic circuits are organized in a standard two-dimensional structure. The FPGA arrays differ significantly from CPLDs as they do not contain AND and OR gates. On the contrary, FPGAs contain logic tiers for the implementation of the required functions.

14.6 Xilinx's Ise Design Environment

The integrated software environment (ISE) design environment of Xilinx [20,21] is an integrated interface consisting of a set of programs through which it is possible to create, simulate, and implement digital circuits in devices with FPGA circuits, or on CPLD circuits. All tools use a graphics user interface (GUI) which allows all programs to be executed using toolbars, menus, or icons.

The Xilinx ISE supports user throughout the implementation phases, from the design up to the programming of a Xilinx device. The ISE environment organizes and conducts a design through the following steps:

■ Design
■ Composition
■ Implementation
■ Control

14.6.1 Xilinx System Generator

System Generator [22,23] is a tool for the design of digital signal processing systems (DSP) through the Simulink environment of MATLAB aiming at implementing design in FPGA. Designs are depicted in Simulink and are modeled using the special functions blocks provided by the libraries

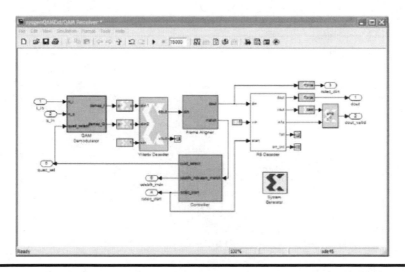

Figure 14.12   System generator screenshot environment.

of the System Generator and Simulink. The tool provides the ability to fully simulate the designed system, to control, evaluate its results, and generate HDL code (VHDL or Verilog) which fully describes it. System Generator (Figure 14.12) uses the Xilinx DSP blockset for Simulink and activates automatically the Xilinx Core Generator for the generation of the optimum netlists for the DSP building blocks. System Generator can execute all implementation applications so as to generate bit flow for FPGA programming.

The System Generator is a DSP design tool of Xilinx, allowing the use of models created in the Simulink environment, on FPGA applications. The applications are integrated in a DSP Simulink environment, using specific Xilinx blocks. The FPGA implementation steps, including the composition, placement, and path are performed automatically, generating an FPGA programming file.

14.6.2 Nallatech XtremeDSP™ Development Kit-IV

The XtremeDSP Development Kit [24] constitutes the ideal development platform for Virtex-4 FPGA technology and was developed by Nallatech (Figure 14.13). It also provides an output for programming and implementation on DIME-II (DSP and imaging processing module for enhanced FPGAs) systems. It contains two independent ADCs—AD6645 ADC channels (14-bits to 105 MSPS) and two independent DAC channels—AD9772 DAC (14-bits to 160 MSPS) of high performance via two input and output channels respectively. It also includes external timing support, an onboard oscillator, as well as Virtex-4 FPGA for user programming and hence constitutes the ideal development platform for signal processing applications.

The development part includes three different Xilinx FPGAs: Virtex-4 that is user-related (Virtex-4 User FPGA: XC4VSX35-10FF668), Virtex II (XC2V80-4CS144) which relates to timing management, and Spartan-II which is related to the peripheral component interconnect (PCI) or universal serial bus (USB) interface for 3.3 V/5 V. User FPGA is a Virtex-4 layout where the master design of the user is transferred. According to the design, the user FPGA interconnects the logic parts appropriately so as to implement the desired processing. The user FPGA inputs are the input channels and the timing signal from the timing FPGA, which constitutes the application implementation clock. The user FPGA outputs are the output channels. The timing FPGA is a Virtex II device. It receives an appropriate design based on which it appropriately interconnects the logic parts so that through the exploitation of the master clock of the developmental part (105 MHz oscillator) it is able to generate timing signals to the user FPGA, the ADCs, DACs, etc. The inputs at the arrays are connected with five MCX connectors and terminate at 50 Ω, both for the input channels and the output channels. The proposed voltage value at the connectors for maximum performance is 2 V P–P, or +/− 1 V.

14.6.3 Nallatech Fuse Software

Software FUSE [25] by Nallatech is based on a Java application that allows the configuration, control, and communication between peripheral systems and FPGA hardware. This allows the development of complex processing systems with seamless software, hardware, and FPGA applications integration. FUSE provides a sufficient number of interfaces, including Tcl Plug-In για το FUSE, FUSE Probe Tool, and FUSE development APIs for C/C++, with optionally available APIs for Java and MATLAB. The key objective of the software operation is FPGA programming.

14.7 Implementation LDPC and Turbo Circuits in FPGA

For the connection of the developmental board of Nallatech with the computer, there was sometimes usage of the Joint Test Action Group (JTAG) interface via parallel cable supplied by Xilinx, sometimes of the PCI interface on a desktop PC and other times of the USB interface. The PCI and USB interface functions with the help of FUSE software by Nallatech while the JTAG communication between FPGA and the computer is implemented initially with the ISE and then with IMPACT software.

To implement the turbo coding circuits, there was usage of System Generator to generate the VHDL code, as stated in detail, through the Xilinx Blockset in a Simulink environment, and there was compilation of source code in VHDL language for LDPC coding. The master clock of the developmental part of the transmitter has a frequency of 105 MHz, and hence, the time unit of the circuit is tclk = 9.5 ns. ADC performs sampling of the input data sequence over a period of 9.5 ns. There was usage of QPSK configuration at a rate of 6 Mbps. Following simulation, the source code created was compiled for FPGA.

The 802.11p PHY FEC turbo coding scheme (Figure 14.13) is presented below and conclusions are drawn from the software on various issues. Specifically, conclusions are drawn regarding the circuit's critical path period in ns, the number of LUTs, the number of FFs used, the number of slices required, the size of memory required by the blocks and, finally, the number of DSP blocks used. The results demonstrate that though the implementation of LDPC coding provides significant benefits for the transmission and retrieval of the signal in adverse environments, there is a drop in performance during programming of the FPGA. In the opposite case, turbo coding consumes fewer resources and has a smaller period, which means it is faster to compile. For the programming of the FPGA circuit, there was usage of Impact software from Xilinx through a parallel connection over a JTAG interface for the receiver, while for the transmitter there was usage of FUSE software via a USB interface. Table 14.2 shows the composition results. The results demonstrate that the transmitter consumes fewer resources, compared to the receiver, both in LDPC and turbo coding. Comparing the two schemes, it arises that LDPC

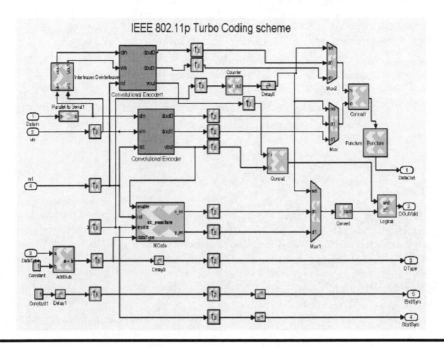

Figure 14.13   IEEE 802.11p turbo coding implementation in Xilinx SG.

Table 14.2   FPGA Implementation Details

Tranceiver Type	Period	LUT	FF	Slices
802.11P LDPC Transmitter	10.175 ns	1279	907	901
802.11P TURBO Transmitter	9.828 ns	1236	887	811
802.11P LDPC Receiver	11.059 ns	1428	1011	988
802.11P TURBO Receiver	10.191 ns	1314	945	903

coding consumes more resources to implement. However, it performs better in simulations, which provides it with an overall advantage at the final comparison.

14.8 Conclusions

This chapter presents the performance evaluation results of a comparative study for IEEE 802.11p PHY employing two different coding schemes, concatenated Reed Solomon convolutional coding and turbo coding. From the obtained simulation results, the BER versus SNR for different kinds of modulation schemes in different channels are calculated. In general wireless communication systems, convolutional codes provide powerful error correction capability, especially in mobile environments with low SNR. The results show that in all environmental cases, both coding schemes achieved significant improvement in our propagation conditions. However, as observed from the presented plots, it must be noted that there is an error floor associated with turbo codes. More specific, when there is low BER the curve flattens a little. Finally, implementation of the FEC coding chain with Xilinx FPGAs using ESL as a fast prototyping approach are presented. The experiments conducted so far show that the implementation of the 802.11p turbo coding does not impose any significant performance or resource usage overhead compared to the 802.11p RS-CC. Comparing the two schemes, it arises that LDPC coding consumes more resources to implement. However, it performs better in simulations, which provides it with an overall advantage at the final comparison.

References

1. Taub H., Schilling D.L., and Saha G., *Principles of Communication Systems*, 3rd Edition, Mcgraw-Hill Education, India, 2008.
2. Gallager R.G., *Low-Density Parity-Check Codes,* Cambridge, MA: MIT Press, 1963.
3. Tanner R , A recursive approach to low complexity codes, *IEEE Trans. Inform. Theor.,* vol. 27, pp. 533–547, September 1981.
4. Margulis G.A., Explicit constructions of graphs without short cycles and low density codes, *Combinatorica*, vol. 2, no. 1, pp. 71–78, 1982.
5. Berrou C., Glavieux A., and Thitimajshima P., Near Shannon limit error-correcting coding and decoding, *Proceedings of the IEEE International Communication Conference (ICC)*, Geneva, Switzerland, pp. 1064–1070, May 1993.
6. Sklar B., Fundamentals of turbo codes, http://www.informit.com/content/images/art_sklar3_turbo-codes/elementLinks/art_sklar3_turbocodes.pdf
7. Sklar B. 'Fundamentals of Turbo Codes', Available from http://ptgmedia.pearsoncmg.com/images/art_sklar3_turboc odes/elementLinks/art_sklar3_turbocodes.pdf, Accessed January 10, 2008.
8. Bahl L., Cocke J., Jelinek F., and Raviv J., Optimal decoding of linear codes for minimizing symbol error rate, *IEEE Trans. Inform. Theor.,* vol. 20, pp. 284–287, March 1974.
9. IEEE Std 802.11-1997 Information Technology—Telecommunications and Information Exchange between Systems-Local and Metropolitan Area Networks-Specific Requirements-Part 11: Wireless Lan Medium Access Control (MAC) And Physical Layer (PHY) Specifications, November 18, 1997, http://ieeexplore.ieee.org/document/654749/
10. Part 11: Wireless LAN Medium Access Control (MAC) and Physical Layer (PHY) specifications: Higher-Speed Physical Layer Extension in the 2.4 GHz band. IEEE Std 802.11b-1999(R2003), http://standards.ieee.org/getieee802/download/802.11b- 1999.pdf
11. Part 11: Wireless LAN Medium Access Control (MAC) and Physical Layer (PHY) specifications High-speed Physical Layer in the 5 GHz Band. IEEE Std 802.11a-1999(R2003), http://standards.ieee.org/getieee802/download/802.11a-1999.pdf
12. Part 11: Wireless LAN Medium Access Control (MAC) and Physical Layer (PHY) specifications Amendment 4: Further Higher Data Rate Extension in the 2.4 GHz Band IEEE Std 802.11g™-2003, http://standards.ieee.org/getieee802/download/802.11g-2003.pdf
13. IEEE 802.11p/D3.0, "Draft Amendment to Standard for Information Technology-Telecommunications and Information Exchange Between Systems—Local and Metropolitan Area Networks Specific Requirements—Part 11: Wireless LAN Medium Access Control (MAC) and Physical Layer Specifications—Amendment 7: Wireless Access in Vehicular Environment" 2007, https://www.ietf.org/mail-archive/web/its/current/pdfqf992dHy9x.pdf
14. Nguyen V.D. and Kuchenbecker H.-P., Intercarrier and Intersymbol Interference Analysis Of OFDM Systems on Time-Varying Channel, University of Hannover, Institut fur Allgemeine Nachrichtentechnik Appelstr. 9A, D-30167 Hannover, Germany.
15. Yucek T., 'Self-interference Handling in OFDM Based Wireless Communication Systems', University of South Florida, Master of Science in Electrical Engineering Department of Electrical Engineering, College of Engineering-University of South Florida.

16. Kato A., Sato K., and Fujise M., ITS Wireless Transmission Technology. Technologies of Millimeter-Wave Inter-Vehicle Communications: Propagation Characteristics. Journal of the Communications Research Laboratory, Vol. 48, No. 4, pp. 99–109, November 2001.

17. ASTM international standards worldwide, http://www.astm.org/.

18. Wymeersch H., Steendam H., and Moeneclaey M., Log-domain decoding of LDPC codes Over GF(q), *Proceedings of the IEEE International Communication Conference (ICC),* Paris, France, pp. 772–776, June 20–24, 2004.

19. Brown S., and Vranesic Z., *Fundamentals of Digital Logic with VHDL Design with CD-ROM,* 3rd Edition, Irwin Electronics & Computer Engineering, Maidenhead, UK: McGraw-Hill Education EMEA.

20. University of Pennsylvania Digital Design Laboratory, Introduction to Xilinx ISE 8.2i, http://www.seas.uppen.edu/-ese201/ISEIntroduction.pdf

21. Xilinx, Project Navigator Overview, http://www.xilinx.xom /itp/Xilinx8/help/iseguide/isegude.htm

22. Xilinx System Generator, http://www.xilinx.com

23. System Generator for DSP Getting Guide, http://www.xilinx.com/support/sw_manuals/sysgen_gs.pdf

24. Nallatech XtremeDSP Development kit-iv user guide, http://www.xilinx.com/support/dicumentation/boards_and _kits/ug_xtremedsp_devkitIV.pdf

25. Fuse System Software User Guide, http://www.nallatech.com/indexx.php/product-briefs.html

FUTURE
DEVELOPMENTS

Chapter 15

Vehicular Network Simulation via ns-3 with Software-Defined Networking Paradigm

Ke Bao and Fei Hu

University of Alabama
Tuscaloosa, Alabama

Contents

15.1 Introduction

As a subcategory of the mobile ad hoc network (MANET), a vehicular ad hoc network (VANET) is a wireless network constructed by vehicles and roadside units (RSUs). A vehicle can forward data with other vehicles and RSUs to realize vehicle-to-vehicle (V2V) and vehicle-to-infrastructure (V2I) communication, respectively. VANETs are expected to provide plenty of services, such as collision avoidance [1], route management [2], surveillance services, and mobile vehicular cloud services [3]. Nevertheless, the implementation of these services is still not fully feasible due to the lack of a dedicated mechanism for resources and connectivity on vehicle mobility and heterogeneous devices. Thus, several VANET scenarios with the software-defined networking (SDN) paradigm have been proposed [3,4,5]. Different than the traditional distributed network structure,

SDN achieves a centralized knowledge and control scenario through logically decoupling the control plane and the data plane. A centralized control system can efficiently simplify the network management, especially when the network has a huge number of nodes. Meanwhile, the intelligent management strategies enabled by a centralized control system would improve the reliability and efficiency of the network data transmission, which will further enhance the performance of VANET services [6–8].

Although most SDN research has been addressed to wired networks, increasingly studies are concentrating on incorporating SDN into a mobile wireless network [9]. As the preliminary studies of SDN in wireless networks show, only a few basic prototypes have been built so far. These prototypes can be classified into two main categories (based on the transmission pattern of control/data traffic [53]), that is, in-band and out-of-band. The out-of-band structure is consistent with the policy of conventional SDNs used for the wired network, in which the packets of control traffic are transmitted separately with the data traffic through an independent channel. Dely et al. [10] proposed the SD-WMN prototype with an out-of-band style. The separation of control/data traffic is realized in [11] by assigning different service set identifiers (SSIDs) to the control/data traffic. Due to the limited resources in WMNs, an in-band-based SD-WMN structure, that is, control/data traffic is transmitted in the same channel, have been realized as well. Both Dely et al. [10] and Yang et al. [12] have designed an SD-WMN platform with an in-band fashion. Particularly, Dely et al. [10] employed OLSR as the routing protocol for both control and data traffic, while Yang et al. [12] established an in-band OpenFlow-based WMN platform for traffic balancing. OpenRoads [13] envisions that users will move between wireless infrastructures. CloudMAC [14] proposes virtualized access points. The Wireless & Mobile Working Group (WMWG) [15] in ONF focuses on wireless backhaul, cellular evolved packet core (EPC), and unified access and management across enterprise wireless and fixed networks (*e.g.*, campus Wi-Fi).

All these studies target limited applications of SDN-based wireless networks and cannot be used as a general performance validation tool of VANET research. To build a more flexible and functional platform, an ns-3–based VANET simulation platform with SDN paradigm is presented in this chapter. The ns-3 simulator is a discrete-event network simulator targeted primarily for research and educational use. Its library covers most of the network structures, such as Ethernet, mesh network, Wi-Fi, and WiMAX. A set of wireless network models are employed to achieve a SDN-based VANET platform. These ns-3 models include a wireless channel model, a mobility model, a UDP model, a MAC layer model, and a wireless channel lost model.

Since the SDN structure is a relative new concept in network standards, ns-3 doesn't provide a complete SDN model. As such, an external SDN library is introduced to integrate with the ns-3 platform. The SDN library is originally based on the Ericsson TrafficLab 1.1 softswitch implementation and then modified in the forwarding plane to support the OpenFlow 1.3 standard. The interface of the SDN library is further revised to establish the interconnection between the SDN library and the ns-3 platform. On the contrary, most SDN simulation platforms are mainly concentrating on a wired network paradigm. To expand the SDN features on the wireless network environment, we also modified some models of ns-3 and the SDN library.

15.2 Background of Simulation

The communication networks can be huge and complex; traditional analytical methods are not enough to estimate the performance and the behavior accurately. Thus, network simulators have been introduced to improve the network evaluation. Normally, a network simulator contains various models to achieve the functionality of computer networks, such as a channel model, a mobility model, a routing model, and an application model. Since there are a huge number of functionalities in different layers (*e.g.*, physical layer, data link layer, routing layer, application layer) of a computer network, a network simulator could be a large and complex software system with the coverage of all these functionalities. Besides various network functionalities, a network simulator also needs to logically integrate the models and make them work together as one whole system. This procedure involves some critical aspects, such as network synchronization, parameter statistics, thread management, model integration, and internal simulator clock. In a network simulator, all the APIs of the models need to be well designed to cooperate with other relative models. Simulating a new protocol in a simulator would incur lots of work, since a huge number of models must be modified to adapt the new model. Based on this consideration, it is preferred to select a professional network simulator that covers as many of the functionalities we need as possible. Nevertheless, the SDN paradigm hasn't been well built in most network simulators. Especially, the current research of SDN is focusing more on wired networks and none of the simulation tools have combined the SDN framework with a wireless network environment. On the contrary, there is also a specialized network simulator for the SDN structure, called Mininet. Mininet is mainly focused on wired network structures. It is short of the functionalities of WMN, in which the ad hoc protocol and mobility functions cannot be achieved. Furthermore, as a lightweight SDN simulator, Mininet is not able to simulate any traditional distributed networking structures either. Users cannot make comparisons between an SDN structure and a traditional distributed network structure by Mininet.

In brief, we decided to employ an ns-3 simulator due to the following advantages.

- *Open source simulator.* As an open and free network simulation tool, ns-3 has a better flexibility than other simulators. Thus, it is relatively easier to add new components to ns-3 for the performance evaluation of the proposed network protocol.
- *Reliability.* Ns-3 has been developed and widely used for more than 10 years. Because it has been widely used and maintained for a long time, it provides a more reliable network simulation system.
- *Optimized simulation structure.* As the inheritor of ns-2, ns-3 has been optimized to obtain more efficient accurate simulation.
- *Closer to realism.* The protocol models in ns-3 are designed to be closer to real computers, where some hardware can be integrated with ns-3 to further achieve emulation.
- *Complete tracing system.* Multiple tracing methods are developed in the ns-3. These tracing methods enable an easier way to trace the network behaviors and parameter statistics.

The ns-3 simulator is a discrete-event network simulator targeted primarily for research and educational use. It is written in C++ and mainly supported by the Unix operating system. Like its ancestor ns-2, ns-3 is developed to provide an open, extensible network simulation platform. However, it is not backward-compatible with ns-2. Basically, the ns-3 is a set of libraries of networking models to simulate how the packet data is operated in various protocols. These models can be integrated by a simulation engine to perform various network simulation tasks. Thanks to the open structure, ns-3 can work with other external software libraries.

Figure 15.1 shows a set of typical models of ns-3. The models of ns-3 cover the most network standards in both wired and wireless networks, and can be classified into three categories as in Figure 15.1: devices (left), utilities (middle) and protocol (right). Device models of ns-3 contain a series of network devices in different standards. Particularly, the CSMA device model is a simple

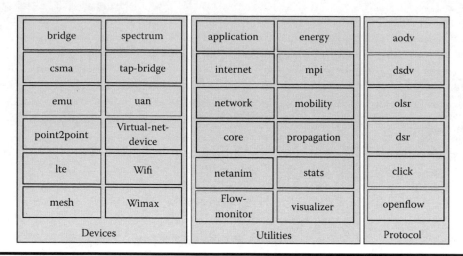

Devices		Utilities		Protocol
bridge	spectrum	application	energy	aodv
csma	tap-bridge	internet	mpi	dsdv
emu	uan	network	mobility	olsr
point2point	Virtual-net-device	core	propagation	dsr
lte	Wifi	netanim	stats	click
mesh	Wimax	Flow-monitor	visualizer	openflow

Figure 15.1 Ns-3 models.

bus network in the spirit of the Ethernet. The LTE model is designed to evaluate the LTE system, such as radio resource management, QoS-aware packet scheduling, intercell interference coordination, and dynamic spectrum access. WiMAX aims to create models of 802:16-based networks. The Wi-Fi models have a wireless network interface based on the IEEE 802:11 standard. The mesh module extends the Wi-Fi module to provide mesh networking capabilities based on the 802:11s standards.

The models located on the right-hand side of Figure 15.1 are some protocol models; aodv, olsr, dsdv, and olsr are routing protocols for ad hoc wireless networks or wired networks. Click is a protocol of pipeline-based routing systems. OpenFlow is a popular standard for SDN. This OpenFlow model of ns-3 only provides simplified features on single OpenFlow node simulation, so it cannot accomplish the simulation task for SDN networks.

Utility models are used to build a platform by integrating various models; this category contains an events scheduler, time arithmetic, tracing, logging, random variables, etc. The packet formats are defined by the network module, and the mobility component provides different mathematical mobility models for wireless network simulation.

15.3 Proposed SDN-Based VANET in ns-3

Since the SDN implementation in ns-3 is outdated and has limited functionalities of SDN, we introduce an external SDN library to enable a complete SDN protocol on an ns-3 platform. The SDN library is based on the OpenFlow protocol version 1.3. It provides an SDN-based switch module and a controller application in OpenFlow. The main structure of the SDN library is shown in Figure 15.2.

The SDN-based network includes a data plane and a control plane. The data plane consists of SDN-based switches, while a control plane includes one or multiple central controllers. In the simulation, both controller and switches stand for network devices. As such, the basis of these two kinds of devices are existing models of ns-3, called NetDevice. In terms of C++, the NetDevice is an important class defining a set of basic functionalities of network devices. All the other network devices in ns3 are the inheritance of the NetDevice class. They inherit the interface

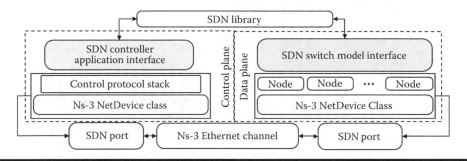

Figure 15.2   SDN-based WMN platform in ns-3.

and functionalities of the NetDevice class and other functions to accomplish different network devices, such as an Ethernet device, Wi-Fi device, and mesh network device. In the simulation, the central controller also stands as a network device, so the external SDN library enables a set of SDN features based on the NetDevice class to obtain a SDN controller. Similarly, an SDN-based switch is also based on the ns3 NetDevice class, while the external SDN library is equipping it with corresponding SDN features.

15.4 SDN-Switch Structure of the SDN Library

The OpenFlow components provided by the SDN library are established upon the NetDevice class of ns-3. There are proper interfaces between OpenFlow devices and the ns-3 NetDevice. As such, the OpenFlow device can be used to interconnect ns-3 nodes using the existing network devices and channels. Figure 15.3 shows the internal switch device structure.

The SDN library provides an OpenFlow port model replacing the port model from ns-3, each port associated with an ns-3 underlying NetDevice. The switch device acts as the intermediary between the ports, receiving a packet from one port and forwarding it to another. The other OpenFlow switch functionalities (flow tables, group table, and meter table) are also provided by the SDN library. Basically, packets entering the switch are sent to the flow table pipeline processing before being forwarded to the correct output port(s). Similarly, the control traffic between the OpenFlow switch and the controller is also processed by the flow table. Based on the category of the control message, the flow table will act upon the message with corresponding actions.

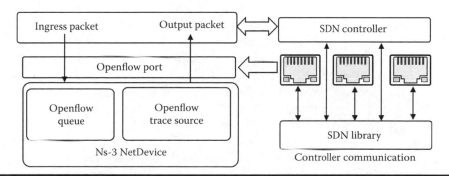

Figure 15.3   SDN-switch structure, SDN library.

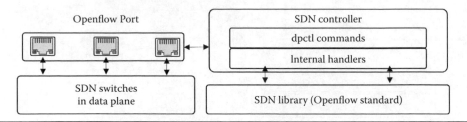

Figure 15.4   SDN controller application interface.

15.5 SDN Controller Application Interface

Regarding the SDN controller application interface, the SDN library provides the basic features of OpenFlow-based controller implementation. It can handle a collection of OpenFlow switches, as illustrated in Figure 15.4. For constructing OpenFlow configuration messages and sending them to the switches, the controller interface relies on the dpctl utility provided by the SDN library. With a simple command-line syntax, this utility can be used to add flows to the pipeline, query for switch features and status, and change other configurations. For control traffic from SDN-based switches, the controller interface provides a collection of internal handlers to deal with the different types of messages. Some handlers cannot be modified by derived class, as they must behave as already implemented. Other handlers can be overridden to implement the desired control logic. In addition, the SDN library is also equipped with a table in the central controller to cache the network topology information. The controller monitors the network by collecting control messages from switches and based on the collected information, the central controller updates the table in real time. Meanwhile, the central controller will send the control messages to switches based on the data in the table.

15.6 A Closer Look at ns-3 with SDN Components

In an ns-3 simulator, a network node is based on a collection of NetDevice objects. The structure of the node is pretty much like an actual computer and contains separate interface cards for Ethernet, Wi-Fi, Bluetooth, etc. [ns3 document wifi]. Figure 15.5 illustrates the framework of each node in ns-3 during the packet transmission. Generally, ns-3 follows the TCP/IP 5-layer model, in which a packet would go through each layer of a node for a forwarding or receiving process. Each of the layers deals with different functionalities as demonstrated in the TCP/IP 5-layer model.

Based on this framework, the main task of ns-3 users is to select corresponding models for different layers to obtain different nodes in the network (*e.g.*, wired CSMA/CA Ethernet, WIFI, WIMAX, MANET). As a widely used network simulator, ns-3 can simulate most categories of network environment by plenty of models in the ns-3 library. A set of these models are enumerated in Figure 15.5. A comprehensive ns-3 library with many models plays an important role for simulation in the purpose of network research. As such, a user can test and compare the performance of various network protocols via flexibly combining different models.

To provide an easier way for node constructing, ns-3 gives us an ns3::Node class and an ns3::NetDevice class. The ns3::Node class can be seen as a container, into which users can aggregate or insert different models or objects to accomplish a comprehensive functionality. In these models and objects, the NetDevice class plays a critical role. The NetDevice is a class defining basic

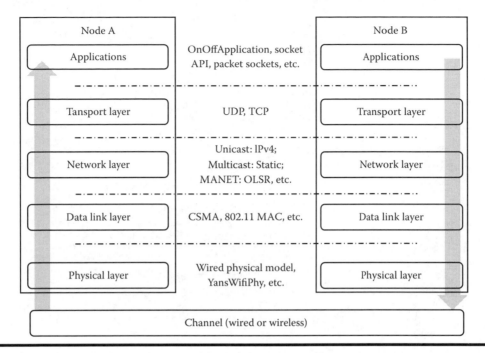

Figure 15.5   Architecture of ns-3 networking simulator.

functionalities of each node, such as packet receiving or forwarding. Based on the class NetDevice, ns3 provides several derived classes to classify a node into different categories. These derived classes include CSMANetDevice, WifiNetDevice, WiMaxNetDevice, and PointToPointNetDevice. Based on the names of these derived classes, it is easy to guess that these devices define different categories of nodes. For example, the CSMANetDevice is used for wired Ethernet networks, in which the transmission media uses the 802.11 CSMA/CA protocol as a MAC layer protocol. In addition, the WifiNetDevice is mainly concentrating on the nodes in a wireless network. Since this chapter aims to explain the VANET simulation, the WifiNetDevice will be discussed in the following sections.

Figure 15.6 shows the architecture of a WifiNetDevice in ns-3 and corresponding API for the models in lower layers. Since the WifiNetDevice is focusing on a wireless network protocol, corresponding models of the WifiNetDevice class in the MAC layer and physical layer are all related to the wireless network features. Specifically, we can define different subcategories of wireless networks in MAC high class, such as ad hoc node and access point node. Meanwhile, the 802.11 DCF protocol is realized in DcfManager class cooperating with DcaTxOp and MacRxMiddle classes. MacLow class is responsible for the four-way handshaking process, while WifiPhy class is modeling the reception of packets and for tracking energy consumption. In brief, ns-3 provides a convenient way to build different categories of networks by classifying the models under different derived categories of ns3::NetDevice class.

As aforementioned in Figure 15.2, the SDN/OpenFlow components of the SDN library are built upon the ns3::NetDevice class. In another words, the realization of the SDN component generally would not affect the lower structure of an ns3::NetDevice class. In this manner, a user can flexibly combine various wireless and wired network protocols with SDN models to realize an SDN-based network. One exception is the network topology discovered by the layer 2 routing protocol, since

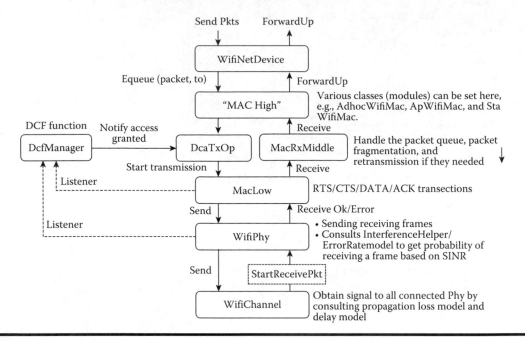

Figure 15.6   WifiNetDevice system in ns-3.

the SDN library is designed to work with a wired Ethernet network in which the central controller would obtain the network topology. Thus, the table located in the central controller only caches the connection between nodes with port numbers by default. On the contrary, VANET requires the MAC address for the packet forwarding. Thus, this is a difference between wireless and wired networks that we must fix. Specifically, we need to add the MAC layer address information and forward it to the SDN component of each node. In ns-3, the layer 2 routing protocol is realized by the arp-l3-protocol class. Thus, after the arp-l3-protocol broadcasts the RREQ message and gets the RREP from the neighbors, the MAC address of neighbors should be forwarded to the SDN component which is locating at the upper layer of the WifiNetDevice class. Then, the SDN component of each node would forward this network topology information to the central controller. In this fashion, the central controller can construct a MAC address-based table with network topology information. This is one of the simplest solutions to transplant a wired network-based SDN library on a wireless network protocol of an ns-3 simulator. Furthermore, traditional wireless network routing protocols, such as OLSR, AODV, and DSR can be used for the SDN-based VANET routing protocol as well. The upper layer SDN component can have the network topology discovered directly by these routing protocols. However, the difficulty is to modify the API between modules. As C++ based software, most of the modules in ns-3 are derived classes. In this fashion, a modification on the API of a module would incur a set of the classes that are relevant to the base class.

15.7 Case Study of SDN-Based VANET

A case study of SDN-based VANET simulation is presented in this section to validate the performance of the proposed ns-3 based simulation platform. Besides the modules we introduced in previous sections, we employ the mobility modules in the ns-3 library to achieve vehicle mobility in the simulation. In addition, the ns-3–based simulator also enables the performance comparison between the proposed SDN-based VANET and traditional VANET (distributed network structure with traditional routing protocol). This VANET case study contains 50 nodes in a 300 m × 1500 m area. The transceiver capability of nodes is 11 Mbps provided by the DsssRate11 Mbps model of the ns-3. The packet size is 1200 bytes.

Figure 15.7 demonstrates the performance of SDN-based VANET and traditional VANET in terms of delivery rate (left-hand side) and delay (right-hand side). In this comparison, we offered different traffic generating rates as shown on the *x*-axis in Figure 15.7. The simulation results show that the SDN-based VANET is outperforming the traditional VANET in both delivered rate and delay.

The mobility is a critical factor for the performance of VANET. In the last simulation, we only considered the performance of VANET in different traffic generating rates. On the contrary, the following figures illustrate the delivered rate and delay in different moving speeds in both SDN-based VANET and traditional VANET.

The left-hand side of Figure 15.8 shows the delivered rate, while the right-hand side shows delay. The x-axis gives the moving speed. As shown in the figures, the increasing of the moving speed would

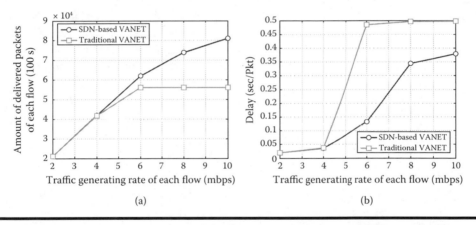

(a)   (b)

Figure 15.7 Performance comparison between PLUS-SW and conventional routing. (a) Delivered packets in different offered load. (b) Average delay in different offered load.

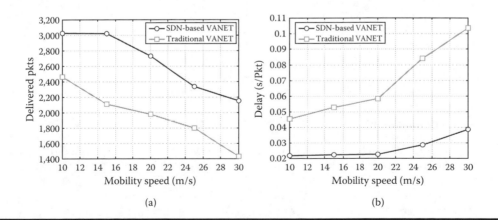

(a)   (b)

Figure 15.8 Performance comparison under different moving speeds. (a) Delivered packets in different moving speed. (b) Average delay in different moving speed.

lower the communication quality of both VANETs. However, the SDN-based VANET possesses a better delivered rate and lower delay than the traditional VANET at all different speeds.

In brief, the above simulation results show a simple case study of the VANET on an ns-3 simulator. Thanks to the powerful library of ns-3, we not only conveniently employ the wireless and mobility modules to achieve an SDN-based VANET, but also the simulation results of the SDN-based VANET can be flexibly compared with other traditional network protocols provided by ns-3.

15.8 Conclusion

This chapter gives the framework of an SDN-based VANET simulation platform on an ns-3 simulator. Since the combination between ad hoc networks and SDN is a relatively new concept, the simulation of SDN-based VANET is still in a preliminary stage. The transplantation of the external SDN library on the ns-3 simulator provides a huge flexibility for network simulation in the purpose of research. Plenty of network modules of the ns-3 can be used to combine with an SDN framework, such as mobility module, ad hoc modules, WiMAX modules. In this manner, we can establish a VAMET platform with heterogeneous physical layer techniques. On the other contrary, the difficulties of the transplantation are concentrating on the API modification between modules in different layers. The proposed SDN-based VANET simulation platform is based on the layer 2 routing. Theoretically, the more advanced ad hoc routing protocols, such as OLSR, can be used for network topology discovery to improve the performance of the VANET. This could be the work for the next step.

References

1. S.K. Gehrig and F.J. Stein, Collision avoidance for vehicle-following systems, *IEEE Trans. Intell. Transport. Syst.*, vol. 8, no. 2, pp. 233–244, June 2007.
2. A. Nandan, S. Das, G. Pau, and M. Gerla, Cooperative downloading in vehicular ad hoc wireless networks, *IEEE Conference WONS*, St. Moritz, Switzerland, pp. 32–41, January 19–21, 2005.
3. I. Ku, Y. Lu, M. Gerla, F. Ongaro, R.L. Gomes, and E. Cerqueira. Towards software-defined VANET: Architecture and services, *13th Annual Mediterranean Ad Hoc Networking Workshop (MED-HOC-NET)*, Piran, Slovenia: IEEE, pp. 103–110, June 2–4, 2014.
4. N.B. Truong, G.M. Lee, and Y. Ghamri-Doudane, Software defined networking-based vehicular ad hoc network with fog computing, *IFIP/IEEE International Symposium on Integrated Network Management (IM)*, Ottawa, ON, Canada: IEEE, pp. 1202–1207, May 11–15, 2015.
5. K. Liu, J.K. Ng, V.C. Lee, S.H. Son, and I. Stojmenovic, Cooperative data scheduling in hybrid vehicular ad hoc networks: VANET as a software defined network, *IEEE/ACM Transac. Netw.*, vol. 24, no. 3, pp. 1759–1773, June 2016.
6. H. Kim and N. Feamster, Improving network management with software defined networking, *IEEE Commun. Mag.*, vol. 51, no. 2, pp. 114–119, February 2013.
7. C.J. Bernardos, A. de La Oliva, P. Serrano, A.Banchs, L.M. Contreras, H. Jin, J.Carlos Zúniga et al., An architecture for software defined wireless networking, *IEEE Wireless Commun. Mag.*, vol. 21, no. 3, pp. 52–61, June 2014.
8. M. Mendonca, K. Obraczka, and T. Turletti, The case for software–Defined networking in heterogeneous networked environments, Nice, France: ACM conference on CoNEXT student workshop (CoNEXT Student '12). ACM, pp. 59–60, December 10, 2012.
9. ONF Solution Brief, Open Flow-Enabled Mobile and Wireless Networks, September 2013.

10. P. Dely, A. Kassler, and N. Bayer, Openflow for wireless mesh networks, *Proceedings of 20th International Conference on Computer Communications and Networks (ICCCN),* Maui, HI: IEEE, pp. 1–6, July 31– August 04, 2011.
11. S. Salsano, G. Siracusano, A. Detti, C. Pisa, P.L. Ventre, and N. Blefari-Melazzi, Controller selection in a wireless mesh sdn under network partitioning and merging scenarios. arXiv preprint arXiv:1406.2470, 2014.
12. M. Yang, Y. Li, D. Jin, L. Zeng, X. Wu, and A.V. Vasilakos, Software-defined and virtualized future mobile and wireless networks: A survey, *Mobile Netw. Appl.,* vol. 20, no. 1, pp. 4–18, February 1, 2015.
13. K.-K. Yap, M. Kobayashi, R. Sherwood, T.-Y. Huang, M. Chan, N. Handigol, and N. McKeown, OpenRoads: Empowering research in mobile networks, *SIGCOMM Comput. Commun. Rev.,* vol. 40, no. 1, pp. 125–126, January 2010.
14. J. Vestin, P. Dely, A. Kassler, N. Bayer, H. Einsiedler, and C. Peylo, CloudMAC: Towards software defined WLANs, *ACM SIGMOBILE Mobile Comput. Commun. Rev.,* vol. 16, no. 4, pp. 42–45, 2013.
15. Wireless & Mobile Working Group (WMWG), [Online]. Available: https://www.opennetworking. org/images/stories/downloads/workinggroups/charter-wireless-mobile.pdf

Chapter 16

WWW: World Wide Wheels—A Paradigm Shift for Transportation Systems via xG

Ali Boyaci and Serhan Yarkan
Istanbul Commerce University
Istanbul, Turkey

Ali Riza Ekti
Balikesir University
Istanbul, Turkey

Muhammed Ali Aydin
Istanbul University
Istanbul, Turkey

Contents

Wireless communications have become an essential part of modern daily life. Starting with traditional voice communications, currently, cellular mobile operators provide high-speed Internet access, file exchange, multimedia communications, video conference services, and online gaming almost anywhere at anytime. It is believed that all wireless nodes/terminals will soon be connected to each other via a common platform such as Internet protocol (IP) network leading to novel concepts and paradigms such as Internet of things (IoT), Internet of everything (IoE), and cyber-physical systems including smart cities, infrastructures, and vehicle-to-everything (V2X) platforms [1].

Vehicle-to-vehicle (V2V) and vehicle-to-infrastructure (V2I) communications are two prominent platforms of a broader concept called V2X for next generation wireless networks (NGWNs) from the perspective of both transportation safety/security and green infrastructures [2]. In case V2V and V2I communications are established, transportation will become safer by the collaborative effort of vehicles and infrastructures via collision avoidance systems, emergency warning protocols, autonomous cars, and so on. Transportation will become greener via V2V and V2I communications as well. Carbon footprints will be reduced by routing optimizations, traffic congestion avoiding navigation, platooning, and traffic information systems [3]. As can be inferred, NGWNs will be the ultimate enabling technology for such envisions.

In this chapter, V2X communications and their relationship with NGWNs will be discussed in detail. Both theoretical and practical aspects along with critical problems and proposed solutions will be provided. Future horizons will be given as well. The chapter is organized as follows: First, NGWNs will be outlined as an enabling technology for V2X. Recently emerging third generation long-term evolution–advanced (3GLTE-A) will constitute the major instance of the NGWN along with methods and protocols appropriate for V2X. Next, key points of physical layer (PHY) aspects of NGWNs will be reviewed from the perspective of V2X. Several scenarios and corresponding solutions proposed in the literature will be given from 3GLTE-A networks perspective. In what follows, an upper–medium access control (MAC) point of view will be adopted and crucial functionalities for V2X in NGWNs such as handoff/handover, routing, networking protocols along with the cloud overview will be summarized. Last, security and privacy will be surveyed. Concluding remarks will then be provided along with details of contemporary studies and future horizons.

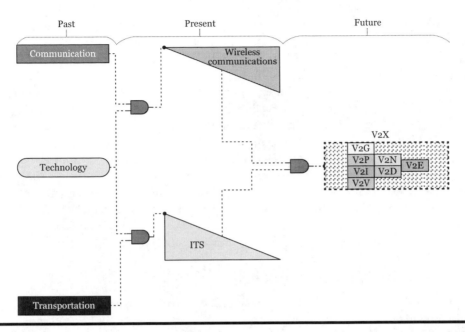

Figure 16.1 Convergence of communication and transportation endeavors and evolution of V2X.

16.1 From Transportation to V2X

16.1.1 Convergence of Wireless Communications and Transportation: ITS

Transportation has already been one of the most prominent necessities for humankind and civilizations. Several millenia, based on the written history, manifest the importance of transportation in terms of several aspects including expediting the dissemination of knowledge and information, trade, relocation, and globalization. Communication can also be considered to be another type of necessity, which serves on bases similar to those of transportation. Even though both transportation and communication supply different forms of solutions, they were regarded as congruent modes of endeavors up until the last two decades. In parallel with the advances especially in digital technology (Figure 16.1), both transportation and communication converge as distinct from the situation in the past. This convergence gives rise to a concept called intelligent transportation systems (ITS) around the 1980s*. As technology becomes cheaper and more capable, both information processing and communication systems start to dominate all parts of modern life. Finally, wireless communications start reverberating across the globe with impressive success. Pervasive presence of wireless communications in daily life pushes ITS to an advanced level: V2X [4]. Ubiquitous connectivity and computing render V2X a comprehensive platform including telematics, interconnected vehicles, infrastructures, pedestrian, grid, and even environment. It is evident that such a vast and diverse interconnectivity could only be established in the presence of a substantial support given by information technology and high-capacity wireless communication networks.

Keeping its historical development in mind, it should be noted that the definition of ITS steadily expands with the technological advances and new concepts. In other words, ITS has

---

* As a historical note, it should be stated that predecessor term for ITS is intelligent vehicle highway system (IVHS) before the 1980s. IVHS includes electronics, control systems, information processing, and sensing which forms an early version of simplex transmissions (not communications!) [25,28].

become a generic "all-in-one" term which includes both already existing and recently emerging technological improvements that are and/or could be applied to transportation endeavor. Two cases exemplifying this fact are discussed in here. The first case extends back to pre-1980s [5,6,7 and references therein]. It is reported that the major focus of ITS was navigation for the driver in a vehicle within that period [8]. However, upon commercial/public release of the global positioning system (GPS), navigation gained a new dimension with several types of information feed such as traffic detours, alerts, and warnings. The second one could be traced back to 1983 when California's fuel-efficient traffic signal management (FETSIM) was introduced for retiming the traffic signals in order to achieve overall decrease in total travel time, redundant stops and delays, and fuel consumption [9]. Note that contemporary understanding of ITS embraces these two cases as default objectives. In light of these two cases and contemporary understanding, one could safely conclude that content of ITS is continuously updated by technological advances and new concepts. From this standpoint, the term ITS will be used in the sense of contemporary, comprehensive understanding throughout this chapter. Clarification will be provided whenever necessary.

16.1.2 Bottom-Up Approach in Analysis of ITS

Broadly speaking, ITS aims to enhance travel safety, provide green transportation by reducing the adverse environmental impact, enable better traffic management strategies, and extend the advantages of transportation to both the commercial and public domain. Based on these observations, one could state that ITS is stimulated heavily by the following prominent factors: (1) fatalities and injuries due to accidents, (2) waste of valuable natural resources such as time, fuel, and money, because of frequent braking and acceleration leading to incessant replacement and repair, and so on due to congestion and nonoptimized travel, and (3) the adverse impact of fuel emissions on the environment. Considering the fact that more than 1.2 million people die each year on roads globally [10], it is apparent that any technological and infrastructural improvement in ITS should first focus on the life-critical aspect of the solution. Next, resource optimization should be pursued. Finally, all aspects of the proposed solution should be revised from the perspective of environment friendly design*. Combining all of the prominent factors together, one could conclude that ITS leverages and positively contributes to economical and societal development both locally and globally.

As discussed previously, ITS converges wireless communications and information technology naturally. The idea of having vehicles, which can be seen as mobile terminals, that are "talking" to each other directly points out wireless communications, whereas the intelligence portion of ITS is established by various forms of solutions embodied in information technology. Evidently, analysis, design, and realization of ITS are all challenging tasks both individually and jointly. Therefore, starting from an analysis, it is wiser to tackle each task individually rather than jointly. However, each and every individual task consists of sophisticated subtasks, modules, and issues. Coupled with interconnections and interrelations between subtasks and modules, even analysis involves intricate details. Hence, a systematic, bottom-up approach is inevitable to study them. Well-established layered organizations such as open systems interconnection reference (OSI) are prominent candidate approaches. This way, both problems and possible solutions along with alternative approaches could be identified easily. The physical layer in this regard should be the natural starting point.

---

* In this regard, one should take into account the regulatory issues of the design and/or solution as well. However, this aspect is outside the scope of this chapter.

16.2 Physical Layer Aspects

In this section, a brief introduction to critical aspects of ITS especially from the perspective of V2X communications is provided. Channel models and transceiver structures, bandwidth, power, several quality of service (QoS) requirements (*e.g.*, delay and priority), signaling schemes, along with several access methodologies are all outlined as well.

16.2.1 Propagation Environment for V2X

16.2.1.1 Importance of Propagation Channel for Wireless Communications

Wireless communication systems can be characterized in various ways based on different criteria such as mobility (*e.g.*, fixed, mobile, and nomadic), range (*e.g.*, body area, short range, and long range), bandwidth (*e.g.*, narrowband, broadband, wideband, and ultrawideband), services provided (*e.g.*, voice, data, multimedia, and gaming), multiple access methodology (*e.g.*, frequency division multiple access [FDMA], time division multiple access [TDMA], code division multiple access [CDMA], transmit message structure (*e.g.*, analog and digital), carrier (*e.g.*, single carrier and multicarrier), antenna configuration (*e.g.*, single–input single–output [SISO], multiinput multioutput [MIMO]), transmission modes (*e.g.*, simplex, half duplex, and full duplex), network formation (*e.g.*, single hop, relay, mesh, and cellular), and radiation energy (*e.g.*, electromagnetic, acoustic, and seismic)*.

Having a plethora of classifications and categorizations for wireless communication systems is not surprising since wireless signals behave quite differently in almost every different scenario and/or category. For instance, the requirements of a fixed wireless communication system are dramatically different from those of a mobile wireless communication system, even though both classifications belong to wireless communication systems. Another example could be the classification based on range. It is known that body area wireless systems, due to several other reasons beside impact of electromagnetic field exposure on human health, operate on a low power regime, whereas long-range systems mandate extremely high levels of power for reliable communication.

Despite the fact that there are various classifications present in the literature, such a vast diversity of categories could be brought together with the aid of a single phenomenon: a propagation channel. Because of the intrinsic propagation mechanism of wireless signals, the physical transmission medium and corresponding communication channel play an important role together in characterizing wireless communication systems. A propagation channel, from this standpoint, should be seen as a mathematical abstraction which encapsulates the differences that lead to the aforementioned diversified classifications. For instance, mobility could be attributed to the channel rather than transceiver and/or receiver so that it could be investigated under a more general abstraction called mobile propagation channels. Range could also be studied from the perspective of propagation channel aspects such as short-range propagation channels as well as long-range propagation channels [11]. The same reasoning applies to bandwidth and to the rest of the aforementioned categories and/or classifications.

---

* In spite of the varieties for radiation energy, this chapter solely focuses on electromagnetic-energy-based transmission.

Electromagnetic waves undergo different mechanisms through the propagation environment. These mechanisms are reflection, refraction, diffraction, and scattering [12]*. The transmitted signal arrives at the receiver by passing through the propagation channel which introduces several types of distortions and degradations. The most plausible way of analyzing the impact of the wireless propagation channel is to start from the received signal since a wireless link could only be established in case the transmitted signal reaches the receiver with certain qualities. The prominent qualitative measure of the signal at the receiver side is the received signal power. Having knowledge about the transmitted signal power, received signal power indicates how much loss is introduced by the wireless channel. Such an approach brings about a different perspective in terms of examining the received signal at the receiver. It is worth recalling that exact values of the received signal power (solutions) at any point in space are attained by wave equation analysis. Yet, such an analysis is cumbersome. Therefore, if loss introduced by the wireless propagation channel could be modeled appropriately, different from wave equation analysis, then a more practical way of analyzing its impact would be possible. Use of stochastic processes along with linear system theory is proven to be one of the most successful ways of modeling the loss introduced by the wireless propagation channel.

In the literature, the loss introduced by the wireless propagation channel is expressed in two different spatial scales: large scale and small scale†. In parallel with the spatial scale discussion, the terminology used for loss in wireless communications takes two different forms as well. In this regard, it is said that wireless propagation channel introduces both large-scale fading and small-scale fading. Here, the spatial scale is quantified by the wavelength of the transmitted waves. In this regard, large-scale fading refers to fluctuations in the received signal power level with displacements of a spatial windows on the order of $>5\lambda$, whereas small-scale fading indicates the variation of the received signal power level with displacements of spatial windows in the order of $<5\lambda$, where λ is the wavelength of the transmitted waveform.

Large-scale fading manifests itself in two submechanisms: path loss and shadowing. Because of such a distinction, in the literature, two different subscales are used to distinguish one from the other. Roughly speaking, path loss is directly related to the area mean which reveals itself with displacements in the order of $>50\lambda$, whereas shadowing implies local mean, which is experienced with displacements in the range of $5\lambda–50\lambda$. Small-scale fading, on the contrary, points out dramatic variations around the local mean due to a multipath effect. Furthermore, combined with mobility, signal power level variations exhibit changes at different rates in a temporal domain within small-scale displacements‡.

All of these remarks point out the importance of the propagation channel for wireless communications. In this regard, comprehensive understanding of V2X communication channels should be the first step to establish reliable, safe, and high-capacity vehicular ad hoc networks (VANETs).

---

* Note here that *absorption* and *polarization* are also reckoned to be two other propagation mechanisms. However, absorption of electromagnetic waves over the air below 10 GHz is generally ignored [26]. When mobile propagation is of interest, vertical polarization is the default choice because signal strength near the ground is statistically higher compared to horizontal polarization at very high frequency (VHF) bands [27]. Although polarization diversity provides expansion in capacity, it is considered to be an extension to the conventional statistical propagation channel characterizations. Because of these two aspects, absorption and polarization are generally not incorporated into the first-tier propagation channel characterizations.

† Sometimes these scales are referred to as "macroscopic" and "microscopic" scales, respectively.

‡ Because rate of change of signal power/amplitude in time could be analyzed better in the frequency domain, the impact of mobility is expressed in the literature as spectral broadening, which is known as "Doppler spread."

16.2.2 V2X Propagation Environment

V2X is a collection of several modes of wireless communication including V2I, V2V, vehicle-to-pedestrian (V2P), vehicle-to-device (V2D), vehicle-to-cloud (V2C), vehicle-to-grid (V2G), vehicle-to-environment (V2E), and vehicle-to-network (V2N). These modes exhibit different channel characteristics; therefore, it is appropriate to first identify peculiar key points to each of them individually. This way, a more comprehensive overview could be obtained.

Before proceeding further, a plausible reasoning is to check validity of the already existing models for relatively well-established cellular channel models and their applicability to the V2X scenarios. This reasoning is not an unrealistic one, since all of the aforementioned keywords for categorization discussed in Section 16.2.1 are seen in both traditional cellular network design stages and V2X scenarios. However, there are critical differences. These differences will be identified clearly when each mode of V2X is examined.

16.2.2.1 V2I Propagation Channels

V2I communication, in the broadest sense, covers the wireless transmission–reception flow of vital safety and operational information between vehicles and different types of infrastructures located in/around the highways. The major focus of V2I is to improve and enhance the safety in transportation so that accidents are totally prevented or minimized[*]. This is achieved by exchanging several types of information between vehicles and roadside units (RSUs). RSUs represent the communication interface of infrastructure which can be a base station or an access point, depending on the scenario.

At first glance, one could conclude that the V2I communication scenario is identical or similar to a traditional wireless cellular network system layout due to the mobile-to-fixed communication organization. However, there are significant differences in terms of corresponding propagation channel abstractions. For instance, a traditional wireless cellular network setup requires antenna heights around 30 m, whereas V2I scenarios require antenna heights to be in the range of 1–6 m. Of course, antenna heights are selected carefully to meet the coverage demands, which automatically implies that the cellular setup is planned to operate—on average—in the range of 1–3 km, whereas the V2I scenario is expected to cover distances in the range of 1–100 m [13]. Another major difference between the traditional cellular setup and the V2I scenario is the mobile speeds. It is clear that V2I scenario consists of higher speed mobile units as compared to those in the traditional cellular setup. Mobility causes differences not only from the perspective of mobile speed but also from the motion patterns of the mobile units. In a traditional cellular setup, directions of motion of the majority of mobile terminals exhibit a high level of randomness. On the contrary, directions of motion of almost all of the mobile terminals in a V2I scenario are deterministic or quasideterministic due to the road structures and traffic flow organization. This creates an interesting transmission–reception mechanism from the perspective of both intended and interfered signals. Sizes, dimensions, and mobility behavior of the objects causing signals to scatter are also different in the two layouts. There are several other differences between the two layouts as well. The prominent parameters are summarized in a comparative way in Table 16.1. As can be seen, even though both layouts fall under the mobile-to-fixed communication category, they exhibit considerably different propagation channel

---

[*] There are some other auxiliary motivations for ITS via V2I in terms of transportation efficiency, optimization, advertisement, and infotainment applications. Some of the examples are localization information broadcast; speed compliance notifications; safe distance warnings; road/weather condition announcements; work zone alerts; electronic toll collection; restricted lane indications; multimedia streaming; local merchant advertisements and so on.

Table 16.1   Comparison between Cellular and V2X Layouts

		Mobile-to-Fixed Scenarios	
		Cellular	V2X
Parameters	Antenna heights	>30 m	1–6 m
	Coverage	1–3 km	1–100 m
	Line of sight (LOS)	Generally absent	Generally present
	Objects in environment	Mainly fixed	Mainly mobile
	Object layout	Almost uniform	Mainly along the road
	Speed	Mainly pedestrian	Mainly high-speed mobile
	Mobility pattern	High level of randomness	quasideterministic

characteristics due to configurations of the fundamental parameters. Therefore, it is not wise to use traditional cellular system propagation models for V2I scenarios.

16.2.2.2 Large-Scale Characterization

Despite the dissimilarities pointed out in here peculiar to V2I channels, large-scale characterization still benefits from well-studied channel models to some extent because of the macro-scale resemblance to cellular communications from mobile-to-fixed transmission aspect. Depending on the scenario of interest, both log-distance and two-ray models are applied to V2I channels in the literature. It is important to note here that extensive field measurement results imply a strong correlation between behaviors of loss in the V2I scenario and predictions of the free-space path loss model. However, this correlation is limited solely to LOS conditions. In other words, the free-space path loss model provides satisfactory results only under LOS conditions. On the contrary, discrepancy between measured data and theoretical predictions becomes dramatic under non-line-of-sight (NLOS) conditions. Therefore, in the literature, NLOS conditions are examined in subcategories.

In V2I environments, NLOS occurs due to different reasons such as obstruction caused by vehicles, foliage, and buildings. Note that these reasons are valid for traditional cellular communications as well. However, the situation is different in V2I environments in the following ways: (1) Coverage is important for trajectories (*i.e.*, highway and roads) rather than areas, (2) mobility follows a quasideterministic pattern (due to the traffic flow organization), and (3) propagation channel changes relatively faster. These items render the NLOS scenario slightly different in terms of analysis even though the causes for obstructions are similar in both V2I and traditional cellular environments.

It is important to keep in mind that RSUs in V2I scenarios are placed in such a way that LOS is generally maintained for the link between any vehicle and infrastructure. Nevertheless, the LOS state could be switched occasionally to an NLOS state due to big-size vehicles (*e.g.*, trucks, vans, and heavy-haul) blocking the direct transmission path between the RSU and the vehicle antenna. As in the general propagation channel modeling studies, there are three mainstream strategies for V2X channels: (1) the deterministic approach, (2) geometry-based approaches, and (3) the stochastic approach.

The deterministic approach relies on having the complete knowledge of the terrain, topography, and contour features. In case such a massive amount of details are given/known, then computational electrodynamics could be employed to obtain the exact characterization. Although there are various techniques available in the literature, computational electrodynamics is known to be a computationally intensive process. Ray tracing (with ray imaging and ray launching) seems to be a promising candidate to alleviate the computational burden. However, ray tracing is still considered to be computationally demanding. Even though computational power and resources available increase day by day, each and every slight change in the configuration causes these calculations to be carried out repetitively. Both being computationally intensive and necessitating the calculations to be carried out repeatedly, computational electrodynamics or ray tracing methods are generally avoided.

Geometry-based approaches depend on a relatively simplified two-dimensional (2D) or three-dimensional (3D) approximations of the communication environment. It is clear that 2D approximation is preferred over 3D approximation. This stems from the fact that analysis for 3D requires an extra dimension and related parameters to be taken into account as compared to that for 2D. It is clear that a simplified and tractable statistical model with sufficient accuracy is always preferable. For that reason, there are numerous 2D models present in the literature. Some of the prominent 2D models are the two-ray flat earth model, the blind bend propagation model, and the distinct signal component model. Among all, the two-ray model is the most frequently encountered/employed 2D propagation model. Besides its simple structure, an important and distinct characteristic of this model is the concept called critical distance (breaking point), which leads to two different decay rates for the received signal power for two different transmitter–receiver separations.

Stochastic approaches could be regarded as natural extensions of geometry-based approaches with certain abstractions. For instance, there exist several straightforward extensions to the two-ray model in large-scale characterizations of V2I environments. The first extension is obtained by parameterizing both before-the-critical-distance and after-the-critical-distance decay. The next stage of extension is acquired by incorporating a random variable into the parameterized version. This way, the deterministic structure of the decay is disturbed by random amplitudes and the model becomes more realistic. Generally, the random amplitude disturbance is selected to be normally distributed with specific parameters depending on the environment and topography. In the literature, it is shown that additional attenuation introduced by big-size vehicles blocking the transmission path could be approximated by a multiple knife-edge diffraction propagation mechanism [14] based on the Fresnel ellipsoid radius [15]. Therefore, further extensions are possible for the two-ray model family.

In light of the discussions above, a generalized version of the macroscopic loss could be given as a function of transmitter–receiver separation d:

$$P(d) = \begin{cases} P(d_0) - 10n_1 \log_{10}\left(\dfrac{d}{d_0}\right) + S_1, & d_0 < d < d_{\mathrm{bp}} \\[2ex] P(d_0) - 10n_1 \log_{10}\left(\dfrac{d_{\mathrm{bp}}}{d_0}\right) - 10n_2 \log_{10}\dfrac{d}{d_{\mathrm{bp}}} + S_2, & d_{\mathrm{bp}} \leq d \end{cases} \tag{16.1}$$

where d_0 is the reference distance; $P(d_0)$ denotes the power level for the reference distance (either known or measured); n_1 stands for the path loss exponent for the distances up to breaking point d_{bp}; n_2 is the path loss exponent for the distances exceeding breaking point; S_1 and S_2 are the random variables which represent the fluctuations around the mean with a specific standard deviation for each scenario, respectively.

It is clear that (16.1) leads to a random variable, $P(d)$, in either scenario. The overall description of $P(d)$ is analogous to that in traditional cellular deployment. In cellular system design, $P(d)$ is assumed to have a log-normal distribution. There are certain justifications in regards to log

Table 16.2 Experimental Values of Parameters for the Dual Slope Model

Link Type	Model Parameters			
	n_1	n_2	σ_{S1}	σ_{S2}
General	(1.83, 3.5)	(2.12, 7.78)	(1.62, 5.6)	(4.4, 8.4)
LOS	(1.66, 3.8)	(2.25, 8.42)	(1.53, 5.38)	(3.95, 4.15)
NLOS	(1.64, 3.56)	(3.52, 4.25)	(6.12, 6.67)	(3.2, 6.26)

normality which mainly capitalize on the central limit theorem. Nevertheless, V2I environments generally maintain LOS; therefore, it is difficult to invoke log normality with the same reasoning adopted for traditional cellular deployment. There are some other problems beside physical suitability of the log normality assumption such as tractability and closed-form analysis of the received signal and link capacity. In the literature, Gamma distribution is proposed to alleviate such problems [16]. In order to give further insight into the model, parameters and corresponding values are tabulated in Table 16.2.

In the sequel, it is worth mentioning that recently emerging options such as communications in THz bands pave the way for extremely wideband transmissions especially for LOS environments. It is known that frequency dependency should be incorporated into path loss models as well [17]. Besides, the impact of antenna polarization and antenna height are also investigated in the literature. Although statistical characterizations of such parameters are crucial in link quality analysis, incorporating them into Equation 16.1 will lead to a more sophisticated version of the model.

As in the LOS scenario, NLOS scenarios exhibit frequency dependency as well due to several causes such as foliage or obstructions. Measurements reveal that as the transmission frequency increases, the impact of foliage leads to a linearly increasing attenuation especially beyond 3 GHz [18]. Losses caused by obstructions due to several other causes are reported in the literature with higher path loss exponents [19,10].

The overall behavior of path loss is given in Equation 16.1 along with the first-order statistics tabulated in Table 16.2. However, a more comprehensive perspective could only be obtained by investigating the higher-order statistics. Measurements report that random fluctuations of $P(d)$ around its mean exhibit spatial correlation. In other words, fluctuations observed in $P(d)$ become decorrelated after a sufficiently long displacement. Spatial coherence can best be captured by a wide-sense stationary (WSS) Gaussian process, $S(\Delta x)$, where Δx represents the spatial separation [21]. Second-order statistics of the process S is obtained by:

$$R_{SS}\{\Delta x\} = E\{S(x)S(x + \Delta x)\} \tag{16.2}$$

where $E\{\cdot\}$ is the expected value operator; $R_{XX}\{\cdot\}$ is the autocorrelation of the stochastic process X, which is assumed to be of WSS form. A normalized version of Equation 16.2 is often used for standardization purposes:

$$\rho_{SS}\{\Delta x\} = \frac{R_{SS}\{\Delta x\}}{\sigma_S^2} \tag{16.3}$$

In the literature, a frequently encountered analytical model for decoherence distance (or sometimes known as decorrelation distance) is expressed in terms of Equation 16.3, the normalized version of Equation 16.2, as:

$$\rho_{SS}\{\Delta x\} = e^{-\frac{|\Delta x|}{d_c}} \tag{16.4}$$

where d_c represents the decorrelation distance where correlation coefficients $\rho_{SS}\{\Delta x\}$ cross below a certain value. Conservative applications adopt d_c satisfying $\rho_{SS}\{\Delta x\} = 0.5^*$, whereas relatively relaxed applications consider $\rho_{SS}\{\Delta x\} \cong 0.368$ or $1/e$ for $\Delta x = d_c$.

16.2.2.3 Small-Scale Characterization

Wireless signals propagate through the transmission medium and reach the receiver antenna as multiple copies. The receiver measures the superposed signal at the antenna and strives to extract the message via signal processing methods. Considering the wave nature of each copy, it is obvious that the measured signal experiences both constructive and destructive interference depending on the propagation mechanisms and geometry. Therefore, any distinct position of the receiver indicates a different combination of superposition of the copies of the transmitted signal. Thus, the receiver observes dramatic fluctuations in the received signal power within the propagation environment. Interestingly, such dramatic fluctuations occur in very small displacements, which are in the order of a couple of wavelengths of the transmitted signal, as exemplified in Figure 16.2. Hence, this phenomenon is known as small-scale fading.

Fading affects the received power dramatically. However, there is a further intriguing observation with regard to the propagation channel: the channel output leads to different interpretations for the same input (*e.g.*, an impulse in a temporal domain) under various parameter settings at the receiver side. The most striking instance of this behavior is the time resolution of the receiver. When time resolution is low, the receiver is unable to distinguish the copies arriving at the antenna from each other. Therefore, the propagation channel consists of a single reflection/copy whose amplitude changes dramatically from the low-time-resolution-receiver perspective[†]. In contrast, when time resolution is high, the receiver gains the capability of distinguishing the individual copies[‡]. In case low time resolution is considered, the channel is said to obey a narrowband assumption, whereas for high time resolution case, the channel is called a wideband channel.

Besides fading and narrow or wideband characterizations, another important factor in describing the small-scale characterization is how fast the behavior of the channel changes in time. Note that this is different from the previous discussions regarding fading and narrow/wideband characterizations. In the previous cases, the channel behavior is assumed to be fixed in time. Such a behavior is known to be time-invariant. But when mobility is introduced, due to steadily

---

[*] This corresponds to the distance approximately 70% of d, that is $\Delta x = 0.7d_c$.

[†] One should keep in mind that the receiver observes a single copy due to its low time resolution even though that observed single copy actually consists of multiple copies arriving almost simultaneously at the receiver antenna.

[‡] Based on the discussion above regarding the time resolution of the receiver, the most pertinent question is: "What sort of transceiver design is required?" There are various ways to answer this question. One possible way is to decide the service type to be carried out by the channel. Voice, data, Internet, online gaming, beaconing, synchronization signaling, relaying are several possible service types to devise an appropriate design. Although service types are diverse, bandwidth required for them is a characteristic property. Due to the time–frequency duality, the higher the bandwidth, the higher the signaling rate or the higher the time resolution.

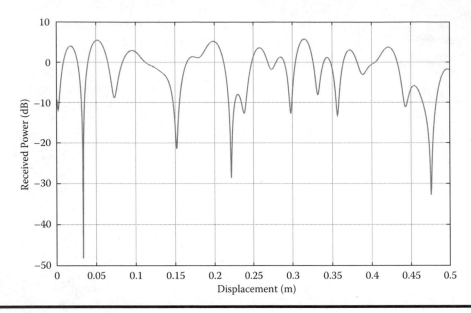

Figure 16.2 **Received power observed by a mobile whose speed is 5 m/s when a pure tone is transmitted at 5.9 GHz. Note that the wavelength of the transmission is ≈0.05 m, whereas the received power varies rapidly in displacements even less than 0.05 m.**

changing propagation geometry, copies of the transmitted signal arrive at the receiver with delays that change in time. Hence, especially for V2I channels, this time-varying nature of the channel behavior should be taken into account.

Considering fading, narrow/wideband characterization and the time-varying nature of the channel altogether, a linear complex baseband equivalent model could be employed to describe these intrinsic behaviors of the V2I channel:

$$h(t,\tau) = \sum_{l=0}^{L-1} h_l(t,\tau) e^{\left(-j2\pi f_c \tau_l(t) + \Theta_l(t,\tau)\right)} \delta\left(\tau - \tau_l(t)\right) \tag{16.5}$$

where $j = \sqrt{-1}$, f_c is the transmitted signal frequency, $h_l(t, \tau)$ stands for the possibly time-varying magnitude; $\tau_l(t)$ is the delay, and $\Theta_l(t, \tau)$ denotes the phase of the l–th component.

A complete description of $h(t, \tau)$ is possible by providing with statistical characterizations of $h_l(t, \tau)$, $\Theta_l(t, \tau)$, and $\tau_l(t)$. As discussed previously, each resolvable copy of the received signal is actually a cluster consisting of superposition of multiple (due to the time resolution of the receiver) indistinguishable sinusoidals such that:

$$h_l(t,\tau) = \sum_{n=1}^{N_l} a_n(t) e^{\left(-j\phi_n(t)\right)} \tag{16.6}$$

where N_l is the number of components in the l-th cluster; $a_n t$ and $\phi_n(t)$ are the time-varying magnitude and phase of the n-th component, respectively. Assuming that N_l is sufficiently large, the central limit theorem could be invoked for Equation 16.6 such that $h_l(t, \tau)$ is assumed to be complex Gaussian distributed. This automatically implies that the magnitude of $h_l(t, \tau)$, that is $|h_l(t, \tau)|$, is Rayleigh distributed any time instant t with:

$$f_X(x) = 2\alpha x e^{-\alpha x^2}, \quad x \geq 0 \tag{16.7}$$

where α is the shape parameter for distribution which corresponds to $E\{|h_l(t, \tau)|2\}$.

Previously, LOS is stated to be the dominant propagation mode in V2I environments. Therefore, LOS should be incorporated into the stochastic description of $h_l(t, \tau)$ An important extension of Equation 16.6 for the V2I scenario is obtained by introducing a specular component which may correspond to LOS path as:

$$h_l(t, \tau) = a_0 e^{(-j\phi_0)} + \sum_{n=1}^{N_l} a_n(t) e^{(-j\phi_n(t))} \tag{16.8}$$

where $(\cdot)_0$ represents the index for the LOS path.

An important detail regarding the statistical characterization of V2I propagation channel is the impact of high mobility present in the environment. When mobility exists in the propagation environment, delays of resolvable paths stemming from scatterers prolong or shorten depending on the direction of the relative motion. Such changes in the delays of the path are perceived as smearing in the frequency domain, which is known to be a Doppler effect. Random motion in the environment leads to a random shift in the frequency domain, which is known to be Doppler spread. Motion causes the propagation channel to be time selective. It is clear that Doppler spread is limited to maximum shift in the frequency domain.

Impact of random, relative motion in the environment, mathematically, has already been included into the channel characterization by the presence of $h_l(t, \cdot)$. The propagation channel is assumed to be of a linear, time-variant form. Hence, the rate of selectivity in time should be evaluated in the statistical sense, which brings about the notion of coherence time. Second-order statistics of $h_l(t, \tau)$ are carried out the same way in the literature as in shadowing analysis.

16.2.2.4 V2V Propagation Channels

V2V propagation channels are considered as a modified version of V2I channels from the following perspectives: (1) antenna heights and (2) maximum Doppler shift. It should be noted that none of these modifications necessitates a drastic change in the models discussed up until this point. Accurate representations are obtained by appropriate selection of suitable parameters in the model. For instance, in case a dual-slope model is adopted, critical distance will be different for V2V from that for V2I. Similarly, decorrelation distance will be different as well. Even though the propagation mechanism is the same, magnitude distribution of the channel coefficients might be chosen to be different from Rayleigh or Rice distributions. Although the Doppler effect is modeled the same way, consequences are more dramatic since relative motion becomes extreme especially for the scenarios on the highway when vehicles are traveling in opposite directions.

16.2.2.5 Other V2X Propagation Channels

As discussed in Section 16.2.2, the propagation environment for other V2X scenarios such as V2P, V2D, V2C, V2G, V2E, and V2N could all be considered under the umbrella of V2I and

V2V propagation models with different settings and parameters. The topography of the environment, terrain elevation, bridges, tunnels, over- and underpasses, trees and vegetation, roundabouts, heavy and/or long vehicles, crosses, pedestrian presence, and traffic density are some of the descriptive aspects which should be taken into account during the parameter selection process.

16.3 MAC Layer Aspects

As in the PHY layer, there are critical differences and issues for V2X communications at the MAC layer in comparison with traditional communication systems and infrastructures. In this section, prominent differences and issues will be investigated.

16.3.1 Peculiar Properties of V2X MAC Layer

V2X environments exhibit high mobility. High mobility points out several issues for the contemporary wireless communications systems. Propagation and related issues are discussed in Section 16.2.1. Of course, some of the points are directly related to the V2X PHY layer. However, there are peculiar characteristics for the MAC layer, which are driven by distinct applications. First and foremost, V2X environments are eminently dynamic. Such dynamism gives rise to network formations' sojourn across a local geographical area. Naturally, network entry and handover should be obvious design extensions. Relaying, which is directly related to coverage, is another aspect. Processing delay requirements and mechanisms (*e.g.,* decode-and-forward and amplify-and-forward) to be adopted shape the MAC layer design as well.

Scalability of multiple access strategies under severe interference with various traffic densities (including traffic congestions) is another interesting issue for V2X environments. Prioritization, over-the-air signaling, resource allocation, reliability, fairness, and decentralized topology along with the compatibility of already existing and recently emerging technologies within harsh propagation environments are some other important topics.

16.3.2 Current Organization of V2X MAC Layer

16.3.2.1 Dominance of the Institute of Electrical and Electronics Engineers (IEEE) 802.11

Initial deployment of V2X, which covers mainly V2I and V2V, relies on two mainstreams: European and United States (US) layout[*]. However, both of them are based on IEEE 802.11. It is important to distinguish here that IEEE 802.11p operating at 5.9 GHz is the selected member of the IEEE 802.11 family suite. IEEE 802.11p allows direct communication and data exchange in short latency by enabling an ad hoc mode in the absence of an infrastructure. As a final point, IEEE 802.11p operates on an unlicensed (but regulated) spectrum.

DSRC adopts mainly IEEE 802.11a with several functional changes. As stated before, some of the changes are directly to PHY layer-related issues. For instance, DSRC employs a half-rate clock in comparison with the traditional IEEE 802.11 layout which reduces the channel spacing to 10 MHz with prolonged temporal parameters. This way, intercarrier interference stemming

---

[*] The European version is called cooperative intelligent transportation system (C-ITS), whereas the US version is known as dedicated short range communications (DSRC).

from high Doppler shifts could be eliminated [22]. Network formation, as touched upon in Section 16.3, is another aspect in DSRC. Considering very stringent requirements of V2X environments such as immediate data exchange without prior handshaking mechanisms, a novel basic service set (BSS), namely outside the context of basic service set (OCB), is defined by disabling all procedures pertaining to common control flow in traditional BSS. Furthermore, with the use of enhanced distributed channel access (EDCA), several modes of operations are allowed with different access categories. Hence, prioritization, which is a crucial property for V2X, is allowed. Also, multihop multichannel operations are enabled in the presence of IEEE 1609.X support. There are various proposals for switching patterns between control and service channels in the literature for multichannel mode of operation.

Different from DSRC, C-ITS is planned to be deployed at the same spectrum band (*i.e.*, 5.9 GHz) with further subdivisions*. In C-ITS, subdivisions are classified based on their functionalities. Traffic efficiency, transportation safety, nonsafety, and reserved-for-future-use are the main categories for the subdivisions. Another critical distinction between DSRC and C-ITS is observed in the routing protocol in the ad hoc mode. In C-ITS, GeoNetworking is introduced and placed in the standardization text, which employs the geographical/physical coordinates in addressing and relaying [23]. Note that the V2X network formation sojourns around and reestablishes very frequently. Therefore, GeoNetworking will provide a very efficient multihop topology at the expense of a slightly more sophisticated protocol flow and signaling overhead.

16.3.2.2 Acknowledgment, Addressing, and Operating on Channels

Before proceeding with the control channel (CCH) and service channel (SCH), it is appropriate to discuss MAC-level acknowledgments since it will shed more light on the distinctions about V2X environments in regards to the IEEE 802.11 suite.

The IEEE 802.11p suite is based on a distributed control function (DCF) with carrier sense multiple access/collision avoidance (CSMA/CA). In a unicast scenario, MAC-level acknowledgments are present. However, there is no MAC-level acknowledgment in the broadcast scenario. This renders the broadcast scenario slightly unreliable. Also, both slot durations and short interframe space (SIFS) are prolonged due to the range concerns. Addressing for both RSUs and onboard units (OBUs) are different too. RSUs have a fixed MAC address, whereas OBUs initialize a random MAC and strive to enter the network. In case a collision occurs, OBUs switch to another MAC address.

Use of channels are critical in V2X. There are two prominent channels in V2X: CCH and SCH. All nodes have to check on CCH, whereas they could monitor a single SCH. Synchronization for CCH and SCH monitoring is performed via GPS. Hence, when the CCH interval starts, all nodes have to listen to CCH. On the contrary, strategies for network formation and selection of SCH are still studied in the literature.

16.3.3 Design Concerns for V2X MAC

There are various design concerns regarding V2X MAC. In this part, prominent design concerns and relevant discussions are given.

---

* C-ITS, the European version, resembles DSRC in many aspects. Therefore, for the sake of clarity, only the major differences will be emphasized in here.

16.3.3.1 Mobility-Related Concerns

V2X environments include different mobility classes. Bearing in mind the velocities of mobile nodes and traffic congestions together, two extreme scenarios manifest themselves automatically. In case high mobility scenario is considered, network lifetimes are very short, the number of nodes involved are few, and message priority is high (due to the increased possibility of accident and/or dangerous situations). On the contrary, traffic congestions yield situations and conditions almost reverted. In traffic congestions, network lifetimes might be long, the number of nodes involved are many, and message priority is relatively very low. As emphasized here, V2X should operate optimally within these extreme ranges of situations and conditions, almost all the time. One should note that "optimality" should take into account multiaccess interference (MAI) as well. Individual behaviors of these parameters and their relationship with mobility is given in Figure 16.3.

In the literature, traffic flow and distribution of vehicles are assumed to be among direct parameters of mobility-related concerns. However, they are second-tier parameters. This is best observed from the perspective of mobility. Low or no mobility, in regular traffic, points out a high density vehicle distribution, which could almost be assumed to be deterministic for the sake of simplicity in analysis. On the contrary, a high mobility scenario implies very sparse distribution. Traffic flow is essentially a deterministic concept due to the road and infrastructure organizations. Hence, it is not recognized among the first-tier parameters.

16.3.3.2 Message Prioritization

V2X communication systems inherently mandate message and traffic prioritization due to time-critical safety requirements. Regular broadcast messages, emergency signals, delay-sensitive data exchange, and acknowledgment-required communications all need to be prioritized and

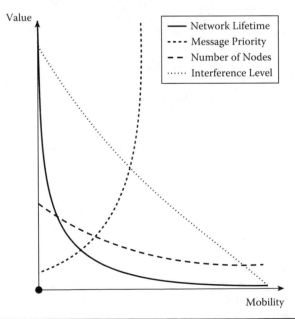

Figure 16.3 Impact of mobility on critical parameters of V2X communications. It is important to keep in mind that V2X communication system designs require to operate in an optimal sense in any mobility scenario.

disseminated accordingly. Note that prioritization should take into account the aforementioned concerns in regards to mobility.

16.3.3.3 Channel Coordination

As discussed above, V2X communication systems should deploy some sort of channel coordination mechanism because of concurrent use of safety-related and nonsafety-related data exchanges; SCH and CCH, along with the requirement of a high level of spectrum utilization. Multichannel operation in V2X improves the overall system performance especially when dumping data from CCH and SCH; under dynamically changing network topography, traffic congestion and flow conditions, and in the presence of cochannel interference (CCI) or adjacent channel interference (ACI).

Channel coordination implies a reliable channel switching mechanism; appropriate protocol design and suite selection; and suitable hardware. Bearing in mind that the possibility of use of both single- and dual-radio solutions, channel coordination becomes one of the most important functionalities of V2X. In this regard, reliability, efficiency, and cost are the fundamental design parameters for channel coordination in V2X networks.

16.3.3.4 HetNet Operations

Heterogeneous networks have already been implemented to some extent especially in metropolitan areas. Therefore, HetNet support should definitely be included in V2X MAC. More rigorously, versatile communication modes as unicast, multicast, and broadcast need to be provided; stringent QoS requirements such as delay-sensitive and high-priority message delivery should be met, with ubiquitous coverage and amalgamation of distributed architectures with centralized networks.

16.4 Upper Layer Aspects

In coordination with both PHY and MAC, there are certain points that need to be investigated in detail for V2X different from traditional mobile and/or ad hoc networks. In order to gain more insight into this, application- and service-oriented perspectives could be adopted.

The first distinction is observed for V2X among its modes such as V2I and V2V. Versatility of the impact zone or region, which defines the geographical range of the target(s), is another important difference. Type of transmission initiation such as event driven, periodic, or node directed is another important aspect. Message dissemination mode such as unicast, multicast, or broadcast in a dynamic topology becomes prominent in V2X networks. Message/frame format and routing strategies are different from those in traditional network architectures as well.

It is obvious that security and privacy related issues should be treated in a special way due to the peculiar structure of V2X networks.

16.5 Security and Privacy for V2X

Security and privacy concepts for V2X systems possess slightly different implications from traditional wireless networks. This stems from the fact that V2X systems carry critical messages whose failures, interruptions, and/or delays might end up injuries, casualties/fatalities with serious material damage. Besides its life-critical aspect, V2X systems have commercial projections which include location-based advertisements, digital content delivery on demand, high-speed toll collection, travel guidances, distributed gaming, and Internet access. These two perspectives enforce the design of V2X systems to reconsider and improve the traditional approaches in the field of security and privacy.

Security encompasses several aspects such as message authenticity, integrity, validity, accuracy, credibility, reliability, nonrepudiation, and even timeliness. Cryptographic methods could be considered to be mature enough to provide necessary and sufficient bases for V2X security. Symmetrical, asymmetrical, and hybrid approaches could be used depending on the security problem and its requirements. Of course, when V2X is of interest, all of these keywords gain an extra dimension due to its life-critical aspect. There is a domino effect in case a security breach takes place because of key exchanges over unsecured channels. Furthermore, timeliness and cost (bandwidth and over-the-air signaling) are two contending criteria, which should be satisfied to some extent while a domino effect is avoided. One common solution to this intricate problem is hardware modules, which lead to very high-speed solutions with optimized operational behaviors. However, security-specific hardware modules increase the cost. There are some other approaches such as prioritizing the message exchange in such a way that only life-critical messages are delivered via a special upper layer filter. This way, computational resources are kept at a minimum and cost is reduced.

One of the interesting aspects of hardware-oriented security is the hardware integrity. Any malicious attack, such as injecting a false emergency alert on the network, could lead to fatalities. Such a possibility could only be prevented by secure hardware controllers, modules, and onboard units.

Privacy implies anonymity and confidentiality. However, V2X systems are heavily location dependent such that physical coordinates are crucial. Furthermore, V2X systems from a network via some sort of beacon signals, which, in turn, could be jointly used to identify the vehicle with digital certificates—and therefore the driver. Hence, privacy and security for V2X systems should be examined more carefully [24].

16.6 Relationship between xG and V2X

In order to better understand the relationship between xG and V2X, use cases and scenario-dependent analysis of V2X are required. This is critical, since standardization efforts of the so-called 5G have been evolving in such a way that some of the requirements of V2X imply extended capabilities for already matured standard items.

In a nutshell, 5G is expected to support: (1) high-velocity terminals, (2) ultralow latency within the network, (3) very high (on the order of multiGbps) capacity data exchange, (4) extended reliability and service availability, and (5) improved security and privacy in all types and modes of communications. On the contrary, V2X comes with several modes such as V2V, V2I so requirements and relevant implications are different. For instance, V2V mode requires direct communications with adjustable sensitivity based on speed information. When there is traffic jam, V2V causes a very high density message exchange in the vicinity of a local region. Furthermore, high-speed interactions lead to extremely high Doppler spread. In addition, it is expected to have V2V interactions always available even though infrastructure support is not present. The situation is slightly different for V2I scenarios. For V2I, broadcast messages are quite important and they are delay sensitive. Multifrequency communications should be supported as well. As can be inferred, V2I is only possible in the presence of infrastructure support.

High-velocity terminals especially at highway traffic bring up significant Doppler shifts. Furthermore, networks should respond to any changes extremely fast due to the life-critical aspect of the vehicular environments. It is known that long-term evolution (LTE) Release 14 supports up to 280 Kmph and 5G is envisioned to support 500 Kmph. Note that in a highway environment, 150 Kmph of absolute speed at each direction of a two-way scenario surpasses the physical limits of LTE Release 14. Furthermore, such high speeds cause the network to expand the coverage within a couple of seconds. Of course, this comes at the expense of a higher link budget due to higher transmit power, improved coding gain, and extended transmission periods. As discussed previously, significant Doppler shifts challenge the already established synchronization schemes such as increasing the number of pilots.

Ultralow latency requirement is one of the keys of V2X environments because of life-critical message exchanges such as precrash warnings and emergency brake notifications. Considering the heterogeneous structure of NGWNs, different vendors and operators should meet this ultralow latency messaging protocol for a successful V2X deployment. In this regard, PC5 interface extension, which is included in LTE Release 14, provides some sort of a solution for V2X requirements. However, as touched upon previously, direct communication mode should be supported for both inter- and intraoperator scenarios within and outside the coverage zones. Furthermore, LTE Release 14 supports 100 ms end-to-end delay, whereas 5G is expected to reduce it down to less than 1 ms. Note that all of the PHY and MAC layer algorithms should be revisited because of the ultralow latency requirement of V2X along with the high-speed terminals communicating with each other.

Another aspect of the challenges is the extended capacity requirement posed by particular V2X scenarios. Specific life-critical use cases such as video assisted overtaking or see-through functionality yield a serious challenge in NGWNs since high-volume data should be exchanged via network at high-speed terminals with extremely low latencies. Coupled with varying traffic density, handling high volumes of life-critical data transmission requires sophisticated PHY and MAC layer techniques.

A peculiar property of V2X scenarios is that the message exchange solely for the V2V mode of operation should stay in the local geographical region. In other words, only relevant vehicles should be informed regarding the content of the message delivery rather than all of the vehicles within a broader region. This imposes a strict control plane and user plane separation with appropriate scaling. As a direct consequence, advanced adaptive antenna techniques such as adaptive beamforming and tracking mechanisms should be adopted.

As discussed above, when infrastructural coverage is not available, life-critical message and data delivery should still be carried out in a synchronous fashion. This implies the presence of two strategies: an external synchronism support such as GPS, and link redundancy for failsafe connectivity. Hence, NGWN should provide enhanced PHY and MAC layer operations and functionalities such as smart switching between operational modes, and fast acquisition and synchronization techniques.

Last but not least, security and privacy for V2X need further elaboration from the perspective of NGWNs. Subscriber identification module or international mobile subscriber identity provides authentication; however, they cannot guarantee to protect the privacy of the communications. Furthermore, it is desired for V2X that mobile network operators should not be able to reassemble the identity and relevant details such as speed and geographical location. In addition, vehicles should not establish a communication network with preshared security keys. All of these concerns point out a uniform architecture in terms of security, privacy, and performance in NGWNs. Public key infrastructures should be established to distribute and manage the digital certificates.

Authorities such as registration and pseudonym certificates should be distributed across the modules and functionalities.

16.7 Conclusions and Future Directions

Considering socioeconomical, environmental, and financial aspects, transportation could be classified to be one of the most influential endeavors of humankind since the beginning of civlization. Communications, on the contrary, has always been reckoned to be a separate endeavor serving on the same bases as transportation. However, technological advances have forced these two endeavors to converge rapidly in the last couple of decades, leading to a concept called ITS. It is believed that ITS paves the way for more secure transportation, green environments, and extended services. With the emergence of NGWNs, ITS evolves into a more generic concept called V2X.

V2X is envisioned to be a very comprehensive concept including V2I, V2V, V2P, V2D, V2C, V2G, V2E, and V2N. These modes exhibit different characteristics and have various requirements. Starting from the PHY layer, modes of V2X drive and shape the architecture and organization of NGWNs in numerous ways. Waveform design, resource allocation, relaying, handover, latency, range, throughput, applications, reliability, privacy, and security are only some of the well-known topics which should be investigated in detail for V2X, especially with the emerging NGWNs such as 5G and beyond.

References

1. J. Jin, J. Gubbi, S. Marusic, and M. Palaniswami. An information framework for creating a smart city through internet of things. *IEEE Internet of Things Journal*, 1(2):112–121, 2014.
2. A. Ozpinar and S. Yarkan. Vehicle to cloud: Big data for environmental sustainability. *Effective Big Data Management and Opportunities for Implementation*, Singh, Manoj Kumar, and Dileep Kumar G. (eds.), p. 182, Hershey, PA: IGI Global, 2016, DOI: 10.4018/978-1-5225-0182-4.
3. A. Boyacı, H. Zaim, and C. Sönmez. A cross-layer adaptive channel selection mechanism for ieee 802.11 p suite. *EURASIP Journal on Wireless Communications and Networking*, 2015(1):214, 2015.
4. H. Rakouth, P. Alexander, Andrew Jr. Brown, W. Kosiak, M. Fukushima, L. Ghosh, C. Hedges, H. Kong, S. Kopetzki, R. Siripurapu, and J. Shen. *V2X Communication Technology: Field Experience and Comparative Analysis.* pp. 113–129. Springer, Berlin Heidelberg, 2013.
5. R.E. Fenton. Automated highway system technology. In *30th IEEE Vehicular Technology Conference*, vol. 30, Dearborn, Michigan, pp. 457–460, September 15–17, 1980.
6. J.E. Gibson. National goals in transportation. *Proceedings of the IEEE*, 56(4):380–384, 1968.
7. T. Ishii, M. Iguchi, T. Nakahara, Y. Kohsaka, and Y. Doi. Computer-controlled minicar system in expo'70: An experiment in a new personal urban transportation system. *IEEE Transactions on Vehicular Technology*, 21(3):77–91, 1972.
8. R.L. French. Historical overview of automobile navigation technology. In *36th IEEE Vehicular Technology Conference*, vol. 36, Dallas, Texas, pp. 350–358, May 20–22, 1986.
9. A. Auer, S. Feese, and S. Lockwood. *History of Intelligent Transportation Systems.* Technical report, Intelligent Transportation Systems Joint Program Office, Washington DC: U.S. Department of Transportation, 2016.
10. World Health Organization. *Global Status Report on Road Safety 2015.* Nonserial Publication. World Health Organization, Geneva, Switzerland, 2015.

11. A. Domazetovic, L.J. Greenstein, N.B. Mandayam, and I. Seskar. Propagation models for short-range wireless channels with predictable path geometries. *IEEE Transactions Communications*, 53(7):1123–1126, 2005.

12. T.S. Rappaport. *Wireless Communications: Principles and Practice.* Prentice Hall Communications Engineering and emerging Technologies Series. Prentice–Hall, NJ: Upper Saddle River, second edition, 2002.

13. P. Belanovic, D. Valerio, A. Paier, T. Zemen, F. Ricciato, and C.F. Mecklenbrauker. On wireless links for vehicle-to-infrastructure communications. *IEEE Transactions on Vehicular Technology*, 59(1):269–282, 2010.

14. M. Boban, T. Tiago, V. Vinhoza, M. Ferreira, and O.K. Tonguz. Impact of vehicles as obstacles in vehicular ad hoc networks. *IEEE Journal on Selected Areas in Communications*, pp. 15–28, 2011.

15. M.A. Qureshi, R.M. Noor, A. Shamim, S. Shamshirband, and K.-K. Raymond Choo. A lightweight radio propagation model for vehicular communication in road tunnels. *PLoS One*, 11(3):e0152727, 2016.

16. A. Abdi and M. Kaveh. On the utility of gamma pdf in modeling shadow fading (slow fading). In *1999 IEEE 49th Vehicular Technology Conference (Cat. No.99CH36363)*, Houston, TX, vol. 3, pp. 2308–2312, May 16–20, 1999.

17. J.M. Jornet and I.F. Akyildiz. Channel modeling and capacity analysis for electromagnetic wireless nanonetworks in the terahertz band. *IEEE Transactions on Wireless Communications*, 10(10):3211–3221, 2011.

18. B. Benzair, H. Smith, and J.R. Norbury. Tree attenuation measurements at 1-4 ghz for mobile radio systems. In *1991 Sixth International Conference on Mobile Radio and Personal Communications*, Coventry, UK, pp. 16–20, December 09–11, 1991.

19. M. Boban, J. Barros, and O.K. Tonguz. Geometry-based vehicle-to-vehicle channel modeling for large-scale simulation. *IEEE Transactions on Vehicular Technology*, 63(9):4146–4164, 2014.

20. J. Gozalvez, M. Sepulcre, and R. Bauza. Ieee 802.11p vehicle to infrastructure communications in urban environments. *IEEE Communications Magazine*, 50(5):176–183, 2012.

21. M. Gudmundson. Correlation model for shadow fading in mobile radio systems. *Electronics Letters*, 27(23):2145–2146, November 07, 1991.

22. C.F. Mecklenbrauker, A.F. Molisch, J. Karedal, F. Tufvesson, A. Paier, L. Bernado, T. Zemen, O. Klemp, and N. Czink. Vehicular channel characterization and its implications for wireless system design and performance. *Proceedings of the IEEE*, 99(7):1189–1212, 2011.

23. K. Sjoberg, P. Andres, T. Buburuzan, and A. Brakemeier. Cooperative intelligent transport systems in Europe: Current deployment status and outlook. *IEEE Vehicular Technology Magazine*, 12(2):89–97, 2017.

24. J. Petit, F. Schaub, M. Feiri, and F. Kargl. Pseudonym schemes in vehicular networks: A survey. *IEEE Communications Surveys Tutorials*, 17(1):228–255, Firstquarter 2015.

25. K. Chen and B.A. Galler. An overview of intelligent vehicle-highway systems (ivhs) activities in North America. In *Information Technology, 1990. 'Next Decade in Information Technology', Proceedings of the 5th Jerusalem Conference on (Cat. No. 90TH0326-9)*, Jerusalem, Israel: IEEE, October 22–25, pp. 694–701, 1990.

26. P. Series. *Propagation data and prediction methods required for the design of earth-space telecommunication systems.* Recommendation ITU-R, Geneva, Switzerland, pp. 618–12, 2015.

27. G.L. Stüber. *Principles of Mobile Communications.* Norwell, MA: Kluwer Academic, 1996. Fourth Printing.

28. P. Varaiya and S.E. Shladover. Sketch of an ivhs systems architecture. In *Vehicle Navigation and Information Systems Conference,* Troy, MI, vol. 2, pp. 909–922, October 20–23, 1991.

Chapter 17

Exploiting Infrastructure-Coordinated Multichannel Scheme for Improving Channel Efficiency in V2V Communications

Xiaohuan Li

Beihang University
Beijing, China

Chunhai Li, Xin Tang, and Rongbin Yao

Guilin University of Electronic Technology
Guilin, China

Xiang Chen

Sun Yat-sen University
Guangzhou, China

Contents

In vehicle ad hoc networks (VANETs), channel coordination scheme is a key technology related to the utilization of channels, which to some extent increases the system throughput. But the available multichannel scheme still suffers from some obvious drawbacks, such as control channel (CCH) congestion, poor channel utilization, multichannel hidden terminal, and missing receiver problems. In this chapter, we study the infrastructure-coordinated multichannel schemes to solve the aforementioned problems with regard to VANETs. Generally, the multichannel schemes are divided into two categories: synchronous multichannel and asynchronous multichannel. For the synchronous scheme, we present a novel synchronous roadside unit (RSU) coordinated multichannel MAC scheme (SRCMC) to allow channel rendezvous on the CCH during the whole synchronization intervals for alleviating the CCH congestion problem. SRCMC also supports simultaneous transmissions on different service channels (SCHs) to increase SCHs utilization. Moreover, we utilize the RSU to record the rendezvous information and broadcast it to the surrounding vehicles on the CCH. Thus, the collision problem on the SCHs can be further avoided. For the asynchronous scheme, we propose a novel asynchronous RSU-coordinated multichannel MAC scheme (ARCMC) to allow RSU switching between different SCHs and retransmitting messages to OBUs which miss rendezvous information. It solves the message loss and rendezvous conflict problems. It is noted that an analytical model is developed to evaluate the aggregate throughput of V2V communications in these different scenes and the performances are analyzed in terms of multichannel hidden terminal, missing receiver, and CCH congestion problems. Simulation results demonstrate that the proposed schemes can provide better performance in channel utilization, system throughput, and avoiding multichannel hidden terminal and missing receiver problem, than that of the existing approaches in the literature.

17.1 Introduction

VANETs have been considered a key component for providing safety and comfort in intelligent transportation systems (ITS) and receive a lot of attention in the research community [1–5]. Its aims are to provide many application services for vehicles and passengers, one of which is traffic safety applications [6]. This type of application is mainly to provide driver safety information to avoid collision with other vehicles; for example, it reminds the drivers to slow down at the crossroads and informs the follower vehicles to avoid rear-end collisions in the case of congestion. The information can be used to reduce the probability of traffic accidents and the number of deaths in accidents by vehicle-to-vehicle (V2V) and vehicle-to-infrastructure (V2I) communication [7–10]. Other applications that are not related to road safety, that is, nonsafety applications, include traffic management and efficiency applications, as well as multimedia entertainment applications. Traffic efficiency and management applications focus on improving the vehicle traffic flow (*i.e.*, traffic coordination or traffic assistance) as well as providing updated local information and messages of relevance bounded in space and/or time. Speed management and cooperative navigation are two typical groups of this type of application [6]. Meanwhile, multimedia information applications focus on improving the experience of driving and providing comfort to the traveling vehicle by VANETs, from which we can obtain locally based services such as point of interest notification, local electronic commerce, and media downloading [11–12]. In the near future, the must-have option for vehicles would no longer be the leather seat, it will be the ultrahigh-speed Internet connectivity that provides the passengers with remarkable multimedia services. Therefore, more and more attention has been paid to vehicular multimedia services in recent years. However, how to realize the nonsafety message transmission between the vehicle nodes, vehicles, and infrastructure (*i.e.*, RSU) to obtain the high-quality infotainment, the low latency, and packet reception rate of service experiences while safety applications must be guaranteed is a great challenge to the limited channel resource carrying capacity, and the channel utilization needs to be further improved.

In VANETs, the safety and nonsafety information is transmitted by wireless access in vehicular environments (WAVE). The WAVE adopts dedicated short-range communication (DSRC) [3] which is based on IEEE 802.11p [4] and IEEE 1609.4 [5], to assist drivers with information and manage vehicle traffic. IEEE 1609.4 defines a multichannel media access control (MAC) scheme, which stipulates multichannel cooperative operation, CCH and SCH parameters, channel priority access parameters, and a channel switching mechanism. The scheme is an access mechanism that relies on universal coordinated time (UTC) [5] for synchronization of channels. Channel synchronization switching is shown in Figure 17.1, in which the channel access time is divided into multiple synchronous intervals (sync intervals) with a fixed length of 100 ms. Each sync interval consists of a CCH interval and an SCH interval. Durations of both the CCH interval and the SCH interval last 50 ms which include a 4 ms guard interval. During the CCH interval, all on board units (OBUs) must tune to the CCH for safety and system control message exchange. Similarly, during the SCH interval, OBUs listen to the CCH or switch to a specific SCH, which is used for nonsafety services.

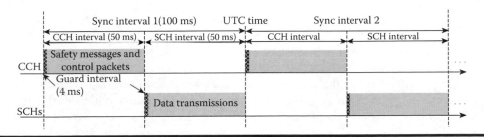

Figure 17.1 Multichannel synchronization switching of IEEE 1609.4.

Although the goal of WAVE is to guarantee transmissions of safety applications and maximize the system throughput for nonsafety services, its multichannel MAC scheme suffers from some obvious drawbacks. First, the channel utilization cannot inherently exceed 50% because of equal-length alternating between CCH and SCH intervals. Second, the switching scheme between CCH and SCHs prohibits intelligent assistance of time intervals in response to variable traffic demands. Moreover, the channel competition mechanism may cause a decrease in the performance of the system due to possible collisions, especially under a crowded urban traffic environment as well as in the case of high-speed moving vehicles. Therefore, an efficient multichannel MAC scheme is a necessity to improve the channel utilization and reduce the contention on CCH.

In this chapter, first, we analyze and compare the characteristics and multichannel efficiency problems of the earlier works for VANETs. Then, we study RSU-coordinated multichannel schemes for improving channel efficiency. Meanwhile, we have the RSU participate in the multichannel coordination process to play a coordinated role in channel access and allocation. Without loss of generality, we consider the synchronous mechanism and asynchronous mechanism respectively. On the one hand, we propose a novel synchronous RSU-coordinated multichannel MAC scheme for VANETs to allow channel rendezvous on the CCH during the whole synchronization intervals and support simultaneous transmission on different SCHs, which can reduce the CCH congestion. On the other hand, we propose a novel asynchronous RSU-coordinated multichannel MAC for VANETs to solve the multichannel hidden terminal and missing receiver problems [2]. In short, the above two RSU-coordinated approaches improve the efficiency of CCH and SCH accessing via different channel rendezvous schemes as well as increase the utilization of channels.

The remainder of this chapter is organized as follows. In Section 17.2, a comprehensive survey of the multichannel coordinate scheme is provided. In Section 17.3, the V2I and V2V communication model is briefly described. In Section 17.4, a novel RSU-coordinated synchronous multichannel MAC scheme is provided. The RSU-coordinated asynchronous multichannel MAC scheme is proposed in Section 17.5. In Section 17.6, we perform simulations to verify the performances of the proposed schemes. Finally, Section 17.7 provides some concluding remarks.

17.2 Related Works

The traditional wireless communication system uses the single-channel communication mode, which makes the whole authorized frequency band a single channel. In the process of data transmission, the intersignal interference will reduce the system throughput. Therefore, a multichannel communication scheme to improve the system throughput is proposed. The basic idea of the multichannel scheme is to divide the authorized spectrum resources into multiple subchannels that are equal and nonoverlapping, and allow multiple nodes in the same coverage area to be able to simultaneously transmit data on different channels, which can improve the channel utilization and system aggregate throughput effectively. Therefore, it is very important to propose an effective multichannel scheme for VANETs.

The design of the channel protocol needs to meet the following requirements: First, the protocol is able to provide a message exchange method for the source node and the destination node. Second, the protocol can allocate available free channels for the source and destination nodes, while avoiding collision of messages. In the actual multichannel system, because the workload of each subchannel, the distribution of nodes, and the channel negotiation mechanism are greatly affected by the network state compared with the traditional single channel system, the multichannel communication system suffers from the problems of multichannel hidden terminal, missing receiver, and CCH congestion.

In some earlier works, different multichannel MAC schemes have been proposed for VANETs, which can be categorized into synchronous multichannel MAC protocols and asynchronous multichannel MAC protocols.

The synchronous multichannel MAC protocols require a strict time synchronization between the CCH interval and the SCH interval. In Refs. [13–14], a multichannel MAC protocol dynamically adjusting the duration of the CCH interval and its analytical model was proposed. This scheme can adjust the length of the CCH interval according to the density of vehicles on the road. It tries to provide reliable transmission for safety messages and to improve the saturation throughput of service data. However, the SCH resources are still wasted during the CCH interval. In Ref. [15], the authors designed a RSU-coordinated multichannel MAC protocol which is based on 802.11 PCF. The original 1CCH+6SCH structure is modified into 1CCH+1ECH+5SCH to guarantee the reliability of safety message transmission on the ECH, which increases the probability of receiving safety messages at the cost of degradation of system throughput on the SCHs. A dedicated multichannel MAC (DMMAC) was proposed in Ref. [16]. It employs the hybrid channel access to provide collision-free and delay-bounded transmission for safety messages. The synchronous algorithm may use one designated service channel in a synchronous interval. It leads to the other five channels wasted. In Ref. [17], the authors proposed a vehicular enhanced multichannel MAC protocol (VEMMAC) for vehicular ad hoc networks. It allows nodes to access the CCH to reserve one of the SCHs to exchange nonsafety messages only during the CCH interval, and broadcast safety messages twice with each in the CCH and SCH interval. In Ref. [18], it allows OBUs to send request for service (RFS) packets to make an agreement with SCH providers. However, making service channel rendezvous only on CCH may lead to a high contention level on the CCH and reduce the success rate of reservation.

The asynchronous multichannel MAC protocols need no strict time synchronization and allow OBUs to switch to any SCH channel through rendezvous. In Ref. [19], the authors proposed an asynchronous multichannel protocol. The protocol did not need tight time synchronization, but it required two rounds of message interaction during the selection of the SCH, which caused additional message load. A multichannel selection strategy through neighbor nodes assistance was analyzed in Ref. [20], but the paper also pointed out that the rapid movement of nodes will have a great impact on the performance. In Ref. [21], the authors proposed an asynchronous multichannel MAC (AMCMAC). It supports simultaneous transmissions on different SCHs, as well as allows other OBUs to make rendezvous with their providers and receivers, or broadcast emergency messages on the CCH. However, this scheme may suffer from serious collision on the CCH when the density of vehicles is high. In Ref. [22], the authors designed a novel distributed asynchronous multichannel medium access control scheme for large-scale vehicular ad hoc networks, named asynchronous multichannel medium access control with a distributed time-division multiple-access mechanism (AMCMAC-D). The proposed scheme supports simultaneous transmissions on different SCH channels while allowing rendezvous and broadcast of emergency messages on the CCH channel. A dynamic service channels allocation (DSCA) method was proposed in Ref. [23] to maximize throughput by dynamically assigning different service channels for users. In Ref. [24], the authors proposed a multichannel medium access control protocol, called multichannel MAC protocol with hopping reservation (MMAC-HR), to resolve the multichannel exposed terminal problem. In Ref. [25], the OBUs share the control message to realize the collaborative asynchronous multichannel MAC. In Ref. [26], the authors proposed a novel multichannel multiple-access protocol (MMA). For ensuring nonsafety messages transmission, two CCHs are utilized in MMA to rendezvous SCH and acknowledgments (ACKs) separately. To the best of our knowledge, these works may suffer from a multichannel hidden terminal problem, because OBU-pair simultaneous transmissions on different SCH channels will miss the rendezvous message on the CCH.

17.3 System Model

In this section, we present a popular system model in VANETs, as shown in Figure 17.2. Consider a VANETs consisting of OBUs driving through the coverage of an RSU. When an OBU arrives in the range of a designated RSU, it could transmit and receive messages from the RSU. Moreover, OBUs could communicate with each other in a one-hop range. In addition, the vehicle can interact with the RSU to form a WAVE basic service set (WBSS) to acquire and update the application that interests the user. The data of the safety and nonsafety applications are transmitted on the CCHs and SCHs, respectively. Based on such a model, the details of an RSU-coordinated synchronous and asynchronous multichannel MAC scheme were studied in the subsequent sections.

17.4 RSU-Coordinated Synchronous Multichannel MAC Scheme

First, we assume that all OBUs in the VANETs are equipped with a half-duplex transceiver which is able to switch between CCH and SCHs to exchange safety and nonsafety messages. It can either transmit or listen but cannot perform both simultaneously. Second, we assume the RSU is equipped with two half-duplex transceivers (transceiver A and transceiver B). Transceiver A always listens to the CCH. It would record the rendezvous information in a current SCH interval. Transceiver B is capable of switching between different SCHs and CCH. It broadcasts the rendezvous information in the next RSU interval to avoid reserved SCHs selected by other OBUs.

Based on the above-mentioned assumptions, in this section, we propose a novel RSU-coordinated synchronous multichannel MAC protocol for VANETs, which not only solves the

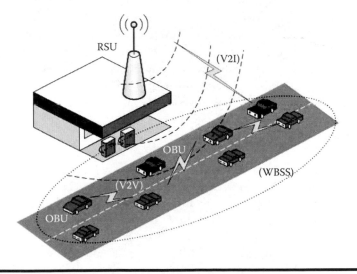

Figure 17.2   System model in VANETs.

multichannel hidden terminal problem but also provides channel rendezvous in the whole synchronous interval and simultaneous transmissions on different SCHs.

In our proposed multichannel protocol, the CCH interval is further divided into RSU interval and RFS interval, as shown in Figure 17.3. The CCH interval starts from the RSU interval, which broadcasts an RSU packet. The RSU interval lasts for a fixed length of 5 ms, which includes a 4 ms guard interval and 1 ms transmission delay of the RSU packet. The value of transmission delay is based on an observation in which a typical safety broadcast message takes up to 1 ms to transmit. Meanwhile, the CCH can be used during the whole sync interval, that is, it can work either in the CCH interval or in the SCH interval. The SCHs can be used in the SCH interval. Besides, we design all OBUs such that they have to tune to the CCH in the CCH interval, which can ensure safety of message exchange and the use of the RSU to broadcast the RSU package in the RFS interval. Moreover, the OBUs which rendezvous with the SCH successfully in the last sync interval (*i.e.*, CCH interval and SCH interval) tune to the selected SCH to exchange nonsafety messages in the current SCH interval, while others are allowed to rendezvous SCH in the CCH or just to listen to CCH.

The operation of the proposed scheme is shown in Figure 17.3. It inherits the multichannel access framework and all nodes (OBUs and RSUs) needing tight time synchronization. In this figure, S stands for safety message and CTR stands for clear to rendezvous message. For the safety messages transmission, OBU tries to access the CCH and contends the channel to broadcast a safety message in the RFS interval. For the nonsafety messages transmission, OBU tries to access the CCH to rendezvous one of SCHs by exchanging RFS/CTR messages during the RFS and SCH interval. Then, successful rendezvous OBUs can realize simultaneous transmissions on different SCHs. To record the rendezvous information, we assume each node keeps two local SCH rendezvous tables: local SCH rendezvous table-1 (LSRT-1) and local SCH rendezvous table-2 (LSRT-2). LSRT-1 records the rendezvous information which is reserved in the CCH interval. It contains the state of each SCH and the time periods which are used to exchange nonsafety messages on the selected SCH, while LSRT-2 records the rendezvous information which is reserved in the SCH interval. The following subsections provide more detailed information on the proposed scheme.

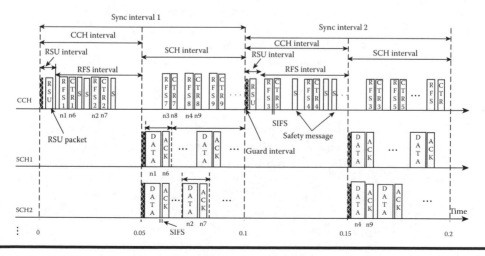

Figure 17.3 RSU-coordinated synchronous multichannel MAC scheme.

17.4.1 The Operation of Time Interval

In the RSU interval, the RSU broadcasts a RSU packet, which includes the accurate rendezvous information that occurred during the last SCH interval. Then, OBUs which have received the RSU packet update LSRT-2 immediately and copy the rendezvous information from LSRT-2 to LSRT-1. In the RFS interval, whenever an OBU wants to transmit a safety message or non-safety message, it has to compete with other OBUs to access the CCH. When an OBU wants to exchange a nonsafety message, it reserves one of the SCHs in a current synchronous interval by exchanging RFS/CTR messages. Moreover, the successful rendezvous information is recorded in LSRT-1. In an SCH interval, successfully reserved OBU pairs tune to the selected SCH channel in an orderly manner to exchange nonsafety messages according to the LSRT-1 and the others tune to the CCH channel to reserve one of the SCHs in the next synchronous interval by exchanging RFS/CTR messages. Similarly, the successful rendezvous information is saved in LSRT-2.

17.4.2 The Operation of Channel Rendezvous

The proposed scheme adopts a RSU-coordinated channel rendezvous scheme and a novel channel selection strategy, aiming to perform contention-free channel access and increased channel utilization.

When a vehicle sends a nonsafety message, the OBU computes the message transmission time, the cut-off time of the current synchronous interval, and the minimum occupied time of the six SCHs according to LSRT-1 or LSRT-2. If the selected channel idle time is enough to transmit the nonsafety message, it will send the RFS packet and initialize the CTR timer. The receive rendezvous OBU decodes the RFS packet. Then, it checks the selected channel idle time and sends the CTR packet. Moreover, the successful rendezvous information will be saved into LSRT-1 or LSRT-2 in different time intervals. With the coordination of the RSU, the proposed scheme could obtain more chances to rendezvous successfully and avoid a multichannel hidden terminal problem. The detail of the channel rendezvous operation is shown in Table 17.1.

17.4.3 Performance Analysis

In this section, we present a performance analysis of the proposed scheme. The motivation of the proposed multichannel MAC scheme to improve the aggregate throughput on SCHs. Generally speaking, the aggregate throughput is defined as the channel utilization multiplied by the channel data rate. In the following, we propose a new analytical model for aggregate throughput on SCHs, which takes into account the following factors: nonsafety message transmission probability in each synchronous interval, average size of the nonsafety-related message, and the number of OBUs.

To facilitate the analysis, some assumptions are made. First, we assume there are N OBUs driving through the coverage of an RSU. Moreover, at each synchronous interval, the nonsafety message transmission probability is S_n. Second, the synchronous interval is further divided into small time slots. Each time slot is just long enough to exchange SCH rendezvous messages (RFS/CTR messages) to make an agreement within OBU pairs. The occupation time of nonsafety message is integer multiples of the time slots. This assumption enables us to use the SCH occupancy time to analyze SCH average utilization. We assume T_p is a random variable representing the SCH occupancy time slots of a nonsafety message which is sent successfully. For simplicity, we assume the length of nonsafety messages for all OBUs is a constant.

Table 17.1 Flowchart of the Channel Rendezvous

// Executed by initiating rendezvous node when a vehicle experiences a nonsafety message
// L: the transmission time of current nonsafety message
// $Ends1$: the cutoff time of the current synchronization interval
// $Endm1$: the minimum occupied time of the six SCH channels according to the LSRT-1
// $RF\ S$: rendezvous information $C\ T\ R$: clear to rendezvous
// $Ends2$: the cutoff time of the next synchronization interval
// $Endm2$: the minimum occupied time of the six SCH channels according to the LSRT-2
(1)   Compute L.
(2)   **If** the channel interval is RFS
(3)   Compute $Ends1$.
(4)   Compute $Endm1$.
(5)   **If** $L + Endm1 < Ends1$
Send the $RF\ S$ packet and initialize the $C\ T\ R$ timer.
(6)   **If** $C\ T\ R < 0$.
Go to step (2)
(7)   **Else**
Update the successful rendezvous information into LSRT-1.
End
(8)   **Else**
Give up the rendezvous.
End
(9)   **Else**
Go to step (8).
End
(10)   Compute $Endm2$.
(11)   **If** $L + Endm2 < Ends2$.
Send the $RF\ S$ packet and initialize the $C\ T\ R$.
(12)   **If** $CTR < 0$.
Give up the rendezvous.
(13)   **Else**
Update the successful rendezvous information into LSRT-2.
End
(14)   **Else**
Give up the rendezvous.
End
// Executed by receive rendezvous node when a vehicle received RFS
(15)   **If** the channel interval is RFS
(16)   **If** $L + Endm1 < Ends1$
Send the CTR and update the successful rendezvous information into
LSRT-1.
(17)   **Else**
Give up the rendezvous.
End
(18)   **Else**
If $L + Endm2 < Ends2$.
Send the CTR and update the successful rendezvous information into
LSRT-2.
(19)   **Else**
Give up the rendezvous.
End
  End

17.4.4 Aggregate Throughput

In vehicular networks, OBUs could randomly choose an SCH for nonsafety message transmission. It means that each SCH has the same average throughput in statistics. Therefore, we could obtain the channel average throughput through analyzing average channel utilization multiplied by the channel data rate.

In this subsection, we use channel occupation time to compute the channel utilization. To analyze the SCH occupation time, we assume that there are i random number of successful rendezvous made during the whole synchronous interval. Therefore, the probability of i successful rendezvous in the whole synchronous interval can be expressed as follows:

$$Y_i = P(Y = i), i \in [0, r] \tag{17.1}$$

where $r = \min\left(d, \dfrac{M \times s}{T_D}\right)$, and d is the number of the slots which could be used for making a rendezvous by the OBU in each synchronous interval. It enables us to know the maximum number of successful rendezvous on the CCH, s is the number of the slots which could be used for transmitting nonsafety messages by the OBU in an SCH interval. M is the number of SCHs. $\dfrac{M \times s}{T_p}$ enables us to know the maximum number of successful nonsafety message transmissions on SCHs. Generally speaking, $d < \dfrac{M \times s}{T_p}$ in a real system, because we could adjust the data rate of SCH to make the T_p small.

Then, Y_i could be rewritten as follows:

$$Y_i = C_d^i \times p^i \times (1 - p)^{d-1} \tag{17.2}$$

where p is a probability that a successful rendezvous is made in each time slot.

Then, p is analyzed and the process of a successful rendezvous is tracked. There are N OBUs driving through the coverage of an RSU and the nonsafety message transmission probability is S_n at each synchronous interval. Therefore, there will be N_n ($N_n = N \times S_n$) SCH rendezvous messages transmitted on the CCH.

Moreover, the OBU randomly selects the time slot to transmit a rendezvous message. If there are no other OBUs to select the same slot to transmit, the rendezvous would be successful. This is similar to an example of we have N_n balls (OBUs) that are thrown into d bins (slots), and there is no more than one ball (OBU) in the same bin (slot). Thus, the probability of a successful rendezvous follows the binomial distribution. So p could be written as follows:

$$p = P(X = 1) = C_{N_n}^1 \times \frac{1}{d} \times \left(1 - \frac{1}{d}\right)^{N_n - 1} \tag{17.3}$$

Furthermore, the average SCH channel utilization ϱ can be obtained as follows:

$$\varrho = \frac{\sum_i Y_i \times Th_i}{c \times M} \tag{17.4}$$

where c is the number of time slots in a synchronous interval, and Th_i denotes the average occupancy time on an SCH channel when i successful rendezvous are made on a CCH. Th_i could be expressed as follows:

$$Th_i = i \times T_p \tag{17.5}$$

By using Equations (17.4) and (17.5), we could obtain the aggregate SCHs throughput as follows:

$$T = \varrho \, R_D M \tag{17.6}$$

where R_D is the SCH data rate.

17.4.5 Maximum Aggregate Throughput

In this subsection, we determine the maximum aggregate throughput on SCHs and the parameters that could affect the achievable SCHs throughput. Within the type of reservation method, the aggregate throughput on SCHs is affected by the number of successful rendezvous and the maximum number of successful nonsafety message transmissions on SCHs $\left(\dfrac{M \times s}{T_p} \right)$. First, we consider the number of successful rendezvous. Equation (17.3) determines the probability of a successful rendezvous in each time slot. Therefore, the number of successful rendezvous can be obtained in the whole synchronous interval, as follows:

$$SP = p \times d = C_{N_n}\frac{1}{} \times \left(1 - \frac{1}{d} \right)^{N_n-1} \tag{17.7}$$

where SP is the number of successful rendezvous. Since d is a constant in a given network, SP depends on N_n. Therefore, we could use the first derivation of Equation (17.7) with respect to N_n:

$$\frac{d(SP)}{d(N_n)} = \left(1 - \frac{1}{d} \right) N_n - 1 + N_n \times \left(1 - \frac{1}{d} \right) N_n - 1 \times \log\left(1 - \frac{1}{d} \right) \tag{17.8}$$

Furthermore, we let Equation 17.8 be equal to 0 and obtain the N_n which produces the aggregate throughput maximum, as follows:

$$N_n = -\frac{1}{\log\left(1 - \dfrac{1}{d} \right)} \tag{17.9}$$

Equation (17.9) indicates that the maximum aggregate throughput on SCHs is affected by the number of SCH rendezvous which is transmitted on the CCH. Within a given network, N_n depends on the number of OBUs N and the nonsafety message transmission probability S_n. Therefore, if N is fixed, we could employ optimization of S_n to make $N_n = -\dfrac{1}{\log\left(1 - \dfrac{1}{d} \right)}$, thus achieving the maximum aggregate throughput on SCHs.

Using Equations (17.3), (17.4), (17.6), and (17.9), we could get the maximum number of successful rendezvous. Moreover, we compare the maximum number of successful rendezvous with the maximum number of successful nonsafety message transmissions on SCHs. The smaller number corresponds to the maximum aggregate throughput on SCHs.

17.5 RSU-Coordinated Asynchronous Multichannel MAC Scheme

The RSU-coordinated asynchronous multichannel MAC scheme needs no strict time synchronization and division interval, so all the channels including CCH and SCHs can be available in the whole time interval. However, due to the fact that the OBU can only use CCH for safety and rendezvous message exchange, it may cause other OBUs to miss the rendezvous message and make the rendezvous records maintained by each OBU incomplete, which will cause the missing receiver and the multichannel hidden terminal problems. From this point of view, we will use the RSU to assist the message exchange of each OBU to solve the above problems. Different to the existing asynchronous multichannel strategy, in our proposed asynchronous multichannel scheme, the RSU is equipped with two half-duplex transceivers (transceiver A and transceiver B). Transceiver A always listens to the CCH to record the safety-related message and rendezvous information. Transceiver B is capable of switching between different SCHs and CCH and retransmitting messages to OBUs which get missed rendezvous information so solve the message loss and rendezvous conflict problems. Besides, in the SCH rendezvous process, an OBU reserves one of SCHs by exchanging RTS/RSU-assisted packet RA/CTS messages to avoid the missing receiver problem.

17.5.1 The Operation of the Proposed Scheme

The research scene and related conditions in this section are consistent with Section 17.3, so we will not repeat here. The RSU-coordinated asynchronous multichannel MAC scheme adopts the RTS/RA/CTS/DATA/ACK packets to rendezvous SCHs and transmit the nonsafety message on the reserved SCH, where the RTS/RA/CTS is completed on the ·CCH and the DATA/ACK is completed on the SCH .

To finish an SCH rendezvous, our proposed scheme designs the RTS packet and CTS packet, which include the SCH *ID*, rendezvous start time *ST*, and rendezvous end time *ET*. Besides, to record the rendezvous information, we assume each node keeps a local SCH rendezvous table, which includes the success rendezvous channel *ID*, *ST*, and *ET* of the channel rendezvous, the source node *ID*, and the destination node *ID* of the rendezvous channel. Most importantly, we design the RSU to send the RA packet to help the SCH rendezvous. The RA packet contains the information of the destination node and the destination SCH, in which the busy node *BN* indicates that the current destination node is not on the CCH, the busy channel *BC* means the SCH selected by the rendezvous node is busy now and the remaining time of channel *RTC* indicates the busy time of the SCHs. The protocol stipulates that the safety-related message transmission and SCHs rendezvous are completed on the CCH, while the transmission of the nonsafety message is completed on the SCHs. The details of our proposed protocol are follows.

When the source node prepares to transmit the nonsafety message to the destination node, it first broadcasts an RTS rendezvous packet on the CCH. In order to avoid the missing rendezvous message problem and the hidden terminal problem due to the uncomplete local rendezvous message, RSU checks the RTS packet from the source node to know whether the destination node is on the CCH or not. If the destination node is not on the CCH, RSU will reply with an RA packet (*BN* = 1) after a unit interval. Otherwise, the RSU further checks whether the SCH is idle during the period of the source node rendezvous time. If the SCH is busy, the RSU responds to the RA packet (*BC* = 1), and if the SCH is idle, RSU does not send an RA packet. Therefore, when a source node receives an RA packet from the RSU after a unit interval, it means that the current rendezvous cannot be completed, and the source node needs to wait for an RTC interval to make a rendezvous again. However, if the source node does not receive any RA packet, it continues to listen to the CCH to wait for the CTS packet from the destination node. At the same time, the destination node waits for two unit intervals when it receives the RTS packet, and if it does not receive the RA packet sent by the RSU, it immediately sends the CTS packet to the source node and switches to the corresponding SCH according to the rendezvous message for the nonsafety data/acknowledgment transmission (DATA/ACK).

In the RSU-coordinated asynchronous multichannel MAC scheme, a node rendezvous an SCH with another node in an RSU coverage area, the detailed RSU-coordinated rendezvous algorithm is elaborated in Table 17.2. The operation of our proposed scheme is shown in Figure 17.4. To better understand, here we take two examples to explain the process of such scheme.

Scene 1: The source node n1 rendezvous the idle SCH1 to the destination node n6. However, due to the destination node n6 being also on the CCH simultaneously, the RSU does not need to send an RA packet, n1 and n6 can make the rendezvous and exchange messages on the CCH and SCH1 respectively.

Table 17.2   Flowchart of RSU-Coordinated Asynchronous Multichannel MAC Scheme

// Executed by an RSU when a source node sends a RTS packet
// *LSRT* : the local SCH rendezvous table.
// *destinationl D*: to receive RTS and send CTS packets
// *channell D*: the field of RTS or CTS, it means one of SCHs, which is chosen by service provider.
// *BC* : the field of RTS or CTS, it means the SCH channel is busy.
// *BN* : the field of RTS or CTS, it means the destination node is absent in CCH.
// *RA*: RSU-assisted packet that is only send by the RSU
// *RT C* : remaining time of channel or node, it means the node or channel will be idle after
  the remaining time.
(1)   Checking *LSRT*.
(2)   **If** there is *destinationID* in *LSRT*
then
(3)   **If** *channelID* is idle in *LSRT* **then**
Go to step (5).
else
BC = 1, go to step (6).
else
BN = 1, go to step (6).
(4)   Do nothing, preparing to receive the CTS from the destination node.
(5)   Send the RA packet.
// Executed by the destination node when the RSU sends and RA packet
(6)   **If** *BN* = 1 or BC = 1, **then**
Initiate rendezvous. again after
t = *RTC*.
End

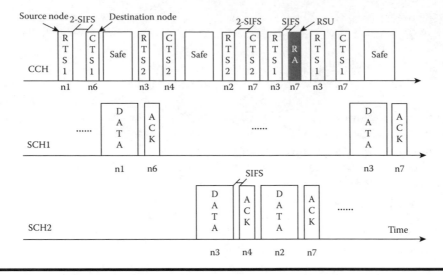

Figure 17.4   RSU-coordinated asynchronous multichannel MAC scheme.

Scene 2: The source node n3 rendezvous SCH1 to the destination node n7 which is exchanging messages with n2 on the SCH2. The RSU waits for a unit interval to send the RA packet and informs the source node n3 that the destination node n7 is not on the CCH due to the n7 exchange message with n2 on the SCH2. After waiting for a *RTC* interval, the source node n3 rendezvous with node n7 again.

17.5.2 The Resolving Methods of Multichannel Hidden Terminal and Missing Receiver Problems

We adopt an RSU-coordinate rendezvous scheme to solve the multichannel hidden terminal problem, as shown in Figure 17.5. After the node n3 sends an RTS (*channelID* = 2) packet to n10, the RSU checks SCH2 is busy, so the RSU sends an RA packet (*BC* = 1) to n3. In this way, the scheme can solve the problem of multichannel hidden terminal effectively.

The RSU-coordinate rendezvous scheme to solve the missing receiver problem is shown in Figure 17.6. After the node n3 sends the RTS packet to n7, the RSU finds the destination node n7 is busy now, so the RSU will send an RA packet (*BN* = 1) to n3. This protocol uses RSU coordination to avoid the missing receiver problem caused by the destination node on the SCH. The n3 will initiate the second rendezvous after *RTC* duration.

17.6 Simulation Results

17.6.1 The Simulation of the RSU-Coordinated Synchronous Multichannel MAC Scheme

The simulation consists of two parts. In the first part, we present simulation results to verify the analytical model. In the second part, we perform simulations to compare the aggregate throughput on SCHs of the proposed scheme with that of the IEEE 1609.4 and VEMMAC (which allows the OBU to try to access the CCH to reserve one of the SCHs only during the CCH interval).

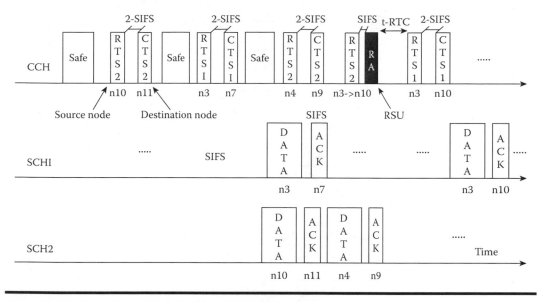

Figure 17.5   Multichannel hidden terminal problem.

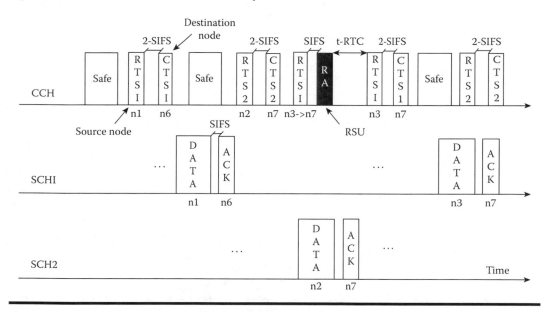

Figure 17.6   Missing receiver problem.

Before performing simulations, we need to choose appropriate values of *d* and *s*. We assume that each of the synchronous intervals is further divided into 100 small time slots ($c = 100$). Therefore, we set $d = 95$ (RFS interval + SCH interval) and $s = 50$ in the proposed scheme. Then, the lowest IEEE 802.11p channel data rate 3 Mbps is chosen to provide robustness and reliability. The other parameters are listed in Table 17.3 for clarity. We perform a 2000 times Monte Carlo simulation.

In this subsection, first we present simulation results to verify the analytical model. Then, the impact of nonsafety message transmission probability in each synchronous interval, the data rate of each SCH, average size of the nonsafety message, and number of OBUs is investigated.

Table 17.3   Simulation Parameters

Parameter	Value
M	6
c	100
d	95
s	50
Data rate	3 Mbps
MAC header	256 bit
PHY header	192 bit
Packet size of CTR	112 bit
Packet size of RFS	160 bit
Packet size of RSU	100 bit
Packet size of safety message	300 bit

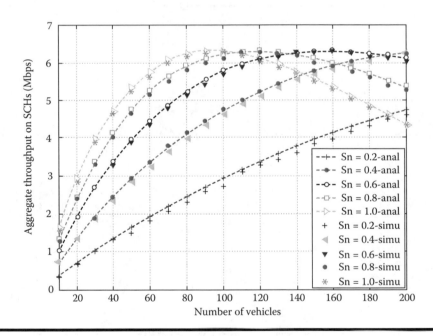

Figure 17.7   Aggregate throughput on SCHs with different S_n.

Figure 17.7 shows the aggregate throughput on SCHs with different nonsafety message transmission probability S_n at each synchronous interval. It is observed that the aggregate throughput on SCHs increases with number of OBUs N for a given S_n, before it reaches the maximum aggregate throughput. However, the aggregate throughput on SCHs decreases with the number of OBUs N for a given S_n. The reason is, after the aggregate throughput on SCHs reaches its maximum, increasing the number of OBUs is equal to adding the load on the CCH. It causes CCH congestion and leads to the decrease of the probability of successful rendezvous. Then, we consider five different values of S_n in the simulation. It is easy to check that greater S_n requires smaller N

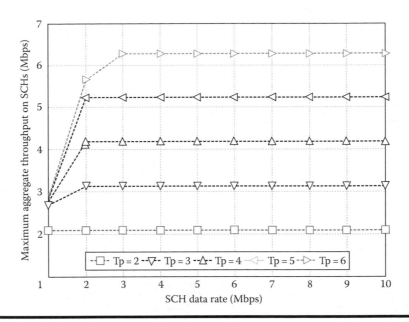

Figure 17.8 Maximum aggregate throughput on SCHs.

to reach the maximum aggregate throughput on SCHs. The reason is that $N_n = -\dfrac{1}{\log\left(1 - \dfrac{1}{d}\right)}$ is

fixed in a given network. The pairs of nonsafety message transmission probability and the number of achieving maximum aggregate throughput OBUs are $(S_n, N_{anal}, N_{sium}) = (1, 94.49, 100)$, $(0.8, 118, 120)$, $(0.6, 158, 160)$. Simulation results have a little fluctuation around the ones by using an analytical model ($<5\%$), which indicates that the analytical model has good performance in computing maximum aggregate throughput.

Next, we present simulation results to verify the impact of SCH data rate and average size of the nonsafety message. We set $T_{pmin} = 2$, $\Delta T_p = 1$, and $T_{pmax} = 6$ in the simulation, when the nonsafety message transmission probability at each synchronous interval is equal to 80%. Corresponding to the simulation parameters in Table 17.3, it is equal to set the average message size $PL_{min} = 6$ kbyte, $\Delta PL = 3$ kbyte, and $PL_{max} = 18$ kbyte. Figure 17.8 shows the maximum aggregate throughput with different SCH data rates. It is observed that the maximum aggregate throughput is unchanged with the same T_p when the SCH data rates are higher than 3 Mbps. The reason is that the maximum number of successful nonsafety message transmissions on SCHs is not the limit of aggregate throughput when the SCH data rate is high. In Figure 17.9, it is easy to check that the aggregate throughput on SCHs increases with T_p for a given N. This is because the aggregate throughput increases with the size of nonsafety message when the probability of successful rendezvous is fixed.

To evaluate the performance of the proposed scheme, we perform simulations and compare the aggregate throughput of the proposed scheme with that of the VEMMAC [17] and IEEE 1609.4.

Figure 17.10 shows the aggregate throughput on SCHs per second using the proposed scheme, VEMMAC and IEEE 1609.4, respectively, when the nonsafety message transmission probability at each synchronous interval is 60%. The ordinate is calculated by the total size of nonsafety messages that are transmitted by the rendezvoused OBU pairs. It is observed that the aggregate throughput of the proposed scheme is higher than that of the aforementioned VEMMAC and

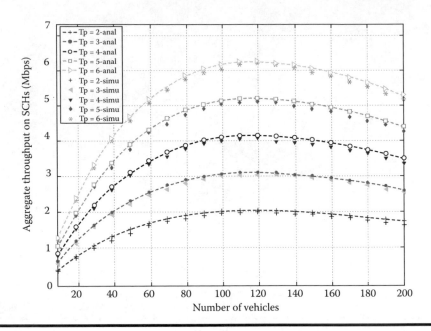

Figure 17.9   Aggregate throughput on SCHs with different T_p.

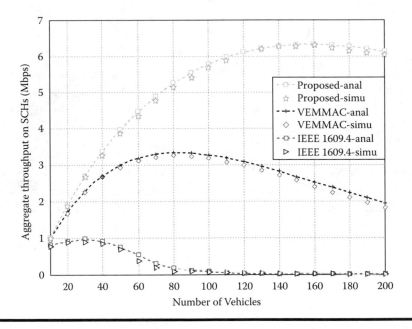

Figure 17.10   Comparison of three schemes in aggregate throughput.

IEEE 1609.4 when the number of OBUs increases. However, when the number of OBUs is less than 20, the gap of the aggregate throughput among the three schemes is very small. While the number of vehicles is more than 20, the two rendezvous schemes have a better performance than IEEE 1609.4. The reason is that the collisions increase with the network load in using one desig-nated SCH of IEEE 1609.4. Furthermore, the proposed scheme could obtain higher maximum

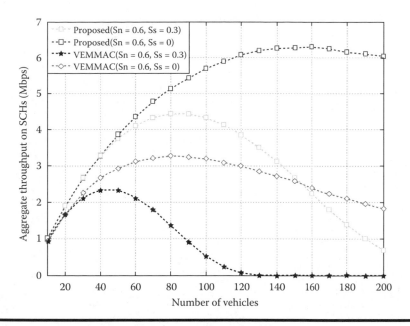

Figure 17.11   Comparison of the proposed scheme and VEMMAC in aggregate throughput when considering safety message transmission on a CCH.

aggregate throughput and allow more OBU pairs to exchange rendezvous messages successfully than that of VEMMAC. The reason is that the proposed scheme allows OBUs to reserve in almost the whole synchronous interval. It could obtain more chances to reserve successfully, when the collisions increase with the network load.

In Figure 17.11, we compare the aggregate throughput on SCHs of the proposed scheme with that of the VEMMAC, when considering safety message transmission on a CCH. Let S_s denote the safety message transmission probability at each synchronous interval. Then, we set the non-safety message transmission probability at each synchronous interval at 60%. To evaluate the performance of the two schemes under different loads on the CCH, we compare the safety message transmission on the CCH with that of no safety message transmission. It is observed that the aggregate throughput of the VEMMAC scheme decreases to 0 faster than the proposed scheme. It demonstrates that the proposed scheme has better performance in CCH congestion control, and it obtains higher channel utilization and reliability, especially under high network load conditions.

17.6.2 The Simulation of the RSU-Coordinated Asynchronous Multichannel MAC Scheme

To show the performance advantages of the scheme, ARCMC is simulated on NS2, compared with the performances of a typical asynchronous multichannel protocol (AMCP) [19] and IEEE 1609.4. In Ref. [27], it is proved that the influence of node movement on channel access is extremely small. Therefore, we consider the simulation is carried out in the transmission range of a single RSU, and each node is in a quiescent state. Half of the nodes participating in the simulation play the role of the service provider, and the other half is the service user. The simulation scenario is shown in Figure 17.2. The main parameters of the simulation analysis are shown in Table 17.4.

Table 17.4   Simulation Parameters

Parameter	Value
Timeslotlength	6
SIFS	100
CWmin	95
CWmax	50
Size of RTS packet	160 bit
Size of RA packet	112 bit
Size of CTS packet	112 bit
Size of safety packet	100 bit
Safety packet broadcast frequency	10 Hz
Channel rendezvous frequency	30 Hz
Size of nonsafety packet	1024 bit
Channel capacity	2 Mb/s
Number of nodes	10–100

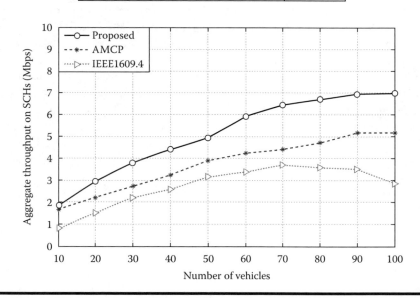

Figure 17.12   Comparison of the proposed scheme, AMCP and IEEE1609.4, in aggregate throughput.

Figure 17.12 shows the aggregate throughput on SCHs of the three different schemes in the case of different numbers of nodes. As the number of nodes rises, the SCH throughput of ARCMC is increasing by 60% to 140% on IEEE 1609.4, while the aggregate SCH throughput of AMCP is increasing by nearly 16% to 100% on IEEE 1609.4. These phenomena can be simply explained as follows. On the contrary, ARCMC and AMCP adopt the asynchronous SCH reservation

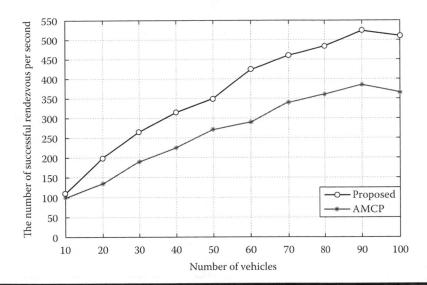

Figure 17.13   The number of successful SCH rendezvous.

mechanism to improve the nonsafety data transmission efficiency on the SCH. On the other hand, when the number of nodes reaches 70, the IEEE 1609.4 has the CCH congestion problem due to the fixed-duration CCH interval which causes the increase of collision probability, so the throughput decreases significantly fast. However, ARCMC and AMCP adopt the asynchronous multichannel mechanism, which alleviates the CCH congestion to a certain extent and effectively reduces the collision probability of the node. Last but not least, ARCMC with RSU-assisted rendezvous mechanism avoids multichannel hidden terminals and misses receiver problems which are not completely solved by AMCP. Consequently, the result shows that the throughput performance of ARCMC is more prominent than AMCP.

Figure 17.13 shows the number of successful SCH rendezvous for ARCMC and AMCP. With the increase in the number of nodes, the number of successful SCH rendezvous of ARCMC is improved and is higher than AMCP. This is because ARCMC uses the RSU to assist the rendezvous mechanism by sending an RA packet to avoid multichannel hidden terminals and missed receiver problems, which increases the number of successful SCH reservations. However, using AMCP to avoid a multichannel hidden terminal problem is at the expense of reducing SCH rendezvous efficiency. Besides, Figure 17.12 shows that the aggregate throughput of ARCMC is higher than AMCP, indicating that the number of successful SCH rendezvous plays a decisive role in the aggregate throughput.

Figure 17.14 shows the safety message reception rate under the three schemes. The reception rate of the safety message under three different schemes decreases slowly as the number of nodes increases. When the number of nodes reaches 100, the reception rate under the IEEE 1609.4 standard still maintains more than 90%, which is significantly better than both ARCMC and AMCP. The reason is that the IEEE 1609.4 standard allows the safety message broadcast only in the CCH interval on the CCH and requires that the nodes must switch to the CCH at the CCH interval. It is clear that a high reception rate of the safety message is ensured. Although ARCMC and AMCP have achieved good system throughput through asynchronous multichannel reservation mechanisms, it reduces the reception rate of the safety message. As a result, the safety message reception rates of ARCMC and AMCP schemes are less than IEEE 1609.4.

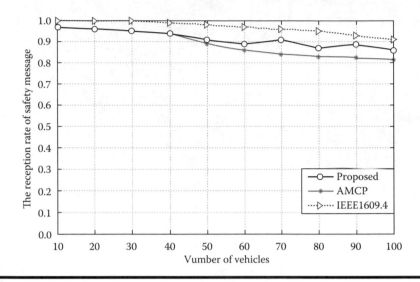

Figure 17.14 The reception rate of a safety message.

In addition, when the number of nodes is less than 50, due to the network traffic under light load, the channel access environment is good. As a result, the safety message reception rate of ARCMC and AMCP with asynchronous multichannel protocols is consistent. However, when the number of nodes is higher than 50, the reception rate of ARCMC is higher than AMCP. It means that when the network load increases gradually, the probability of multichannel hidden terminal and missing receiver problems increases. Consequently, we design RSU-coordinate mechanism that can effectively reduce the waste of channel resources caused by the above two problems and ensure more nodes on the CCH exchange safety messages.

17.7 Conclusion

In this chapter, we analyze and compare the basic ideas, characteristics, and existing problems of the representative multichannel coordination schemes for VANETs. First of all, we propose a novel synchronous RSU-coordinated multichannel MAC for VANETs to allow channel rendezvous on the CCH during the whole synchronization intervals and support simultaneous transmission on different SCHs, which can reduce the control channel congestion. Then, we propose a novel asynchronous RSU-coordinated multichannel MAC for VANETs to solve the multichannel hidden terminal and missing receiver problems. Finally, simulation results demonstrate that the proposed schemes can provide better performance in terms of channel utilization and system throughput than that of the existing approaches in the literature.

Acknowledgments

This work was supported by the National Natural Science Foundation of China (61762030, 61661016,61561014), Natural Science Foundation of Guangxi Province under Grant

(2016GXNSFGA380002), the Key Laboratory of Cognitive Radio and Information Processing, Ministry of Education (Guilin University of Electronic Technology), CRKL160106, and the Guangxi Colleges and Universities Key Laboratory of cloud computing and complex systems.

References

1. L. Atzori, A. Iera, and G. Morabito, The Internet of things: A survey, *Comput Netw.*, vol. 54, no. 15, pp. 2787–2805, Oct. 2010.
2. R. Uzcategui and G. Acosta-Marum. Wave: A tutorial, *Proc. IEEE Commun. Mag.*, pp. 126–133, Mar. 2009.
3. J. B. Kenney, Dedicated short-range communications (DSRC) standards in the United States, *Proc. IEEE*, vol. 99, no. 7, pp. 1162–1182, Jul. 2011.
4. IEEE, *Wireless LAN medium access control (MAC) and physical layer (PHY) specifications amendment 6: Wireless access in vehicular environments*, IEEE Std. P802.11p/D11.0, 2010.
5. S. A. Mohammad, A. Rasheed and Qayyum A, VANET architectures and protocol stacks-A survey, *Proc. LNCS*, pp. 95–105, Mar. 2011.
6. G. Karagiannis, O. Altintas, E. Ekici, G. Heijenk, B. Jarupan, K. Lin, and T. Weil, Vehicular networking: A survey and tutorial on requirements, architectures, challenges, standards and solutions, *IEEE Commun. Surv. Tutorials*, vol. 13, no. 4, pp. 584–616, Fourth Quarter 2011.
7. M. L. Sichitiu and M. Kihl, Inter-Vehicle communication systems: A survey, *IEEE Commun. Surv. Tutorials*, vol. 10, no. 2, pp. 88–105, 2008.
8. CAR 2 CAR, Communication Consortium Manifesto, Car 2 Car Communication Consortium, http://www.car-2-car.org/fileadmin/downloads/C2C-CC_manifesto_v1.1.pdf, 2007.
9. TR102638 E, Intelligent Transport System (ITS), Vehicular Communications, Basic Set of Applications, Definition. Sophia Antipolis: ETSI Std, ETSI ITS Specification TR, 102:638, 2009.
10. Vehicle Safety Applications, US DOT Intelli Drive Project-ITS Joint Program Office, Washington D.C, 2008.
11. PreDrive C2X deliverable D44.1, Technical Performance of DRIVE C2X functions in full-scale FOT operations. Pre Drive C2X project, Sindelfingen, Germany, 2013.
12. H. Hartenstein, K. Laberteaux, VANET: Vehicular Applications and Inter-Networking Technologies, John Wiley & Sons, Hoboken, 2006.
13. Q. Wang, S. Leng, H. Fu, Y. Zhang, and H. Weerasinghe, An enhanced multichannel MAC for the IEEE-1609.4-based vehicular ad hoc networks, *Proc. IEEE INFOCOM*, pp. 2534–2537, Mar. 2010.
14. Q. Wang, S. Leng, H. Fu, and Y. Zhang, An IEEE 802.11p-based multichannel MAC scheme with channel coordination for vehicular ad hoc networks, *IEEE Trans. Intelligent Transportation Syst.*, vol. 13, no. 2, pp. 449–457, Jun. 2012.
15. J. Wang, Y. Ji, X. Wang, and F. Liu, RSU-coordinated multi-channel MAC with multi-criteria channel allocation, *Proc. IEEE ICCVE*, pp. 60–65, Dec. 2012.
16. N. Lu, Y. Ji, F. Liu, and X. Wang, A dedicated multichannel MAC protocol design for VANET with adaptive broadcasting, *Proc. IEEE WCNC*, pp. 1–6, Apr. 2010.
17. D. N. M. Dang, H. N. Dang, C. Do, and C. S. Hong, An enhanced multi-channel MAC for vehicular ad hoc networks, *Proc. IEEE WCNC*, pp. 351–355, Apr. 2013.
18. L. Tang, C. Wang, Q. Chen, and P. Gong, An optimal back-off parameters study for multi-channel MAC protocols in VANETs, *Proc. IEEE WOCC*, pp. 1–5, May 2013.
19. J. Shi, T. Salonidis, E. W. Knightly, Starvation Mitigation Through Multi-channels Coordination in CSMA Multi-hop Wireless Networks, *2006 ACM 7th International Symposium on Mobile Ad Hoc Networking and Computing*, pp. 214–225, May, 2006.
20. T. Luo, M. Motani, V. Srinivasan, CAM-MAC: A Cooperative Asynchronous Multi-channel MAC Protocol for Ad Hoc Networks, *2006 IEEE International Conference on Broadband Communications, Networks and Systems*, pp. 1–10, Nov. 2006.

21. C. Han, M. Dianati, R. Tafazolli, and R. Kernchen, Asynchronous multichannel MAC for vehicular ad hoc networks, *Proc. IEEE VNC*, pp. 170–176, Nov. 2011.
22. C. Han, M. Dianati, R. Tafazolli, and X. Liu, A novel distributed asynchronous multichannel MAC scheme for large-scale vehicular ad hoc networks, *IEEE Trans. Veh. Tech.*, vol. 61, no. 7, pp. 3125–3138, Sep. 2012.
23. S. Park, Y. Chang, F. Khan, and J. A. Copeland, Dynamic service-channels allocation (DSCA) in vehicular ad-hoc networks, *Proc. IEEE CCNC*, pp. 351–357, Jan. 2013.
24. K. H. Almotairi and X. Shen, Multichannel medium access control for ad hoc wireless networks, *Wirel. Commun. Mob. Comput.*, vol. 13, no. 11, pp. 1047–1059, Aug. 2013.
25. T. Luo, M. Motani, and V. Srinivasan, Cooperative Asynchronous Multichannel MAC: Design, analysis, and implementation, *Wirel. Commun. Mob. Comput.*, vol. 8, no. 3, pp. 338–352, Aug. 2009.
26. M. Ni, Z. Zhong, and D. Zhao, A novel multichannel multiple access protocol for vehicular ad hoc networks, *Proc. IEEE ICC*, pp. 528–532, Jun. 2012.
27. J. Hass and J. Deng, Dual Busy Tone Multiple Access (DBTMA) performance evaluation, *Proc. IEEE VTC*, pp. 314–319, May 1999.

Index